Lecture Notes in Physics

Editorial Board

H. Araki, Kyoto, Japan
E. Brézin, Paris, France
J. Ehlers, Potsdam, Germany
U. Frisch, Nice, France
K. Hepp, Zürich, Switzerland
R. L. Jaffe, Cambridge, MA, USA
R. Kippenhahn, Göttingen, Germany
H. A. Weidenmüller, Heidelberg, Germany
J. Wess, München, Germany
J. Zittartz, Köln, Germany

Managing Editor

W. Beiglböck
Assisted by Mrs. Sabine Lehr
c/o Springer-Verlag, Physics Editorial Department II
Tiergartenstrasse 17, D-69121 Heidelberg, Germany

Springer
Berlin
Heidelberg
New York
Barcelona
Budapest
Hong Kong
London
Milan
Paris
Santa Clara
Singapore
Tokyo

The Editorial Policy for Proceedings

The series Lecture Notes in Physics reports new developments in physical research and teaching – quickly, informally, and at a high level. The proceedings to be considered for publication in this series should be limited to only a few areas of research, and these should be closely related to each other. The contributions should be of a high standard and should avoid lengthy redraftings of papers already published or about to be published elsewhere. As a whole, the proceedings should aim for a balanced presentation of the theme of the conference including a description of the techniques used and enough motivation for a broad readership. It should not be assumed that the published proceedings must reflect the conference in its entirety. (A listing or abstracts of papers presented at the meeting but not included in the proceedings could be added as an appendix.)

When applying for publication in the series Lecture Notes in Physics the volume's editor(s) should submit sufficient material to enable the series editors and their referees to make a fairly accurate evaluation (e.g. a complete list of speakers and titles of papers to be presented and abstracts). If, based on this information, the proceedings are (tentatively) accepted, the volume's editor(s), whose name(s) will appear on the title pages, should select the papers suitable for publication and have them refereed (as for a journal) when appropriate. As a rule discussions will not be accepted. The series editors and Springer-Verlag will normally not interfere with the detailed editing except in fairly obvious cases or on technical matters.

Final acceptance is expressed by the series editor in charge, in consultation with Springer-Verlag only after receiving the complete manuscript. It might help to send a copy of the authors' manuscripts in advance to the editor in charge to discuss possible revisions with him. As a general rule, the series editor will confirm his tentative acceptance if the final manuscript corresponds to the original concept discussed, if the quality of the contribution meets the requirements of the series, and if the final size of the manuscript does not greatly exceed the number of pages originally agreed upon. The manuscript should be forwarded to Springer-Verlag shortly after the meeting. In cases of extreme delay (more than six months after the conference) the series editors will check once more the timeliness of the papers. Therefore, the volume's editor(s) should establish strict deadlines, or collect the articles during the conference and have them revised on the spot. If a delay is unavoidable, one should encourage the authors to update their contributions if appropriate. The editors of proceedings are strongly advised to inform contributors about these points at an early stage.

The final manuscript should contain a table of contents and an informative introduction accessible also to readers not particularly familiar with the topic of the conference. The contributions should be in English. The volume's editor(s) should check the contributions for the correct use of language. At Springer-Verlag only the prefaces will be checked by a copy-editor for language and style. Grave linguistic or technical shortcomings may lead to the rejection of contributions by the series editors. A conference report should not exceed a total of 500 pages. Keeping the size within this bound should be achieved by a stricter selection of articles and not by imposing an upper limit to the length of the individual papers. Editors receive jointly 30 complimentary copies of their book. They are entitled to purchase further copies of their book at a reduced rate. As a rule no reprints of individual contributions can be supplied. No royalty is paid on Lecture Notes in Physics volumes. Commitment to publish is made by letter of interest rather than by signing a formal contract. Springer-Verlag secures the copyright for each volume.

The Production Process

The books are hardbound, and the publisher will select quality paper appropriate to the needs of the author(s). Publication time is about ten weeks. More than twenty years of experience guarantee authors the best possible service. To reach the goal of rapid publication at a low price the technique of photographic reproduction from a camera-ready manuscript was chosen. This process shifts the main responsibility for the technical quality considerably from the publisher to the authors. We therefore urge all authors and editors of proceedings to observe very carefully the essentials for the preparation of camera-ready manuscripts, which we will supply on request. This applies especially to the quality of figures and halftones submitted for publication. In addition, it might be useful to look at some of the volumes already published. As a special service, we offer free of charge LaTeX and TeX macro packages to format the text according to Springer-Verlag's quality requirements. We strongly recommend that you make use of this offer, since the result will be a book of considerably improved technical quality. To avoid mistakes and time-consuming correspondence during the production period the conference editors should request special instructions from the publisher well before the beginning of the conference. Manuscripts not meeting the technical standard of the series will have to be returned for improvement.

For further information please contact Springer-Verlag, Physics Editorial Department II, Tiergartenstrasse 17, D-69121 Heidelberg, Germany

Jochen Greiner (Ed.)

Supersoft X-Ray Sources

Proceedings of the International Workshop
Held in Garching, Germany,
28 February – 1 March 1996

 Springer

Editor

Jochen Greiner
Max-Planck-Institut für Extraterrestrische Physik
D-85740 Garching, Germany

Scientific Organising Committee:
J. Greiner (Germany, chair), F. Meyer (Germany), J. Trümper (Germany)

Conference Secretary:
L. Falke

Cataloging-in-Publication Data applied for.

Die Deutsche Bibliothek - CIP-Einheitsaufnahme

Supersoft x-ray sources : proceedings of the international workshop, held in Garching, Germany, 28. February - 1 March 1996 / Jochen Greiner (ed.). - Berlin ; Heidelberg ; New York ; Barcelona ; Budapest ; Hong Kong ; London ; Milan ; Paris ; Santa Clara ; Singapore ; Tokyo : Springer, 1996
  (Lecture notes in physics ; Vol. 472)
  ISBN 3-540-61390-0
NE: Greiner, Jochen [Hrsg.]; GT

ISSN 0075-8450
ISBN 3-540-61390-0 Springer-Verlag Berlin Heidelberg New York

This work is subject to copyright. All rights are reserved, whether the whole or part of the material is concerned, specifically the rights of translation, reprinting, re-use of illustrations, recitation, broadcasting, reproduction on microfilms or in any other way, and storage in data banks. Duplication of this publication or parts thereof is permitted only under the provisions of the German Copyright Law of September 9, 1965, in its current version, and permission for use must always be obtained from Springer-Verlag. Violations are liable for prosecution under the German Copyright Law.

© Springer-Verlag Berlin Heidelberg 1996
Printed in Germany

The use of general descriptive names, registered names, trademarks, etc. in this publication does not imply, even in the absence of a specific statement, that such names are exempt from the relevant protective laws and regulations and therefore free for general use.

Typesetting: Camera-ready by the authors
SPIN: 10520086   55/3142-543210 - Printed on acid-free paper

Preface

The workshop on supersoft X-ray sources (and related objects) was held at the Max-Planck-Institut für extraterrestrische Physik (MPE) in Garching, Germany from February 28 to March 1, 1996. The decision to organise this workshop was made for three reasons. (1) With all of the ROSAT PSPC data being publicly available to every interested scientist, and with most of the relevant data (pointed observations of the Magellanic Clouds and the Andromeda galaxy and the all-sky survey data for a galactic population) being investigated, the number of expected discoveries of new supersoft sources along the previously used search strategies is small. (2) Since the community of scientists working actively in the field of supersoft X-ray sources is rather small, and since the work to be done, especially the optical identification and follow-up work, requires rare time on large telescopes, it seemed worthwhile to bring all researchers in close contact to eventually combine the individual efforts. (3) With the variety of observational facts on supersoft X-ray sources and the establishment of links to rather diverse research areas it seemed appropriate to supply a forum for theoreticians to discuss new results, and to strengthen the connection between observers and theoreticians.

After the discovery of supersoft X-ray sources with Einstein Observatory observations, the ROSAT satellite with the MPE-built PSPC has discovered nearly three dozen new supersoft sources and has thus established luminous supersoft X-ray sources as a class of objects. Though many different classes of objects emit supersoft X-ray radiation (defined here as emission dominantly below 0.5 keV which corresponds to effective temperatures of the emitting objects below 50 eV), such as the Moon, single, nearby early type stars, single white dwarfs, central stars of planetary nebulae, PG 1159 stars, magnetic cataclysmic variables, some symbiotic binaries and a surprisingly large number of active galactic nuclei, the class of supersoft X-ray sources (as meant in these proceedings) is characterised by high bolometric luminosities (for galactic source standards) in the range $10^{36}-10^{38}$ erg/s. Optical observations have revealed the binary nature of several of these objects, supporting the idea that one deals with systems in which a white dwarf accretes matter at high rates from a more massive donor companion, and then burns it steadily on its surface.

The discovery of this class of objects has motivated substantial theoretical research, including population synthesis studies. Work has progressed on the study of the evolution of binary systems in which the donor is more massive than the accretor, and on understanding the role that stellar winds may play in such systems. The unique features of accretion disks in which the luminosity from the central source provides more energy than the accretion itself have also begun to be studied. Understanding hydrogen burning on white dwarfs is of central importance to our understanding of supersoft sources. Theoretical studies have also linked supersoft sources to diverse astrophysical phenomena, including Type Ia supernovae and the planetary nebula luminosity function. All of this led to extensive discussions between the attendees throughout the workshop. Indeed, the principal aim of the workshop was the presentation and discussion of recent observational results, as well as their theoretical implications for basically all the above mentioned areas.

This workshop would not have been possible without the dedication of our secretary Lynn Falke. It is a pleasure to express our thanks to her and the other staff members of MPE for providing the appropriate environment and a friendly atmosphere for effective discussions during the workshop.

The Max-Planck-Institut für extraterrestrische Physik not only provided the logistical support but also funded all the expenses of the workshop including the printing of these proceedings. This put us in the position of not having to charge a conference fee. We thank all the participants for their lively and fruitful contributions which made the three-day workshop so successful.

Garching
April 1996

Scientific Organising Committee
J. Greiner
F. Meyer
J. Trümper

Contents

Part I Population Properties of SSS

**On the Evolution, Numbers and Characteristics
of Close-Binary Supersoft Sources**
R. Di Stefano, L.A. Nelson   3

White Dwarfs with H/He Burning as Supersoft X-Ray Sources
M. Kato 15

**Hot High-Gravity NLTE Model Atmospheres
Applied to Supersoft Sources**
H.W. Hartmann, J. Heise 25

Luminous Supersoft X-Ray Sources in Globular Clusters
R. Di Stefano, M.B. Davies 33

**The Integrated X-Ray Spectrum of Galactic Populations
of Luminous Supersoft X-Ray Sources**
R. Di Stefano, C.M. Becker, G. Fabbiano 37

Neutron Stars Can Do It Too if They Want to
N.D. Kylafis 41

SSSs as Progenitors of the BHCs
W. Kundt 45

Part II Winds and Accretion Disks in SSS

Simulation of the Visual Light Curve of CAL 87
S. Schandl, E. Meyer-Hofmeister, F. Meyer 53

Accretion Disks in Supersoft X-Ray Sources
R. Popham, R. Di Stefano 65

Part III X-Ray and Optical Observations of SSS

Supersoft X-Ray Sources in M 31
J. Greiner, R. Supper, E.A. Magnier 75

X-Ray and Optical Observations of RX J0925.7−4758: Constraints on the Binary Structure
C. Motch 83

X-Ray Energy Spectrum of RX J0925.7−4758 with ASCA
K. Ebisawa, K. Asai, K. Mukai, A. Smale, T. Dotani, H.W. Hartmann, J. Heise 91

Optical Photometry of RX J0019.8+2156 over the Last Three Years
T. Will, H. Barwig 99

Phase Resolved UV Spectroscopy of RX J0019.8+2156
B.T. Gänsicke, K. Beuermann, D. de Martino 107

Optical Spectroscopy of RX J0439.8−6809 and 1E 0035.4−7230
A. van Teeseling, K. Reinsch, K. Beuermann, H.-C. Thomas, M.W. Pakull 115

Photometric Observations of Supersoft Sources in the LMC
P.C. Schmidtke, A.P. Cowley 123

Optical Light Curve and Binary Period of the Supersoft X-Ray Transient RX J0513.9−6951
C. Motch, M.W. Pakull 127

Non-LTE Model Atmosphere Analysis of the Supersoft X-Ray Source RX J0122.9−7521
K. Werner, B. Wolff, M.W. Pakull, A.P. Cowley, P.C. Schmidtke, J.B. Hutchings, D. Crampton 131

Implications of Light Metals (Li - Ca) on NLTE Model Atmospheres for Hot Stars
T. Rauch 139

Part IV Intrinsic Variability of SSS

The Long-Term X-Ray Lightcurve of RX J0527.8−6954
J. Greiner, R. Schwarz, G. Hasinger, M. Orio 145

Interpretation of the Long-Term Optical Variations of RX
J0019.8+2156
F. Meyer, E. Meyer-Hofmeister 153

ROSAT Monitoring of the LMC Supersoft Transient Source
RX J0513.9−6951
S.G. Schaeidt 159

Optical Variability of the LMC Supersoft Source
RX J0513.9−6951
K.A. Southwell, M. Livio, P.A. Charles, W. Sutherland, C. Alcock, R.A.
Allsman, D. Alves, T.S. Axelrod, D.P. Bennett, K.H. Cook, K.C. Freeman,
K. Griest, J. Guern, M.J. Lehner, S.L. Marshall, B.A. Peterson, M.R. Pratt,
P.J. Quinn, A.W. Rodgers, C.W. Stubbs, D.L. Welch 165

Optical and X-Ray Variability of Supersoft X-Ray Sources
K. Reinsch, A. van Teeseling, K. Beuermann, H.-C. Thomas 173

Part V   Supersoft Sources as SN Ia Progenitors

Type Ia Supernovae and Supersoft X-Ray Sources
M. Livio 183

Luminous Supersoft X-Ray Sources as Progenitors
of Type Ia Supernovae
R. Di Stefano 193

A New Model for Progenitors of Type Ia Supernovae and Its
Relation to Supersoft X-Ray Sources
I. Hachisu, M. Kato, K. Nomoto 205

Transient and Recurrent Supersoft Sources as Progenitors
of Type Ia Supernovae and of Accretion Induced Collapse
P. Kahabka 215

Part VI   Symbiotics and Other Related Systems

ROSAT Observations of Symbiotic Binaries
and Related Objects
K.F. Bickert, J. Greiner, R.E. Stencel 225

X-Ray Properties of Symbiotic Stars: I. The Supersoft Symbiotic Novae RR Tel and SMC3 (=RX J0048.4–7332)
U. Mürset, S. Jordan, B. Wolff 251

Multiwavelength Observations of the Symbiotic Star AG Dra During 1979–1995
R. Viotti, R. González-Riestra, F. Montagni, J. Mattei, M. Maesano, J. Greiner, M. Friedjung, A. Altamore 259

UV and X-Ray Monitoring of AG Draconis During the 1994/1995 Outbursts
J. Greiner, K. Bickert, R. Luthardt, R. Viotti, A. Altamore, R. González-Riestra 267

A Candidate Isolated Old Neutron Star
R. Neuhäuser, F.M. Walter, S.J. Wolk 279

A Systematic Search for Supersoft X-Ray Sources in the ROSAT All-Sky Survey
J. Greiner 285

A Search for Optical Counterparts to Supersoft X-Ray Sources in the ROSAT Pointed Database
C.M. Becker, R. Remillard, S.A. Rappaport 289

Part VII   Catalog of Supersoft X-Ray Sources

Catalog of Luminous Supersoft X-Ray Sources
J. Greiner 299

Part VIII   Appendix

Subject and Object Index 341
Author Index 349

List of Participants

Barwig, Heinz   `barwig@usm.uni-muenchen.de`
  Universitätssternwarte München, Scheinerstraße 1, 81679 München, Germany
Becker, Christopher   `cmbecker@space.mit.edu`
  Massachusetts Institute of Technology, Center for Space Research, 77 Massachusetts Ave, Rm 37-432, Cambridge, MA 02139, USA
Beuermann, Klaus   `beuermann@usw050.dnet.gwdg.de`
  Universitäts-Sternwarte Göttingen, Geismarlandstraße 11, 37083 Göttingen, Germany
Bickert, Klaus   `kfb@mpe-garching.mpg.de`
  Max-Planck-Institut für Extraterrestrische Physik, Giessenbachstraße, 85740 Garching, Germany
Di Stefano, Rosanne   `rdistefano@cfa.harvard.edu`
  Harvard-Smithsonian Center for Astrophysics, 60 Garden Street, Cambridge, MA 02138, USA
Ebisawa, Ken   `ebisawa@lheavx.gsfc.nasa.gov`
  USRA, NASA/GSFC, code 668, Greenbelt, MD 2020771, USA
Friedjung, Michael   `friedjung@iap.fr`
  Institut d'Astrophysique (CNRS), 98 bis Boulevard Arago, 75014 Paris, France
Gänsicke, Boris   `boris@uni-sw.gwdg.de`
  Universitäts-Sternwarte Göttingen, Geismarlandstraße 11, 37083 Göttingen, Germany
Gonzalez-Riestra, Rosario   `ch@vilspa.esa.es`
  IUE Observatory - VILSPA, P.O. Box 50727, 28080 Madrid, Spain
Greiner, Jochen   `jcg@mpe-garching.mpg.de`
  Max-Planck-Institut für Extraterrestrische Physik, Giessenbachstraße, 85740 Garching, Germany
Haberl, Frank   `fwh@mpe-garching.mpg.de`
  Max-Planck-Institut für Extraterrestrische Physik, Giessenbachstraße, 85740 Garching, Germany
Hachisu, Izumi   `hachisu@chianti.c.u-tokyo.ac.jp`
  Dept. of Earth Science and Astronomy, University of Tokyo, Komaba 3-8-1, Meguro-ku, Tokyo, Japan

Hartmann, Wouter — w.hartmann@sron.ruu.nl
SRON, Sorbonnelaan 2, 3584 CA Utrecht, The Netherlands

van den Heuvel, Ed — edvdh@astro.uva.nl
Astronomical Institute and Center for High Energy Astrophysics, University of Amsterdam, Kruislaan 403, 1098 SJ Amsterdam, The Netherlands

Jordan, Stefan — supas058@astrophysik.uni-kiel.d400.de
Institut für Astronomie und Astrophysik, Universität Kiel, Olshausenstraße 40, 24098 Kiel, Germany

Kahabka, Peter — ptk@astro.uva.nl
Astronomical Institute and Center for High Energy Astrophysics, University of Amsterdam, Kruislaan 403, 1098 SJ Amsterdam, The Netherlands

Kato, Mariko — mariko@educ.cc.keio.ac.jp
Department of Astronomy, Keio University, 4-1-1, Hiyoshi, Kouhoku, Yokohama 223, Japan

Kundt, Wolfgang — wkundt@astro.uni-bonn.de
Institut für Astrophysik, Universität Bonn, Auf dem Hügel 71, 53121 Bonn, Germany

Kylafis, Nikolaos D. — kylafis@iesl.forth.gr
University of Crete, Physics Department, P.O. Box 2208, 71409 Heraklion, Crete, Greece

Livio, Mario — mlivio@stsci.edu
Space Telescope Science Institute, 3700 San Martin Drive, Baltimore, MD 21218, USA

de Martino, Domitilla — ddm@vilspa.esa.es
IUE Observatory - VILSPA, P.O. Box 50727, 28080 Madrid, Spain

Meyer-Hofmeister, Emmi — emm@mpa-garching.mpg.de
Max-Planck-Institut für Astrophysik, Karl-Schwarzschild-Straße 1, 85740 Garching, Germany

Meyer, Friedrich — emm@mpa-garching.mpg.de
Max-Planck-Institut für Astrophysik, Karl-Schwarzschild-Straße 1, 85740 Garching, Germany

Motch, Christian — motch@cdsxb7.u-strasbg.fr
Observatoire de Strasbourg, 11, rue de l'Université, 67000 Strasbourg, France

Mürset, Urs — muerset@astro.phys.ethz.ch
Institut für Astronomie, ETH Zentrum, 8092 Zürich, Switzerland

Neuhäuser, Ralph — rne@mpe-garching.mpg.de
Max-Planck-Institut für Extraterrestrische Physik, Giessenbachstraße, 85740 Garching, Germany

Pakull, Manfred — pakull@cdsxb7.u-strasbg.fr
Observatoire de Strasbourg, 11, rue de l'Université, 67000 Strasbourg, France

Popham, Robert — rpopham@cfa.harvard.edu
Harvard-Smithsonian Center for Astrophysics, 60 Garden Street, Cambridge, MA 02138, USA

Rappaport, Saul A. — sar@mit.edu

Massachusetts Institute of Technology, Center for Space Research, 77 Massachusetts Ave, Rm. 37-551 Cambridge, MA 02139, USA

Rauch, Thomas   supas074@astrophysik.uni-kiel.d400.de
Institut für Astronomie und Astrophysik, Universität Kiel, Olshausenstraße 40, 24098 Kiel, Germany

Reinsch, Klaus   reinsch@usw050.dnet.gwdg.de
Universitäts-Sternwarte Göttingen, Geismarlandstraße 11, 37083 Göttingen, Germany

Ritter, Hans   hsr@mpa-garching.mpg.de
Max-Planck-Institut für Astrophysik, Karl-Schwarzschild-Straße 1, 85740 Garching, Germany

Schaeidt, Stephan   sgs@isow37.vilspa.esa.es
Max-Planck-Institut für Extraterrestrische Physik, Giessenbachstraße, 85740 Garching, Germany

Schandl, Susanne   suh@mpa-garching.mpg.de
Max-Planck-Institut für Astrophysik, Karl-Schwarzschild-Straße 1, 85740 Garching, Germany

Schmidtke, Paul C.   schmidtke@scorpius.la.asu.edu
Arizona State University, Tempe, AZ 85287 - 1504, USA

Southwell, Karen   k.southwell1@physics.oxford.ac.uk
Oxford University, Dept. of Astrophysics, Nuclear Physics Building, Keble Road, Oxford, England OX13RH

Supper, Rodrigo   ros@mpe-garching.mpg.de
Max-Planck-Institut für Extraterrestrische Physik, Giessenbachstraße, 85740 Garching, Germany

van Teeseling, André   andre@uni-sw.gwdg.de
Universitäts-Sternwarte Göttingen, Geismarlandstraße 11, 37083 Göttingen, Germany

Thomas, Hans-Christoph   hcthomas@mpa-garching.mpg.de
Max-Planck-Institut für Astrophysik, Karl-Schwarzschild-Straße 1, 85740 Garching, Germany

Trümper, Joachim   jtrumper@mpe-garching.mpg.de
Max-Planck-Institut für Extraterrestrische Physik, Giessenbachstraße, 85740 Garching, Germany

Viotti, Roberto   uvspace@saturn.ias.fra.cnr.it
Istituto di Astrofisica Spaziale, CNR, Via Enrico Fermi 21, 00044 Frascati RM, Italy

Werner, Klaus   werner@astrophysik.uni-kiel.d400.de
Universität Potsdam, Lehrstuhl Astrophysik, Postfach 601553, 14415 Potsdam

Will, Tobias   will@usm.uni-muenchen.de
Universitätssternwarte München, Scheinerstraße 1, 81679 München, Germany

Part I

Population Properties of SSS

On the Evolution, Numbers and Characteristics of Close-Binary Supersoft Sources

R. Di Stefano[1], L.A. Nelson[2]

[1] Harvard-Smithsonian Center for Astrophysics, Cambridge, MA 02138
[2] Bishop's University, Lennoxville, QC Canada, J1M 1Z7

Abstract. The ability to perform detailed evolutionary calculations is essential to the development of a well-defined and testable binary model. Unfortunately, traditional evolutionary calculations cannot be used to follow a significant fraction of possible close-binary supersoft sources (CBSSs). It is therefore important to examine the input physics carefully, to be sure that all relevant and potentially important physical processes are included. In this paper we continue a line of research begun last year, and explore the role that winds are expected to play in the evolution of CBSSs. We find that at least a subset of the systems that seemed to be candidates for common envelope evolution may survive, if radiation emitted by the white dwarf drives winds from the system. We study the effects of winds on the binary evolution of CBSSs, and compute the number and characteristics of CBSSs expected to be presently active in galaxies such as our own or M31.

1 Close-Binary Supersoft Sources

Close-binary supersoft sources (CBSSs) are binaries containing white dwarfs which can accrete matter from a more massive and possibly slightly evolved companion. Orbital periods during the epoch of supersoft source (SSS) behavior are typically on the order of one day. These systems are characterized by the fact that mass transfer occurs on a time scale closely related to the thermal time scale of the donor. The name "close-binary supersoft source" has been coined to distinguish these systems from other accreting white dwarf systems which may exhibit episodes of SSS behavior (see Tab. 1). The model was developed (van den Heuvel et al. 1992 [vdHBNR]; Rappaport, Di Stefano, & Smith 1994 [RDS]) as a possible description of systems such as CAL 83, CAL 87, and RX J0527.8–6954. (see Greiner et al. 1996, and Greiner, Hasinger & Kahabka (1991) for references.)

There are several features which make the CBSS model an attractive explanation for a subset of the observed luminous supersoft X-ray sources. Central among these is the fact that thermal-time-scale mass transfer from main-sequence or slightly evolved stars yields an accretion rate compatible with the steady burning of accreting hydrogen on the surface of a C-O white dwarf. The associated luminosity, temperature, and orbital period are in the range observed for $\sim 4-6$ of the sources (see Tab. 2). This is significant because no conventional white dwarf model (e.g., cataclysmic variables, or symbiotics) seems compatible with the systems' properties. Although neutron star (see Greiner et al. 1991,

Table 1. Binaries that may appear as luminous supersoft X-ray sources

(1) System Type	(2) Mass Transfer Mechanism	(3) $\dot{M}$ $M_\odot$/yr	(4) P_{orb} days	(5) Steady/ Recurrent	(6) Winds
CVs	mb &/or gr	$<\sim 10^{-8}$	$<\sim 0.2$	R*	nova**
CBSSs	thermal time scale readjustment of donor	$>\sim 10^{-7}$	$\sim 0.2 - 3.0$	S	yes
		$<\sim 10^{-7}$	$\sim 0.2 - \mathcal{O}(10^2)$	R*	nova**
WBSSs	nuclear evolution of donor	$>\sim 10^{-7}$	$\sim 3.0 - \mathcal{O}(10^2)$	S	yes
		$<\sim 10^{-7}$	$\sim 3.0 - \mathcal{O}(10^2)$	R*	yes
Symbiotics (Wind-Driven)	stellar winds from evolved donor	$>\sim 10^{-7}$	$\mathcal{O}(10^2)$	S	yes
		$<\sim 10^{-7}$	$\mathcal{O}(10^2)$	R*	nova**

(1) Here we consider only systems in which hydrogen-rich material is accreting onto the surface of a C-O white dwarf. CV= cataclysmic variable; CBSS=close-binary supersoft source; WBSS=wide-binary supersoft source; wind-driven symbiotics are the only binaries listed here in which the donor does not fill its Roche lobe. WBSSs are discussed further in Di Stefano (1996a,b). (2) The primary mass transfer mechanism is listed: mb=magnetic braking; gr= gravitational radiation. (3) $\dot{M}$ = rate at which mass impinges on the surface of the white dwarf; approximate values are given. (4) P_{orb} = orbital period; approximate values are given. (5) Steady vs Recurrent SSS activity: when $\dot{M}$ is in the correct range (typically $>\sim 10^{-7} M_\odot$/yr) the source will burn nuclear fuel more-or-less steadily. * Recurrent sources: for smaller values of $\dot{M}$, hydrogen will burn sporadically. (6) Winds: "yes" indicates that the system is likely to emit a steady wind; ** "nova" indicates that mass ejection is likely to be primarily associated with nova explosions. Yungelson et al. (YLTTF; 1996) have performed a new population synthesis study that covers all of the binary systems listed in Tab. 1, save for wide-binary supersoft sources (WBSSs), which have been considered as a class more recently (Di Stefano 1996). In fact, YLTTF included in their analysis some systems not listed in Tab. 1, such as helium-accretors and planetary nebulae.

Kylafis & Xilouris 1993, Hughes 1994, Kylafis 1996) and black hole (see, e.g., Cowley et al. 1990) models have been suggested, there is no definitive evidence in favor of them for sources that emit only soft X-radiation. Thus, while keeping an open mind about the nature of the sources, and especially about the possibility that more than one physical model may be required to describe those systems whose nature has not yet been established, we choose to concentrate here on developing concrete, testable signatures of the white dwarf models.

2 The Importance of Evolution

This paper has two major themes. The first is the importance of developing an ability to carry out detailed evolutionary calculations. The second is the role played by winds in helping us to do this.

No binary model can be said to be well-defined or well-developed unless we know how to evolve the individual systems to which it is meant to apply. Yet, the standard approach to binary evolution proves to be problematic for many CBSSs. The reason for this is that the ratio q of the mass of the donor, m, to that of the white dwarf accretor, M, is typically greater than unity. Thus, the donor's Roche lobe tends to shrink during mass transfer. If the donor cannot shrink at least as quickly, the evolution cannot be followed via the standard Roche-lobe-filling approach; in fact there is a real risk that a common envelope might form. Complicating matters is the fact that the donor in CBSSs is often so evolved that it is less able than a main sequence star of the same mass to shrink in response to mass loss. Thus, it appears that we may not be able to follow the evolution of a significant subset of CBSSs. It is therefore necessary to carefully consider whether all of the relevant physics has been included. As we will see in §3, we find that winds may play an important role in the physics and therefore also in the evolution.

The ability to track the evolution of specific systems transforms the conceptual CBSS model into a concrete model that is predictive and testable. It allows us to answer two types of question: (1) which of the observed systems may be realizations of the CBSS model? (2) what are the ranges of properties that a galactic population of CBSSs should be expected to exhibit?

2.1 Testing the Model for Individual Sources

Of the systems for which we have measured values of the orbital period, $P_{\rm orb}$, there are 6 whose properties are roughly consistent with the CBSS model. These are listed in Tab. 2. Of the systems listed, 3 (RX J0513.9–6951, CAL 83, and CAL 87) are strong candidates for the CBSS model. For each of these systems, the uncertainty boxes inferred for the bolometric luminosity and temperature enclose significant area within the steady-burning region computed by Iben (1982). Furthermore, the orbital periods are within the range computed by RDS for the CBSS model. But these circumstances are only weak arguments in favor of the model. To make a stronger argument, the model must make firm specific predictions for each system. Evolutionary calculations will allow us to compute from first principles $\dot{m}$, $\dot{M}$, and $\dot{m}_{ej}$, i.e., the donor's mass loss rate, the accretion rate of the white dwarf, and the rate of mass ejection from the system, respectively. We can then test whether these values are consistent with the observed luminosity and other properties of the system. Evolutionary calculations also allow us

Table 2. Observed Candidates for the Close-Binary Supersoft Model

SSS	P_{orb}	kT^*	L (ergs/s)
RX J0513.9–6951	10.3 hrs	30 – 40 (bb)	$0.1 - 6. \times 10^{38}$
CAL 83	1.04 days	20 – 50 (bb)	$0.6 - 2. \times 10^{37}$
CAL 87	10.6 hrs	65 – 75 (wd)	$1. - 10. \times 10^{38}$
1E 0035.4–7230	4.1 hrs	40 – 50 (wd)	$0.8 - 2. \times 10^{37}$
RX J0019.8+2156	15.8 hrs	25 – 37 (wd)	$3. - 9. \times 10^{36}$
RX J0925.7–4758	3.5 days	45 – 55 (bb)	$1. - 10. \times 10^{37}$

These are the SSSs whose properties are not well-described by other accreting white dwarf models, but which may be roughly consistent with the CBSS model. *The fits used either a pure thermal model (bb) or a white dwarf model atmosphere model (wd). See Greiner (1996) for references.

to post-dict the prior history of each system, so that we can compute quantities such as the total mass ejected by the system throughout its evolution and the average luminosity. These post-dictions can also be checked for consistency with the data. For example, the study of the nebula surrounding CAL 83 seems to lead to a smaller estimate of the time-averaged luminosity over the past $\sim 10^5$ years than the present most-likely value of L (Remillard, Rappaport & Macri 1994). We can determine if evolutionary calculations predict that the average value of L over the past $\sim 10^5$ years is lower than the present value as determined by ROSAT. In general, evolutionary calculations allow us to better assess the likelihood that each observed system fits the CBSS model.

Three of the systems listed in Tab. 2 seem, on the face if it, either because of the value of their luminosity or of P_{orb}, to be less likely members of the CBSS class. In two cases (particularly RX J0019.8+2156 and, to a lesser extent 1E 0035.4–7230), the inferred bolometric luminosities are low when compared to the position of the steady-burning region. But this does not mean that the systems are not CBSSs. It may mean that they are either on their way into or out of the steady-burning region. Computations of their possible evolutionary histories will help us to better constrain their nature.

2.2 Testing the Model for the Total Population: Past Work

RDS attempted to create, via a population synthesis analysis, the total population of close-binary supersoft sources that should be expected in the disk of our own Galaxy, or in other spirals, such as M31. They computed the properties of all sources whose luminosity and temperature fell squarely within the steady-burning region computed by Iben (1982). Their approach was conservative, and led to a prediction of ~ 1000 presently active CBSSs in the Galaxy. Later work by Di Stefano & Rappaport (1994) established that interstellar absorption would be expected to shield the vast majority of these systems from our view, so that the small number of sources observed in our Galaxy (~ 6) and in M31 (~ 15) is consistent with the large population computed by RDS.

Recently, a more comprehensive population synthesis study has been carried out by YLTTF. The new work is more complete in several ways. First, it includes all classes of SSS, except for wide-binary supersoft sources. Second, YLTTF compute the properties of CBSSs when the systems are not in the steady-burning regime, as well as when they are in it. YLTTF derive numbers that are similar to those computed by RDS. Given the uncertainties inherent in the calculations carried out by both groups, the ranges of numbers they compute are compatible. Since the two groups used somewhat different methods, the compatibility of their results seems to indicate that the results are robust.

Both RDS and YLTTF were hindered by the fact that complete evolutionary tracks could not be computed for a significant fraction of the systems created in their simulated galaxies. The two groups handled this difficulty in different ways, but each may be described as being appropriately conservative, in that they did not make specific predictions for the presence or observability of sources for which the standard evolutionary formalism was likely to fail.

3 A New Population Synthesis Study

Our goal is to perform a population synthesis to track the evolution of all systems which do not experience common envelopes during the phase in which mass is being transferred to the white dwarf. We differ from the previous population synthesis studies of CBSSs primarily in our approach to evolution. We include three new features. These are described sequentially in sections 3.1, 3.2, and 3.3. The evolutionary equations are presented in §3.4.

3.1 The Response of the Donor to Mass Loss

We specifically track, in each time step, the response of the donor to mass loss. In general, this response is described by the adiabatic index ξ_{ad}: $\xi_{ad} = d[log(r)]/d[log(m)]$, where r is the radius of the donor, and m is its mass. ξ_{ad} can be estimated through comparisons with Henyey-like calculations. We find that a formula:

$$\xi_{ad} = \tilde{\xi}_{ad} \left[1 - \left(\frac{m_c}{\tilde{m}_c} \right)^2 \right], \qquad (1)$$

fits the numerical data, with the preferred values for $\tilde{\xi}_{ad}$ and $\tilde{m}_c$ equal to 4.0 and 0.2, respectively. Technically, $\tilde{m}_c$ depends on the value of m, but its dependence is relatively weak. Since the most massive cores found among donors in CBSSs are close to $0.2 M_\odot$, ξ_{ad} can be small. Thus, CBSSs are at high risk for having the Roche-lobe-filling formalism break down.

In addition to this direct response, the donor may also be required to respond to the shrinking of its Roche lobe. In particular, if the donor can shrink to keep up with its Roche lobe, its radius may become smaller than the equilibrium radius of a star of the same mass and state of evolution. The star would therefore like to expand to achieve its equilibrium radius, r_{eq}, but can do so only on a thermal time scale. This thermal-time-scale push toward equilibrium introduces a further change in radius. We have

$$\frac{\dot{r}_{th}}{r} = \frac{r_{eq} - r}{f \, \tau_{KH}}, \qquad (2)$$

where τ_{KH} is the donor's Kelvin-Helmholz time, and the value of f is determined by fitting to the results of Henyey-like calculations.

3.2 Mass Ejection by the White Dwarf

The white dwarf can retain only a fraction, β, of the total mass that falls onto its surface. That is, if the rate of accretion is $\dot{M}$, then only $\beta \dot{M}$ is burned and retained by the white dwarf. Because the accretor is a white dwarf, the study of steady nuclear burning provides some useful guidelines for the value of β (see, for example, Paczyński 1970; Sion, Acierno & Tomcszyk 1979; Taam 1980; Nomoto 1982; Iben 1982; Fujimoto 1982; Fujimoto and Sugimoto 1982; Fujimoto & Taam 1982; Fujimoto and Truran 1982; Sion & Starrfield 1986; Livio, Prialnik & Regev 1989; Prialnik & Kovetz 1995). For each value of the mass (M) of the white dwarf, there is a range $\dot{M}_{min} < \dot{M} < \dot{M}_{max}$ within which matter that is accreted can burn more-or-less steadily as it accretes.

For $\dot{M} < \dot{M}_{min}$, matter accumulates and burns explosively only after a critical amount of mass has been accreted. There are open questions about how much mass is ejected, but we find that these do not affect the numbers of sources dramatically. For $\dot{M}$ within the steady nuclear burning range, it is, in principle, possible that all of the incident mass can be burned. It was, however, pointed out that the start of nuclear burning is expected to turn on a wind, decreasing the maximum possible value of β to something less than unity (Livio 1995). The case that has been least-well understood is that in which $\dot{M}$ is greater than $\dot{M}_{max}$. The key question is whether accreting matter in excess of what can be burned accumulates in an envelope around the white dwarf, or whether it can be ejected "in real time". In the former case, the system risks undergoing a common envelope phase, unless the formation of a giant-like envelope can somehow turn off mass accretion when the white dwarf fills its own Roche lobe. Di Stefano et al. (1996b; DNLWR) made a case that, for some systems, the binary evolution could be stabilized if the excess matter can be ejected. The basic physics behind the argument is that (for $m > M$) the loss of mass from the system tends to allow the stars to move further apart and moderates the shrinkage of the Roche lobe, particularly if not much angular momentum is carried away by matter which leaves the system.

DNLWR showed that reasonable energy considerations alone would allow as much as $\sim 99\%$ of the incident matter to be ejected, if the rest of it was burned. Since then, Hachisu, Kato and Nomoto (1996) have verified that there are steady nuclear burning solutions in which excess matter is ejected in a steady wind. This justifies choosing the functional form of β to fall as $\sim \dot{M}_{\max}/\dot{M}$ as $\dot{M}$ increases significantly above $\dot{M}_{\max}$. In the calculations used to derive the results presented in this paper, we have used the following prescription for β.

$$\dot{M} < \dot{M}_{\min} : \quad \beta = \left(\frac{\dot{M}}{\dot{M}_{\min}}\right)^{a_1}; \tag{3}$$

$$\dot{M}_{\min} < \dot{M} < \dot{M}_{\max} : \quad \beta = \beta_{\max} \tag{4}$$

$$\dot{M} > \dot{M}_{\max} : \quad \beta = \left(\frac{\dot{M}_{\max}}{\dot{M}}\right)^{a_2} \tag{5}$$

For our "standard" runs we chose $a_1 = 2$ and $a_2 = 1$. We also chose $\beta_{\max}$ to be unity; this is a conservative choice, almost certainly leading to an underestimate of the number of CBSSs which survive.

3.3 Radiation-Driven Winds

Because the white dwarf is the less massive of the companions, matter that leaves the system from its vicinity carries a large specific angular momentum. This is why radiation-driven winds, which can exit the system with smaller specific angular momentum, may play an important role. Typically the Roche-lobe geometry dictates that $\sim 5\%$ of the radiation emitted by the white dwarf strikes the donor. This generally represents $\mathcal{O}(100)$ times as much energy as is typically emitted by the donor. The radiation is soft, and does not penetrate deeply. It is therefore likely to generate a wind. Let $\dot{m}_{rd}$ represent the rate of mass loss due to a radiation-driven wind. In reality, only part of such a wind is likely to be emitted directly from the donor's surface, and part will be ejected from the highly irradiated accretion disk (see Popham & Di Stefano 1996). Thus, the angular momentum carried by the wind may be greater than that appropriate to the donor. Therefore, although in the work described here we will assume that the wind emanates from the donor, in ongoing work we relax this assumption.

3.4 The Evolutionary Equations

The Roche-lobe-filling condition, in combination with the principle of conservation of angular momentum leads to an expression for $\dot{m}$, the donor's total mass loss rate.

$$\dot{m}\mathcal{D} = \mathcal{N}. \tag{6}$$

In Eq. (6), $\mathcal{N} = m\left[\frac{\dot{J}_{dis}}{J} - \frac{1}{2}\frac{\dot{r}}{r}\right]$; $\dot{J}_{dis}$ is the rate at which angular momentum is drained from the system through the processes of gravitational radiation and magnetic braking, and $\dot{r}$ is the sum of the thermal term given by Eq. (2) and the donor's rate of change of radius due to nuclear burning. $\mathcal{D}$ is given by $\mathcal{A} + \beta \mathcal{B}$, and

$$\mathcal{A} = \left(1 + \frac{\xi_{ad}}{2}\right) - \frac{q}{2}\left\{\frac{1}{(1+q)} + \frac{f'}{f}\right\} - \frac{[\mathcal{F}q^2 + (1-\mathcal{F})]}{(1+q)} \tag{7}$$

$$\mathcal{B} = \mathcal{F}\left\{-q + \frac{q}{2(1+q)} - \frac{q^2}{2}\frac{f'}{f} + \frac{q^2}{(1+q)}\right\} \tag{8}$$

In these equations $\mathcal{F}$ is the fraction of the mass leaving the donor that is incident on the white dwarf; $\dot{M}$, the mass accretion rate of the white dwarf, is $\beta\mathcal{F}\dot{m}$. A generalization of these equations is considered in Di Stefano 1996b.

4 Number of CBSSs in the Steady-Burning Region

To compute the numbers of CBSSs in the steady-burning region, we have performed a set of Monte-Carlo simulations. Each simulation was characterized by a different set of assumptions about the value of the common envelope ejection parameter, α, and the distribution of properties among primordial binaries. Each simulation yielded a set of systems that might be expected to pass through a phase of mass transfer as a CBSS; we referred to these systems as CBSS candidates. We tracked the evolution of each CBSS candidate, and recorded its properties during the time it spent in the steady nuclear burning region. Because of uncertainties about how to perform the evolutions, each separate set of CBSS candidates was actually subjected to a variety of "treatments" in which different assumptions were made about the evolution (e.g., was there a radiation-driven wind? how efficiently could energy incident from the white dwarf be used to drive such a wind? etc.) Our preliminary results are shown in Tab. 3.

These results can be roughly understood from Fig. 1. The figure indicates that more systems can be evolved if there is a radiation-driven wind. Although this might lead us to think that there could be a significant increase in the number of systems in the steady-burning region, it is also the case that radiation-driven winds tend to moderate the mass accretion rate. Thus, some systems that might have been steady nuclear burners will not now enter the steady nuclear burning region; others may spend less time there. The overall number of SSSs that should be active in a galaxy such as our own at any one time will therefore not be radically different in calculations which include the effects of winds.

5 Properties of CBSSs in the Steady-Burning Region

Although the numbers of systems are not very different from what has been found in previous investigations, there are differences in the distribution of system properties. The most profound difference is that we find that winds will be a prominent feature of many CBSSs. Winds were not included in either the early (RDS) or more recent (YLTTF) population synthesis studies of these systems. Indeed since the donors have evolved to, at most, the base of the giant branch

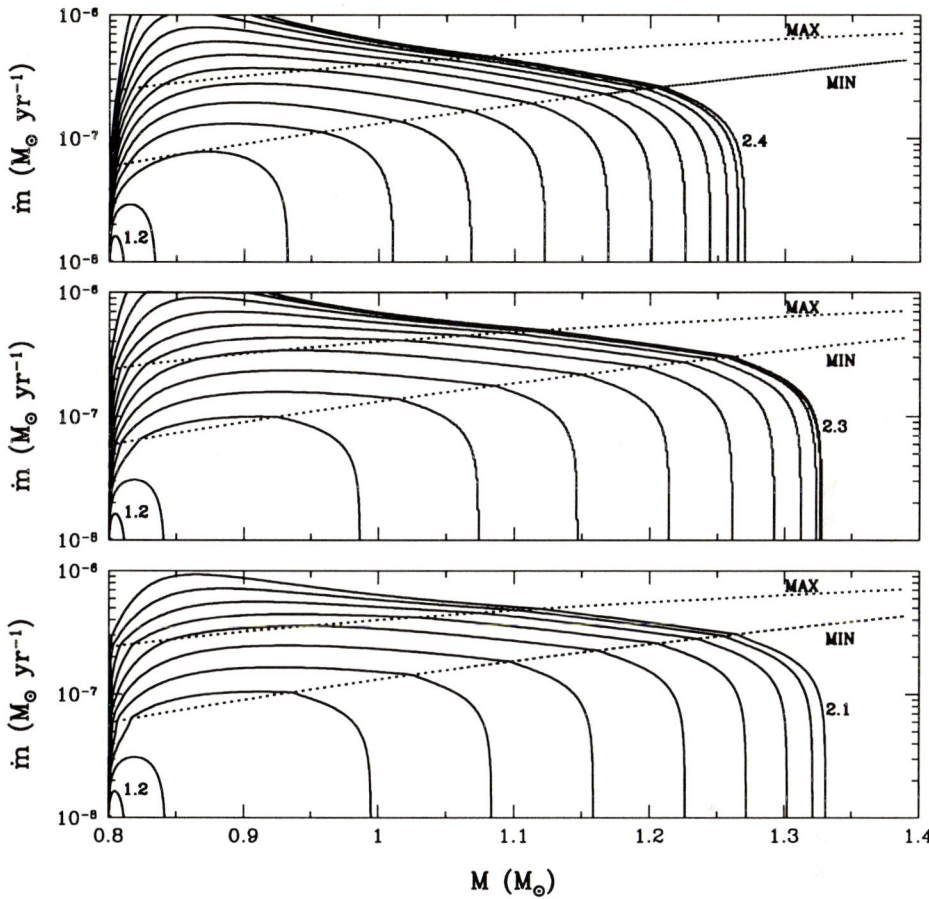

Fig. 1. The mass of the white dwarf is plotted along the horizontal axis; the rate at which mass is falling onto the surface of the white dwarf is plotted along the vertical axis. In all of the evolutionary tracks shown, the initial white-dwarf mass is taken to be $0.8 M_\odot$, and the donor starts on the TAMS. Within each panel, each curve is characterized by the initial value of the donor's mass, $m_d(0)$; the bottom curve has $m_d(0)$ marked; $m_d(0)$ is incremented by 0.1 in each subsequent curve. The value of $m_d(0)$ associated with the upper curve of each panel is the largest value in the sequence for which $\mathcal{D} > 0$ throughout the evolution. No radiation-driven winds were included in the evolutions shown in the bottom panel; radiation-driven winds, with an efficiency factor of 0.001 (0.01) were included in the middle (top) panel. Note that low to moderate efficiency can be expected to maximize both the number of active sources and the number of possible supernovae. As the wind efficiency increases, only slightly more systems can be evolved, but the mass transfer rates tend to be lower than would be necessary for significant mass accretion to occur.

Table 3. Numbers of CBSSs in the Steady-Burning Region

$Treatment^*$	1	2	3	4	5
Data Set 1**					
$f = 0$	~ 2400	~ 3200	~ 2600	~ 2900	~ 2500
$f = 0.001$	~ 1400	~ 1700	~ 1200	~ 4000	–
Data Set 2					
$f = 0.0001$	–	~ 1200	–	–	–
$f = 0.001$	–	~ 1500	–	–	–
$f = 0.01$	~ 500	~ 1500	~ 4400	~ 2500	–

* "Treatment" refers to the values chosen for $\tilde{\xi}_{ad}$, a_1, and a_2. Treatment 2 is our "standard" treatment: $\tilde{\xi}_{ad} = 4.0$, $a_1 = 2$, $a_2 = 1$. Treatment 1 is more conservative in that $\tilde{\xi}_{ad} = 2.0$. Treatment 3 tends to increase the number of system that can be evolved, since the specific angular momentum of all mass leaving the system is chosen to be that appropriate to the donor; however, it also moderates the mass transfer rate, so that some systems that would otherwise be steady nuclear burners do not enter the steady nuclear burning region. In treatment 4, $a_1 = 1$; this allows for more mass to be accreted below the steady nuclear burning region. Treatment 5 is an "optimal treatment; the same parameters are used as for the standard treatment, but $\mathcal{D}$ is set to a minimum value whenever the computed value dips below zero. ** Data set 2 is our "standard" data set; the population synthesis calculation that gave rise to it had α, the common envelope efficiency factor, set equal to 0.8. To derive data set 1, α was taken to be 0.1. The wind efficiency factor, f, is a constant of proportionality that enters into the relationship between $\dot{M}$ and the rate at which matter is lost through the radiation-driven wind. Details can be found in DN; here, the values can be viewed as providing a relative measure of the efficiency of producing the wind; there is no radiation-driven wind for $f = 0$.

during the phase of most-active mass transfer, they would not be expected to emit a large wind if they existed in isolation. However, the white dwarf's ejection of matter that it cannot burn in "real time", coupled with radiation-driven winds, will lead many CBSSs to eject a significant wind. Thus, systems in which there has been significant mass ejection may, in essence, shield themselves from our view. YLTTF have considered this process for wind-driven symbiotics. Investigations of the distribution of properties among the systems we expect to be able to detect must therefore explicitly consider winds ejected from the system, and their ability to shield active systems from our view. Such investigations are underway.

6 Conclusions

It has long been known that the evolution of a binary in which the donor is more massive than the accretor, and also possibly slightly evolved, can be problematic. ROSAT's discovery of the class of luminous supersoft X-ray sources, and the subsequent proposal of the close-binary supersoft (CBSS) model, has forced us to face the difficulties associated with such potentially problematic evolutions. Indeed the motivation for doing so is strong, since detailed evolutionary calculations are required to shape the CBSS model into a well-defined theory which is testable both for individual observed sources and for the range of characteristics one should expect among a galactic population of sources.

If nature has thus presented us with a problem to solve, it has also been kind in choosing to pose the problem in a venue that contains some obvious clues to its resolution. The fact that the accretor is a white dwarf is important, because the physics of nuclear burning places useful constraints on the value of the mass retention factor, β, as a function of the white dwarf mass and accretion rate. Further, if, as recent work suggests, β can become small when the accretion rate exceeds the rate compatible with steady nuclear burning, this helps to stabilize the evolution. The white dwarf nature of the accretor also leads to radiation-driven winds. Indeed, an important feature of the work described here, which will be presented in more detail elsewhere (Di Stefano 1996b; Di Stefano & Nelson 1996), is the inclusion of radiation-driven winds. It is physically reasonable to expect such winds because of (1) the tremendous energy associated with nuclear burning, and (2) the fact that the radiation emitted by the white dwarf is so soft that it cannot penetrate beyond the outer layers of the donor. Mass ejection through radiation-driven winds can help to stabilize the binary evolution. Furthermore, winds are a feature in some of the observed systems which may be close-binary supersoft sources (see Greiner 1996 for references).

We find that, across a broad range of assumptions about the evolution of the binary systems both prior to and during any CBSS phase, the number of steady-nuclear burning sources is in the range of $\sim 1000-4000$ presently active systems. The inclusion of radiation-driven winds does not have a dramatic effect on the number of presently active close-binary supersoft sources. Some systems that would have entered the steady-burning region if winds had not been included, will not do so now; others, which could not be evolved before, will live in the steady-burning region during part of their active mass-transfer phase. What is perhaps more interesting are some subtle shifts in the distribution of system properties; we are presently investigating these. An important observational consequence of the work presented here is that it predicts that winds are likely to be ejected from many, perhaps most, CBSSs during their transit across the steady-burning region. Further, the total amount of mass ejected from the system from the earliest stages of mass transfer up until the present can be calculated.

Results that go beyond those presented here, both in the extent of the parameter space explored and in the analysis of the distribution of system properties, will be described in DN.

Acknowledgement: We would like to thank Scott Kenyon, Saul Rappaport, and J. Craig Wheeler for discussions, and Trevor Wood for assistance with calculations performed during the early phases of the work. This work has been supported in part by NSF under GER-9450087 and by the NSERC (Canada).

References

Cowley A.P., Schmidke P.C., Crampton D., Hutchings J.B., 1990, ApJ 350, 288
Di Stefano R., 1996a, this volume p. 193
Di Stefano R., 1996b (in prep.)
Di Stefano R., Rappaport S., 1994, ApJ 437, 733
Di Stefano R., Paerels F., Rappaport S., 1995, ApJ 450, 705
Di Stefano R., Nelson L.A., Lee W., Wood T., Rappaport S., 1996, in Thermonuclear Supernovae, Proc. of the NATO ASI workshop, eds. R. Canal, P. Lapuente-Ruiz (in press)
Di Stefano R., Nelson L.A., 1996 (DN; in prep.)
Fujimoto M.Y., 1982, ApJ 257, 752
Fujimoto M.Y., Sugimoto D., 1982, ApJ 257, 291
Fujimoto M.Y., Truran J.W., 1982, ApJ 257, 303
Fujimoto M.Y., Taam R., 1982, ApJ 260, 249
Greiner J., 1996, this volume p. 299
Greiner J., Hasinger G., Kahabka P., 1991, A&A 246L, 17
Greiner J., Schwarz R., Hasinger G., Orio M., 1996, this volume p. 145
Hachisu I., Kato M., Nomoto K., 1996, this volume p. 205
Hughes J.P., 1994, ApJ 427L, 25
Iben I. Jr., 1982, ApJ 259, 244
Iben I. Jr., Nomoto K., Tornambe A., Tutukov A.V., 1987, ApJ 313, 727
Kato M., Saio H., Hachisu I., 1989, ApJ 340, 509
Kenyon S.J., Livio M., Mikolajewska J., Tout C.A., 1993, ApJ 407, 81
Kirshner, R.P., Winkler P.F., Chevalier R.A., 1987, ApJ 315, L135
Kylafis N.D., Xilouris E.M., 1993, A&A 278, L43
Kylafis N.D., 1996, this volume p. 41
Livio M., 1995, in Millisecond Pulsars: A Decade of Surprise, eds. A.S. Fruchter, M. Tavani, D.C. Backer (ASP: San Fransisco), p. 72
Livio M., Prialnik D., Regev O., 1989, ApJ 341, 299
Nomoto K., 1982, ApJ 253, 798
Nomoto K., 1996 (priv. comm.)
Paczyński B., 1970, Acta Astr. 20, 287
Popham R., Di Stefano R., 1996, this volume p. 65
Prialnik D., Livio M., 1995, PASP 107, 1201
Prialnik D., Kovetz A., 1995, ApJ 445, 789
Rappaport S., Di Stefano R., Smith J., 1994, ApJ 426, 692
Remillard R., Rappaport S., Macri L., 1994, ApJ 439, 646
Shara M., Prialnik D., 1994, AJ, 107, 1542
Sion E.M., Acierno M.J., Tomcszyk S., 1979, ApJ 230, 832
Sion E.M., Starrfield S.G., 1986, ApJ 303, 130
Smith C.R., Kirshner R.P., Blair W.P, Winkler P.F., 1991, ApJ 375, 652
Taam R.E., 1980, ApJ 242, 749
Tutukov A.V., Yungelson L, Iben I. Jr., 1992, ApJ 384, 580
van den Heuvel E.P.J., Bhattacharya D., Nomoto K., Rappaport S.A., 1992, A&A 262, 97

White Dwarfs with H/He Burning as Supersoft X-Ray Sources

M. Kato

Keio University, Kouhoku-ku, Yokohama, 223 Japan

Abstract. In order to examine the possibility of white dwarfs with surface nuclear burning to be a supersoft X-ray source (SSS), evolution of white dwarfs is followed using the optically thick wind theory. The phenomena examined here are (1) planetary nebular nuclei, (2) the last He shell flash of a star in its post-AGB stage, (3) nova outburst, and (4) He nova (helium shell flash on a white dwarf). All of these objects evolve horizontally leftward in the H-R diagram and reach the stage of strong EUV or soft X-ray emissions. Theoretical light curves in supersoft X-ray, EUV, UV and optical are given, as well as the evolutional time scale and the lifetime as SSS.

1 Introduction

White dwarfs with surface nuclear burning are candidates for supersoft X-ray sources. They follow a horizontal path in the H-R diagram leftward and reach the hot luminous stage before getting dark. There are four types of white dwarfs with surface nuclear burning: planetary nebular nuclei, last He shell flash in a star just evolved into the white dwarf region from red giant, nova and He nova (helium shell flash on a white dwarf). All of these objects evolve horizontally leftward in the H-R diagram and reach the stage of strong EUV or soft X-ray emissions. The wind mass loss commonly occurs in these horizontal paths and the evolutional time scale are obtained using the optically thick wind theory.

Optically thick wind is a continuum-radiation wind in which the acceleration occurs deep inside to the photosphere. The structure of the envelope is obtained by solving the equations of motion, continuity, radiative transfer, and energy conservation. The equations and the numerical method are summarized in detail by Kato & Hachisu (1994).

The evolution of the star is followed by a sequence of the static and steady wind solutions. The evolutional time scale is drastically shortened by the appearance of the new opacity (Rogers & Iglesias 1992, Iglesias & Rogers 1993, Iglesias & Rogers 1996). This new opacity has a large enhancement at the temperature of $\log T$ (K) ~ 5.2 that causes strong acceleration of the winds. The optically thick wind occurs in a wide range of the parameters of the white dwarf mass and the surface temperature, of which large mass loss rate shortens the evolutional time-scale in the H-R diagram.

This paper will present the time scale and the light curves for these four types of stars as well as the lifetime as a supersoft X-ray source.

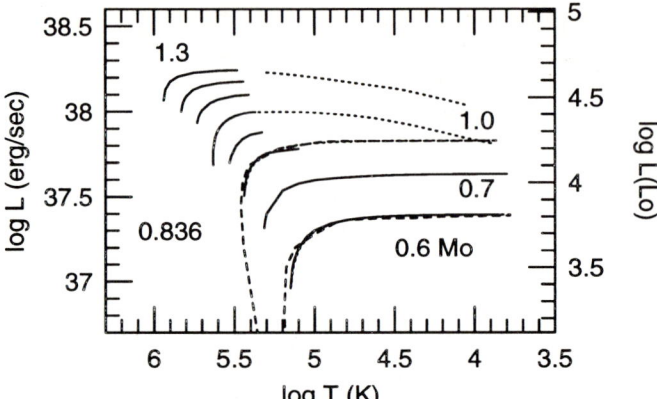

Fig. 1. Evolutional track in the H-R diagram for the central star of planetary nebulae. Optically thick wind occurs in the dotted region. The white dwarf mass is attached to each curve. Dashed curves for 0.6 and 0.836 M$_\odot$ denote the evolutional track obtained by Blöcker & Schönberner (1990) with the old opacity. Dotted curves for 0.6 and 0.836 M$_\odot$ denote the solutions calculated using optically thick wind theory with the old opacity; they are very close to the dotted curve of 0.836 M$_\odot$ and to the solid curve of 0.6 M$_\odot$.

2 Planetary Nebular Nucleus

The evolutional track in the H-R diagram is calculated using the optically thick wind theory. The chemical composition of the envelope is assumed to be uniform with $X=0.7$ and $Z=0.02$ for hydrogen and heavy elements, respectively. The radius of the bottom of the hydrogen-rich envelope is assumed to be that of the hot core in asymptotic giant branch stars (Iben 1982: radius of the center of hydrogen-burning shell in Fig. 1).

Figure 1 shows the evolutional path of planetary nebula nuclei with various core mass. Optically thick wind mass-loss occurs in massive stars (> 0.7 M$_\odot$) in the dotted regions whereas no wind occurs in 0.6 and 0.7 M$_\odot$. Optically thick wind stops at the point of the left edge of the dotted part and the hydrogen nuclear burning extinguishes at the left side of the each track.

The wind mass loss rate strongly depends on the surface temperature and weakly depends on the white dwarf mass; the mass loss rate is obtained as large as 10^{-4}M$_\odot$yr^{-1} for low temperature models ($\log T \sim 4.0$) and decreases to less than 10^{-6}M$_\odot$yr^{-1} in the models just before the wind ceases.

The duration time scales for each track from $\log T(K)=4.0$ to the hydrogen extinguish point are summarized in the third column of Tab. 1. In massive stars the time scale is much reduced due to the strong mass loss and small envelope mass. In less massive stars where no wind occurs the time scale is also shortened due to the effects of the new opacity. The large peak of OPAL opacity reduces envelope mass to 5.3×10^{-4}M$_\odot$ from 8.3×10^{-4}M$_\odot$ of the solution with the old

Table 1. Time scale of evolution, period as supersoft X-ray source

Object	core mass ($M_\odot$)	total duration (yrs)	wind phase[1] (yrs)	SSS ($F_X > -7$) (yrs)	life ($F_X > -8$) (yrs)	time[2] ($F_X > -9$) (yrs)
PNN						
	0.6	3600	no wind	—	—	—
	0.7	700	no wind	—	90	330
	0.836[3]	~140	~40	—	100	120
	1.0	28	7.7	21	24	26
	1.3	1.3	0.63	0.9	1.0	1.1
FINAL He SHELL FLASH						
	0.6	3900	no wind	—	1100	1700
	1.3		8.4	>0.5	>1.2	>2.3
	1.377		4.0	>0.4	>0.8	>1.4
NOVA						
(classical)	0.6	12.8	3.9	8.5	10	11
	1.0	1.3	0.61	0.8	0.9	1.0
	1.33	0.13	0.11	17 d	26 d	32 d
(reccurent)	1.377	0.17	0.14	27 d	38 d	45 d
(slow)[4]	0.8	>22	>5.8	>15	>19	>20
He NOVA						
	1.3		6.6	> 0.5	>1.2	> 2
	1.377		2.2	>0.2	> 0.4	>0.7

[1] Duration from $\log T = 4.0$ to the point where the wind stops.
[2] $F = \log(L_X/4\pi D^2)$, D=1.0 kpc
[3] Accurate value cannot be obtained due to numerical difficulties.
[4] Minimum value of the time scale; see text.

opacity with the same surface temperature of $\log T=4.0$. This effect is small in high surface temperature models because most of the envelope mass lies in the higher temperature side to the opacity peak ($\log T \sim 5.2$).

Therefore the evolutional time scale for a 0.6 $M_\odot$ star to evolve from $\log T=4.0$ to 4.9 is reduced to 3000 years from 4900 years of the models with the old opacity. The old duration time 4900 yrs is consistent with Blöcker & Schönberner's (1990) value of 4000 yrs from $\log T = 3.78$ to 4.87 for a 0.605 $M_\odot$ star obtained with the old opacity, considering their assumption of the given wind mass-loss rate whereas no wind in our case. Their track in the H-R diagram is well reproduced by our solutions with the old opacity as well as those with OPAL opacity as shown in Fig. 1.

Note that the updated OPAL opacity (Iglesias & Rogers 1996) is used to calculate the massive star models ($> 0.836 M_\odot$) in Fig. 1 whereas the first version of OPAL opacity is used for all the other models in this paper. The effect is relatively small; the envelope mass decreases by 8 percent in case of 0.6 $M_\odot$ planetary nebular nuclei with $\log T=4.0$, which reduces the evolutional time scale

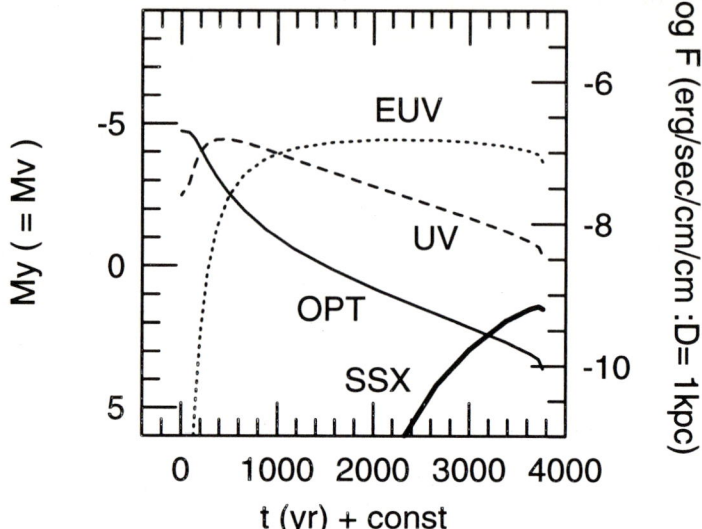

Fig. 2. Light curves for a planetary nebula nucleus of mass 0.6 $M_\odot$. No optically thick wind occurs. The photospheric temperature is $\log T = 4.0$ at $t = 142$ yrs and hydrogen extinguishes at $t = 3760$ yrs. OPT:visual, UV:ultraviolet (912-3250 Å), EUV: extreme ultraviolet (100-912 Å), SSX: supersoft X-ray (30-100 Å).

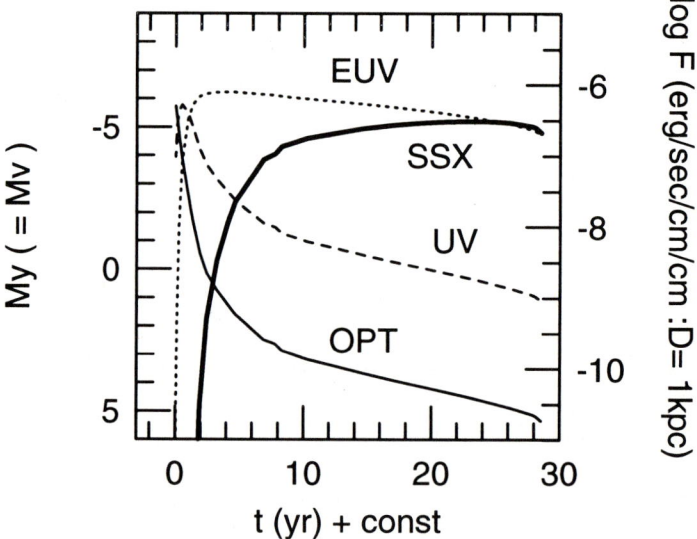

Fig. 3. Same as in Fig. 2 but of core mass 1.0 $M_\odot$. The optically thick wind ceases at $t = 7.8$ yr. The starting point of the light curve $t = 0$ corresponds the stage of photospheric temperature $\log T = 3.9$. Hydrogen burning extinguishes at the end of the curve ($t = 28.6$ yrs).

correspondingly by the similar amount. The basic properties of the occurrence of optically thick winds are unchanged.

The light curves of various wavelength band for 0.6 and 1.0 $M_\odot$ are shown in Figs. 2 and 3, respectively. As the optical magnitude drops earlier, fluxes of short wavelength radiation increase. In most of its life time, the planetary nebula nuclei will be observed as a very bright EUV source with a faint optical counterpart.

In case of 0.6 $M_\odot$, the supersoft X-ray flux is not strong because of the low maximum surface temperature (log $T = 5.15$). In very massive stars, it reaches $F \geq -7$, where $F = \log(L_X/4\pi D^2)$ with the distance to the star D temporary assumed to be 1.0 kpc. The duration of the star being a supersoft X-ray source is summarized in Tab. 1.

3 Final He Shell Flash

A star can undergo a final thermal pulse (He shell flash) at the white dwarf cooling track in the H-R diagram (Iben et al. 1983). Once the He shell flash occurs, the star swells briefly to red giant dimensions. The evolutional track of such a star is recalculated using the OPAL opacity. The chemical composition of the envelope is assumed to be $Y=0.98$ and $Z=0.02$, and the radius of the core is taken from Iben (1982; Edge of CO core in AGB star in Fig. 1). The residual hydrogen above the He layer is neglected partly because of the wind mass loss and of the mixing into the He layer by convection (Iben 1982). The evolutional time scale of final He shell flash is comparable to that of the planetary nebulae nuclei as shown in Table 1. Optically thick wind occurs in massive stars ($\geq 0.8 M_\odot$). The light curves of He shell flash on 0.6 and 1.3 $M_\odot$ white dwarfs are shown in Figs. 4 and 5, respectively.

4 Nova: H Shell Flash on a White Dwarf

Nova occurs on an accreting white dwarf in a close binary system. Theoretical light curves of nova and evolutional courses in the H-R diagram are calculated for the wide parameter range of the white dwarf mass (Kato & Hachisu 1994). The massive white dwarf shows rapid decline in its light curve whereas less massive star shows slow decline. Light curve fitting between theoretical and observational curves gives an estimate of the white dwarf mass. For example, in case of classical novae, in which the heavy element abundance is often observed in ejecta, light curve of Nova Muscae 1983 is fitted by the theoretical curves of mass 0.5 - 0.6 $M_\odot$ (Kato 1995a), and the moderately fast nova, nova Cygni 1978 is fitted by the curve of 1.0 $M_\odot$ (Kato & Hachisu 1994).

The light curves of 1.0 $M_\odot$ white dwarf are shown in Fig. 6. The composition of the envelope is assumed to be $X=0.35$, $Y=0.33$, $C=0.1$, and $O=0.2$. (for the details of the light curve fitting of Nova Cyg 1978, see Kato & Hachisu 1994). This

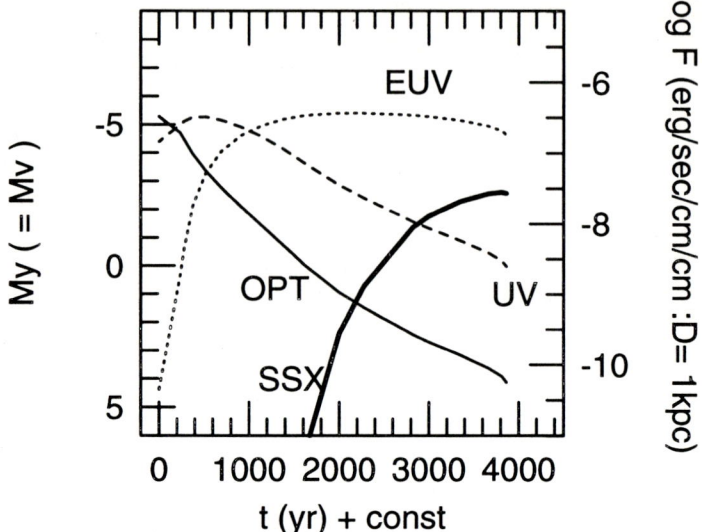

Fig. 4. Light curves of the final He shell flash on a 0.6 $M_\odot$ star. No optically thick wind occurs. At $t = 0$ the photospheric temperature is $\log T = 4.0$. Hydrogen burning extinguishes at $t = 3860$ yr.

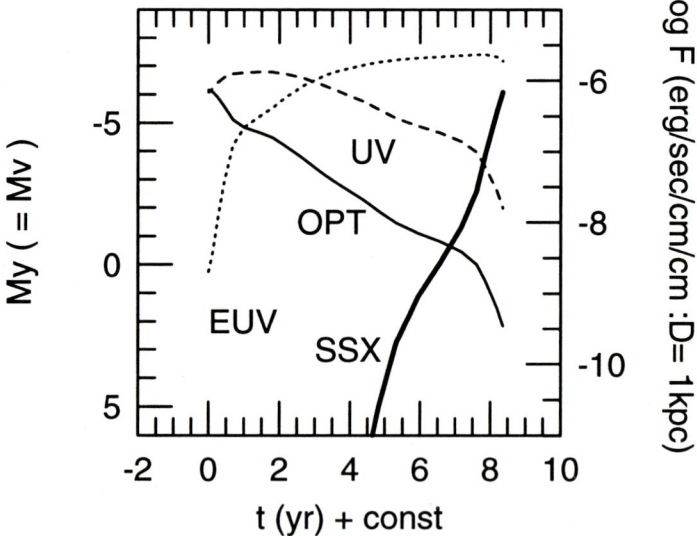

Fig. 5. Same as to Fig. 4 but for 1.3 $M_\odot$ star. At the starting point of the light curve $t = 0$ the photospheric temperature is $\log T = 4.09$. Optically thick wind ceases at $t = 8.36$ yr.

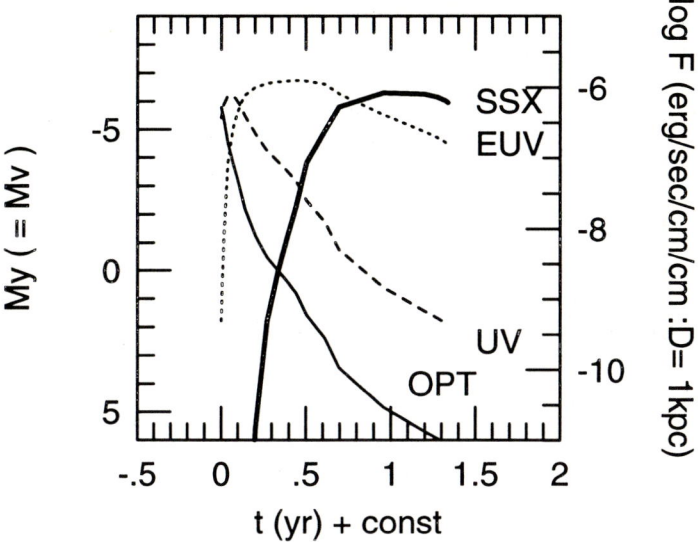

Fig. 6. Theoretical light curve of a classical nova with mass 1.0 $M_\odot$. At $t = 0$ the photospheric temperature is log $T = 4.06$. Optically thick wind ceases at $t = 0.61$ yr. Hydrogen burning extinguishes at $t = 1.34$ yr.

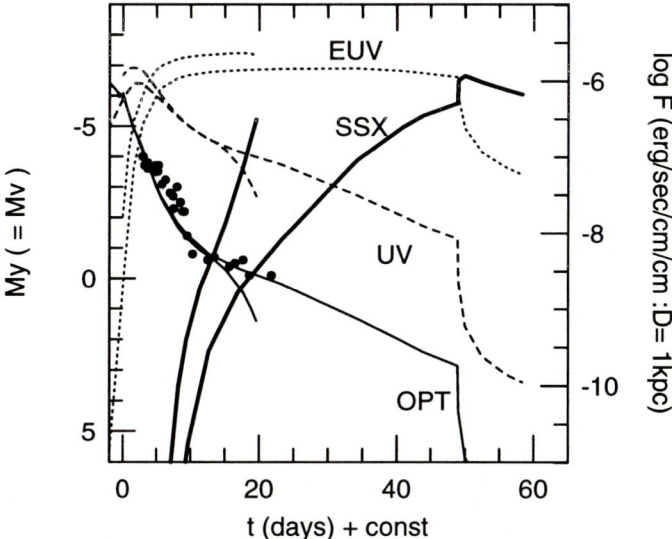

Fig. 7. Theoretical light curves of a recurrent nova with mass 1.377 $M_\odot$. The observational data by Sekiguchi et. al 1988 are denoted by dots. The two sets of the models are shown with different chemical composition; $X = 0.7$ and $Z = 0.02$ (curves finish at $t=58.3$) and $X = 0.1$ and $Z = 0.02$ (finish at $t=19.6$). In the model with $X=0.7$ the photospheric temperature at $t=0$ is log $T = 3.93$. Optically thick wind ceases at the point denoted by dot at $t = 43.9$ days. Hydrogen burning extinguishes at $t = 58.3$ days. In the model with $X=0.1$, the photospheric temperature at $t= 0$ is log $T=4.19$ and only the optically thick wind phase is shown.

figure shows that a classical nova is a bright transient supersoft X-ray source in its later phase when the absolute visual magnitude drops to as low as +4.

Another type of novae, recurrent novae are thought to occur in a very massive white dwarf from their rapid evolution. The light curve fitting (Kato (1995a)) shows the white dwarf to be as massive as 1.35-1.38 $M_\odot$ in four recurrent novae.

Figure 7 shows the observational data of the recurrent nova U Sco in the 1987 outburst (Sekiguchi et. al 1988) as well as the theoretical light curves for the white dwarf mass of 1.377 $M_\odot$. The theoretical curves show a good agreement with the observational data of U Sco. The two sets of theoretical light curves correspond to a different chemical composition of the envelope, i.e., $X=0.7$ and $Z=0.02$ (thick curves) and $X=0.1$ and $Z=0.02$ (thin) (for more details, see Kato 1995a). The light curves of two models are very similar in the first stage, and deviate from each other in the later stage where the optically thick wind weakens to cease. Observational data in such a later phase is valuable to determine the chemical composition. Therefore multiwavelength observations of recurrent novae are very important to determine the white dwarf parameters.

The slow nova shows a slow decline time scale as long as several tens of years. Kato (1995a) shows that the light curve of a slow nova, RR Pic is consistent with the theoretical curve of a 0.8–0.9 $M_\odot$ white dwarf. The total duration of a slow nova is much longer than the optical decline time scale. For example, in case of a 0.8 $M_\odot$ white dwarf (Kato 1995a), the visual magnitude drops from the flat maximum stage within 1 year, whereas the total duration lasts 22 yrs, most of which observed is the supersoft X-ray emission as shown in Tab. 1. Less massive stars than 0.8 $M_\odot$ will be a supersoft X-ray source long after the optical decline. Therefore, the slow nova will be observed as a persistent supersoft X-ray source.

The theoretical light curves and the time-scale of evolution, however, depend on the history of the star. Kato (1995b) suggested that the slow nova has a helium layer under the hydrogen-rich envelope. Helium rich composition of ejecta often observed in slow novae is explained by her assumption that a part of the helium layer is mixed into the hydrogen envelope and eventually lost from the system due to the wind mass loss. The mass of the helium layer, therefore, decreases after each H shell flash, i.e., slow nova outburst. After all of the helium layer will be lost from the white dwarf surface, the system will make a classical nova outburst instead of slow nova.

In this scenario, the mass of the helium layer depends on the stellar history. Moreover, if the star has just evolved into the white dwarf region from a red giant, the stellar core may not be enough cooled down. Therefore, the radius of the bottom of the hydrogen rich envelope cannot be uniquely given but takes various values for the same core mass, which produces a wide variety in the light curves of slow novae.

Table 1 shows the time scales for a slow nova on a 0.8 $M\odot$ cold white dwarf with no He layer. They correspond to the shortest limit of the life time. In the star with massive He layer or a hot core or both, this time scale becomes much longer. The longest limit of this time scale may roughly correspond to that of a planetary nebula nucleus shown in the upper part of the same table.

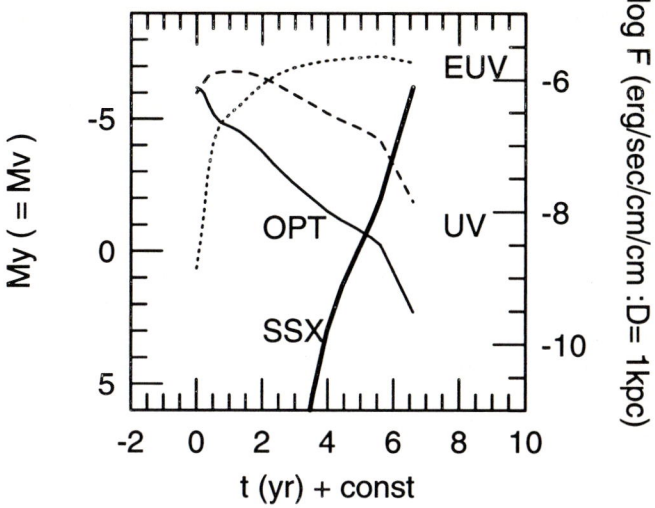

Fig. 8. Theoretical light curves of a He nova with mass 1.3 $M_\odot$. Optically thick wind ceases at $t = 6.60$ yrs. The light curves start at the stage with the photospheric temperature $\log T = 4.09$ at $t=0$.

5 He Nova: He Shell Flash on a White Dwarf

He nova is a nova like object that occurs in a close binary system. The mass accreting white dwarf experiences hydrogen shell flash that causes a nova outburst if the mass accretion rate is smaller than $\leq 10^{-7} M_\odot \text{yr}^{-1}$. If a part of the accreted matter remains after one nova outburst, the white dwarf will develop a helium layer under the hydrogen burning zone. This helium layer will grow every time of the hydrogen shell flash. When the mass of the helium layer reaches a critical value, unstable helium ignition triggers a nova-like phenomenon, i.e., a He nova. The evolutional course of a He nova on a 1.3 $M_\odot$ white dwarf is calculated by Kato, Saio & Hachisu (1989) with the old opacity.

Fig. 8 shows light curves of helium nova on a 1.3 $M_\odot$ white dwarf recalculated with OPAL opacity. The time scale and wind properties are very similar to those in the last helium shell flash (Fig. 5). Slight differences mainly come from the difference of the white dwarf radius; the last helium shell flash occurs on a hot white dwarf that has a slightly larger radius than the cool white dwarf in helium nova. The strength of the shell flash depends on the mass accretion rate; rapid accretion causes earlier ignition that causes a weak shell flash. When the ignition mass is small, the light curve starts from the middle part of Fig. 8 because the star cannot expand to red giant dimensions with small envelope mass.

The relation between the mass accretion rate and the starting point of the light curve is approximated as follows. If we ignore the amount of helium burnt in the rising phase of He nova, we can equate the envelope mass with the ignition mass and use the relation between ignition mass and mass accretion rate. In this approximation, the envelope mass 4.31×10^{-4} $M_\odot$ (at $t=0$ in Fig. 8) [3.5×

$10^{-4} M_\odot$ ($t = 0.22$ yrs), $4.4 \times 10^{-5} M_\odot$ ($t = 4.89$ yrs)] corresponds to mass accretion rates of $7.6 \times 10^{-8} M_\odot \text{yr}^{-1}$, [$10^{-7} M_\odot \text{yr}^{-1}$, $8.4 \times 10^{-6} M_\odot \text{yr}^{-1}$], respectively.

The latter accretion rate is the steady hydrogen-burning rate around the point at $\log T = 5.5$ in the H-R diagram. In this accretion rate, the helium nova begins with a small increase in optical but with a large increase in EUV and supersoft X-ray fluxes that makes the star to be a very bright EUV/supersoft X-ray source with dark optical counterpart from the beginning of the outburst.

When the white dwarf accretes hydrogen-rich matter with high accretion rate ($> 1 \times 10^{-6} M_\odot \text{yr}^{-1}$), the hydrogen burning is stable and the star remains at the same position in the H-R diagram. Such a system is interested in the relation to Type Ia supernovae, because the white dwarf possibly grows to reach the Chandrasekhar limit to make a supernova explosion. In such a system, however, the He burning is unstable ($<$ a few $\times 10^{-7} M_\odot \text{yr}^{-1}$) to cause weak He novae and a part of the accreted matter will be lost from the system. Therefore, in the study of binary evolution or scenario of Type Ia supernova that include an accreting white dwarf, we cannot neglect the effects of He nova outbursts.

6 Summary

The time scale of evolution and light curves are given for the planetary nebular nucleus, last helium shell flash, nova and He nova that evolve in the horizontal path leftward in the H-R diagram. The evolutional time-scale depends on the white dwarf mass, nuclear fuel and the white dwarf radius. Very short evolutional time is obtained in massive white dwarfs due to optically-thick wind mass-loss.

Many of those objects will be observed as a bright supersoft X-ray source when the surface temperature increases to $\log T \sim 5$, at the same time the visual magnitude M_v drops to less than 2.

There is a wide variety in the lifetime of these objects as SSS sources. Some of them are transient supersoft X-ray sources and others are persistent sources.

References

Blöcker T., Schönberner D., 1990, A&A 240, L11
Iben I. Jr., 1982, ApJ 259, 244
Iben I. Jr., Kaler J.B., Truran J.W., Renzini A., 1983, ApJ 264, 605
Iglesias C.A., Rogers F.J., 1993, ApJ 412, 752
Iglesias C.A., Rogers F.J., 1996, ApJ 464 (in press, June 20)
Kato M., 1995a, in Cataclysmic Variables, eds. Bianchini A., Della Valle M. & Orio M. (Dordrecht:Kluwer) p. 243
Kato M., 1995, in Flares and Flashes, eds. Greiner J., Duerbeck H. & Gershberg R.E., Lecture Notes in Physics 454 (Berlin:Springer) p. 237
Kato M., Hachisu I., 1994, ApJ 437, 802
Kato M., Saio H., Hachisu I., 1989, ApJ 340, 509
Rogers F.J., Iglesias C.A., 1992, ApJS 79, 507
Sekiguchi K., Feast M.W., Whitelock P.A., et al. 1988, MNRAS 234, 281

Hot High–Gravity NLTE Model Atmospheres Applied to Supersoft Sources

H.W. Hartmann, J. Heise

SRON Laboratory for Space Research, 3584 CA Utrecht, The Netherlands

Abstract. The spectra of optically thick soft X-ray sources, such as Supersoft Sources, are often fitted with blackbodies. Heise et al. 1994 calculated LTE atmospheres and showed that these are more efficient X-ray emitters than blackbodies. In this paper we study the assumption of LTE and present calculations of hot, high–gravity NLTE model atmospheres in hydrostatic and radiative equilibrium. The range of temperatures is chosen such that they emit significantly in the soft X-ray range (0.1 − 2.0 keV). The gravitational fields of $g = 10^{7.5} - 10^9$ cm s^{-2} are applicable to massive white dwarfs. It appears that NLTE spectra are comparable to LTE spectra if $\log g \geq 9.0$, i.e. for the most massive white dwarfs ($M \geq 0.6$ M$_\odot$). At these gravities the density is sufficiently high to assume that collisional ionizations dominate and that thermal equilibrium (LTE) determines the degree of ionizations and the atomic population levels.

We show a first fit of H, He, C, N, O, Ne NLTE model atmospheres to the observed ROSAT spectrum of the galactic supersoft source RX J0925.7–4758 . The resulting effective temperature strongly depends on the assumed model parameters, such as the gravity and the metallicities. At present X-ray spectral resolution, temperature, gravity and abundances cannot be determined independently.

1 Introduction

Atmospheres of hot white dwarfs are known to emit soft X-rays (0.1 − 2.0 keV). Among the first discoveries are isolated hot DA white dwarfs such as the prototype HZ43, and sources like Sirius B. The spectrum of an optically thick soft X-ray source is often taken as blackbody emission. This is a fair approximation if the opacity as a function of energy is roughly constant. However, in the soft X-ray range the emission spectrum at certain temperatures and densities is often dominated by a few ions, such as the helium and hydrogen–like ions of carbon, nitrogen and oxygen. The atmospheric emission deviates largely from a blackbody at the effective temperature: the atmosphere is transparent for soft X-rays. Optical depth of the order one in the soft X-ray range is reached at a depth where the temperature is much higher than the effective temperature. In such cases it becomes necessary to compute spectra as is generally done in the calculation of stellar atmospheres: a plane–parallel plasma slab in a gravitational field of given strength is assumed to be in hydrostatic and radiative equilibrium. The temperature structure is such that the total flux through the slab is constant.

Heise et al. 1994 have shown that LTE model atmospheres for white dwarfs are more efficient X-ray emitters than blackbodies and consequently the luminosity of those LTE model atmospheres stays below the Eddington limit. For this they had to assume that at surface gravities of white dwarfs, which are of the order of $10^7 - 10^9$ cm s^{-2}, the density of the atmosphere is in thermal equilibrium and that the Saha–Boltzmann law determines the degree of ionization and the atomic level occupation. This assumption may not be valid in the outer low–density region of the atmosphere where NLTE is a better assumption.

We present NLTE models with effective temperatures in the range 3×10^5–1×10^6 K. So far, only NLTE models with temperatures up to 4×10^5 K have been calculated (Rauch 1995). The main purpose of our NLTE grid is to explore how the X-ray emission changes with gravity, effective temperature and abundance, and to apply these models to several ROSAT spectra of supersoft sources.

2 Model Atmospheres

We calculated NLTE models in radiative and hydrostatic equilibrium with cosmic (Allen 1973) LMC and SMC (Dennefeld 1989) abundances for He, C, N, O and Ne. All other metals were ignored.

The gravity ranges from $\log g = 7.5$ to $\log g = 9.0$ in steps of 0.5. The effective temperature is limited by the surface gravity via the Eddington limit, since at higher effective temperatures the model atmosphere would no longer be hydrostatic, and ranges from 3×10^5 K up to 4×10^5 K for $\log g = 7.5$ and up to 1.08×10^6 K for $\log g = 9.0$ in steps of 1×10^4 K.

The model atmospheres are calculated using the computer code TLUSTY178 (Hubený 1988). This code uses the complete linearization technique to solve the coupled set of radiative transfer, radiative equilibrium and statistical equilibrium equations. The difference equation of the radiative transfer equation is represented by the standard Feautrier scheme. Convergence is achieved when the relative changes of the temperature, total number density and electron density are smaller than 5×10^{-3}. For a detailed description of the computer code TLUSTY178 we refer to Hubený (1988).

The models are calculated on 70 depth points and a frequency grid containing 268 points including frequencies just above and below threshold ionization energies. The frequency range runs from 10^{15} (~ 4 eV) to 2×10^{18} (~ 8.3 eV) Hz. We ignored line opacities and line blanketing. These effects are evidently important and will have to be taken into account eventually, but at present we consider them as the next stage in the refinement of the models, which should also take expanding atmospheres into account.

2.1 Model Atoms

In hot white dwarf atmospheres a significant fraction of the ions is in the exited state. However, we restrict ourselves to a limited number of ionization states and atomic levels. It appears that the spectrum of hot, high–gravity atmospheres is often dominated by the lowest levels of one or two ionization stages of a particular element. Therefore, we selected those ionization stages that we expected to be most dominant in the range of parameters of interest.

2.2 Treatment of Continuum Opacities

Photoionization cross–sections of the ground state were determined using data and fitting formula from Verner & Yakovlev 1995:

$$\sigma_{nl}(E) = \sigma_0 F(E/E_0) \text{ Mb},$$

$$F(y) = \left[(y-1)^2 + y_w^2\right] y^{-Q} \left(1 + \sqrt{y/y_a}\right)^{-P}, \quad (1)$$

where n and l are the principal and subshell orbital quantum number respectively, E is the photon energy in eV, $y = E/E_0$. σ_0, E_0, y_w, y_a and P are fit parameters and $Q = 5.5 + l - 0.5P$.

For the photoionization cross–sections of the excited atomic levels we use the interpolation formula (cf. van Teeseling 1994):

$$\sigma_\nu = \sigma_0 \times 10^{-18} \left[\alpha \left(\frac{\nu}{\nu_T}\right)^{-s+1} + (\beta - 2\alpha) \left(\frac{\nu}{\nu_T}\right)^{-s} + (1 + \alpha - \beta) \left(\frac{\nu}{\nu_T}\right)^{-s-1} \right]. \quad (2)$$

Here, ν_T is the threshold frequency of the transition and σ_0, α and β are fit parameters. These cross–sections, when calculated for ground states, do not differ by more than 6 % in the energy range of interest ($\sim 0.1 - 2.0$ keV) as compared to Verner & Yakovlev (1995).

3 Results

3.1 NLTE Models Compared with LTE Models

In Fig. 1 (left panel) we show both a LTE and a NLTE model at $T_{\text{eff}} = 5 \times 10^5$ K for $\log g = 8.0$. The abundances are cosmic for both models. The C VI edge at 0.49 keV is in absorption in the case of LTE whereas this transition is in emission in NLTE. A general known feature of NLTE models (when NLTE is important) is to decrease the depth of the absorption edge, even to reverse it to an emission edge. This effect starts at edges with the lowest energies and builds up towards higher energies when the ions become more highly ionized. Emission edges do not show up in LTE spectra when the temperature is increased. Consequently, the NLTE soft X-ray spectrum is no longer cut off at a certain edge as it is in the LTE case, but contributes considerably to the total flux beyond this edge.

The change from absorption to emission is shown for C VI (0.49 keV), N VII (0.67 keV), O VII (0.74 keV), O VIII (0.87 keV) and Ne IX (1.20 keV) in Fig. 1. The temperature ranges from $T_{\text{eff}} = 6 \times 10^5$ to 1×10^6 K. Fig. 1 (right panel) clearly shows that the edge of O VII vanishes when the temperature rises above 8×10^5 K and O VII becomes completely ionized.

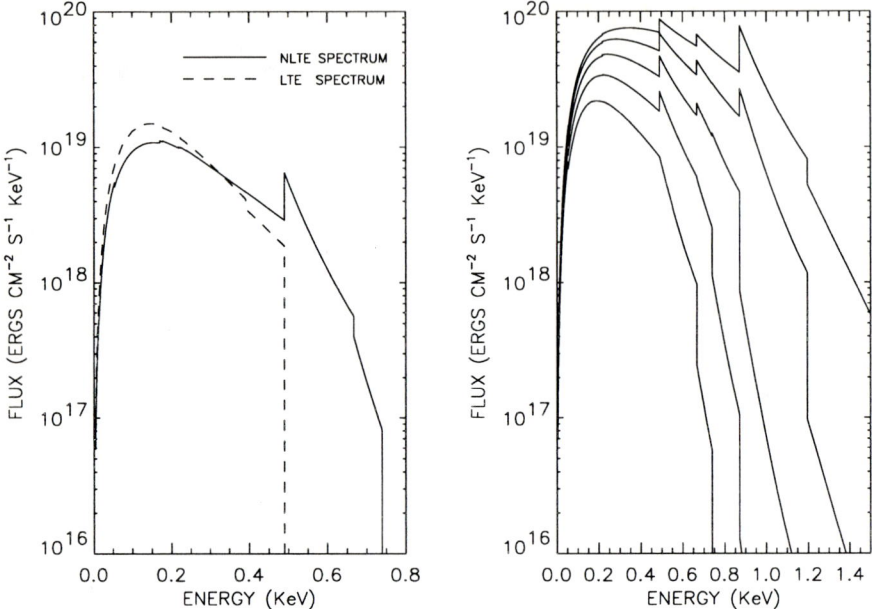

Fig. 1. White dwarf model atmospheres with cosmic abundances. Left panel: LTE and NLTE models are compared for $\log g = 8.0$ and $T_{\text{eff}} = 5 \times 10^5$ K. Right panel: NLTE models for $\log g = 9.0$ and $T_{\text{eff}} = 6 \times 10^5$ to 1×10^6 K in steps of 10^5 K

3.2 Dependence on Gravity

When the gravitational field is increased from $\log g = 7.5$ to $\log g = 9.0$ while the temperature is kept constant, the emission edges gradually disappear (see left panel of Fig. 2). Note that this causes the soft X-ray flux beyond the C VI edge (0.49 keV) to be much higher for $\log g = 7.5$ than for $\log g \geq 8.0$. The departure coefficients for the ground state of C VI are shown in Fig. 2 (right panel). Those departure coefficients (or b–factors) of a certain transition are defined as

$$b = n/n^*, \qquad (3)$$

where n and n^* are the NLTE and LTE population densities of that level respectively. The b–factors give an indication of the departure from thermal equilibrium and thus from LTE at the C VI threshold frequency. It appears that for increasing $\log g$ the departure from LTE becomes smaller, although the outer parts of the atmosphere with $\tau_R \leq 10^{-1}$ are still far from an LTE situation. Increasing $\log g$ even more would be unrealistic, since white dwarfs in general do not have gravities above $\log g = 9.0$. Note that the inner part of the calculated atmosphere with $\tau_R \geq 1$ represents an LTE situation rather well with b close to 1. Higher densities cause the collisional ionizations to dominate and thermal equilibrium to be restored.

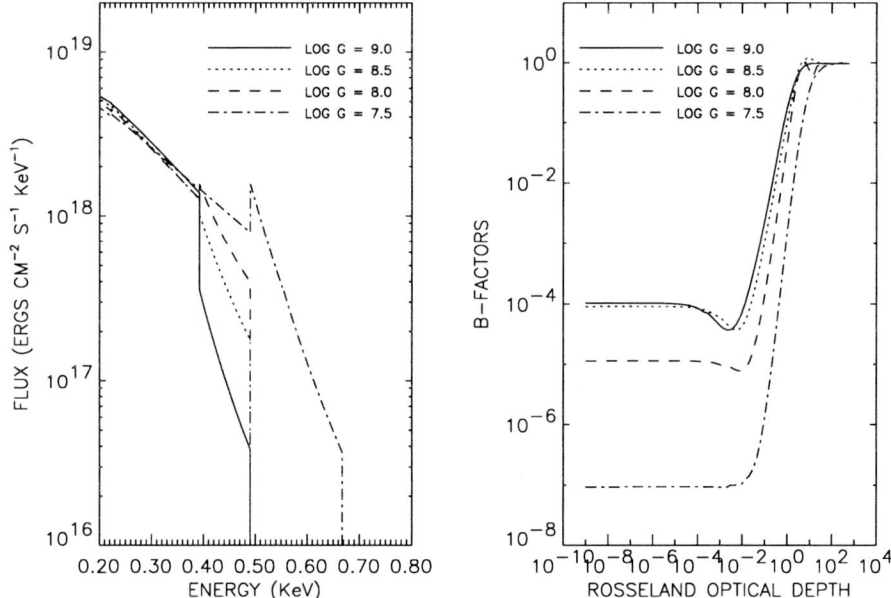

Fig. 2. Left panel: NLTE model atmospheres at $T_{\text{eff}} = 4 \times 10^5$ K at $\log g$ between 7.5 and 9.0. Right panel: b–Factors for the C VI → C VII transition in four model atmospheres at $\log g$ between 7.5 and 9.0. Abundances are cosmic

3.3 Dependence on Abundances: Cosmic, LMC and SMC

In order to check on the effect of abundances we demonstrate NLTE models with cosmic, LMC and SMC abundances rather than calculating an extensive grid of models with a range of abundances for He, C, N, O and Ne. In general, the metallicities of the LMC are roughly 0.25 times cosmic and those of the SMC are roughly 0.15 times cosmic (for more details see Dennefeld 1989). In Fig. 3 we show spectra with $T_{\text{eff}} = 6.5 \times 10^5$ K and $\log g = 8.5$.

Note the influence on the spectrum of the N VII (0.67 keV) and O VII (0.74 keV) edge and the O VIII (0.87 keV) edge going from absorption to emission when the metallicities are decreased. Evidently the soft X-ray flux beyond the O VIII (0.87 keV) edge is much lower for cosmic abundances compared to LMC and SMC abundances.

4 Model Atmosphere Fits

We will fit the NLTE model atmospheres to data of RX J0925.7−4758 and compare the results with earlier blackbody and LTE fits. RX J0925.7−4758 is a galactic supersoft source, discovered in the ROSAT galactic plane survey (Motch et al. 1991). It was reobserved in two pointed observations on November 20 by ROSAT PSPC and on November 23 by ROSAT HRI. The total exposure time

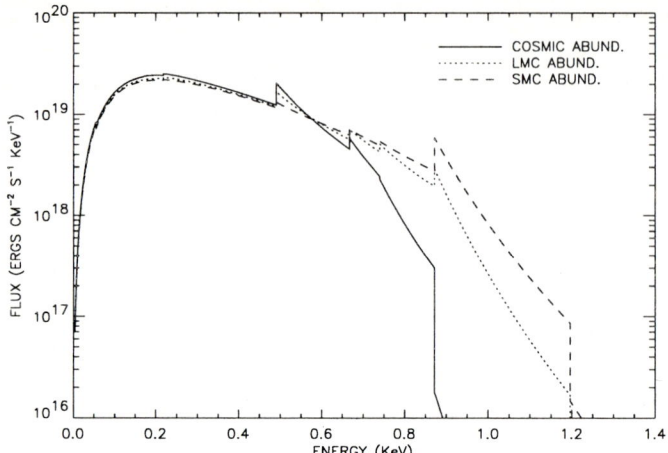

Fig. 3. Spectra for NLTE models with cosmic, LMC and SMC abundances. $T_{\text{eff}} = 6.5 \times 10^5$ K and $\log g = 8.5$

of all three observations was only 7.8 ksec. The PSPC count rate of 1.0 cts s^{-1} showed evidence for variability on timescales of several hours. RX J0925.7–4758 has been spectroscopically identified with a 17^{m} star with spectral features which resemble those of conventional LMXBs. RX J0925.7–4758 has a bolometric luminosity close to the 1 M$_\odot$ Eddington limit. The companion is possibly a $1.5 - 2.5$ M$_\odot$ main sequence star (van den Heuvel et al. 1992). The RX J0925.7–4758 ROSAT data have been fitted with blackbody spectra at temperatures between $30 - 55$ eV and a high column density of $(1.4 - 3.7) \times 10^{22}$ cm^{-2} (Motch et al. 1994), and a (blackbody) luminosity of $\sim 10^{38}$ erg s^{-1}. This luminosity and column density constrain the distance of RX J0925.7–4758 to 425–2000 pc assuming that intrinsic absorption is not a general feature for supersoft sources.

We assumed a distance to RX J0925.7–4758 of 1 kpc. The LTE fit required a temperature of at least 10^6 K ($\chi^2 = 0.86$). The NLTE model resulted in a better fit ($\chi^2 = 0.78$) although there is a significant excess in the data above 1.3 keV which does not show up in the folded model spectrum (see Tab. 1 and Fig. 4).

Table 1. Fit parameters for RX J0925.7–4758

	kT (eV)	$\log N_H$ (cm^{-2})	$\log L$ (erg sec^{-1})	$\log R$ (cm)	$\log g$ (cm s^{-2})
RX J0925.7–4758 :					
NLTE	72	22.0	35.7	7.6	9.0
LTE	>85	22.1	36.0	7.6	9.0
Blackbody[1]	30–55	22.1–22.6	38.3	7.3	–
bb + edges[2]	90	22.0	35.7	7.4	–

[1] Motch et al. (1994), [2] Ebisawa et al. (1995)

RX J0925.7–4758 was observed by ASCA SIS in November 1994. Ebisawa et al. 1995 superimposed four edges to the blackbody spectrum fitted to the ASCA data. The fit could thus be improved. Ebisawa et al. derived a temperature of 90 eV and a radius of the compact object of 230 km, see Tab. 1. It is remarkable that all fits to RX J0925.7–4758 result in a small radius < 370 km.

5 Discussion and Conclusions

For $\log g < 9.0$ it appears that NLTE model atmospheres differ considerably from LTE model atmospheres. LTE model spectra show strong absorption edges at dominant ions, whereas those edges may even be in emission in NLTE, depending on the effective temperature, gravity and abundance. The presence of emission edges is strongly influenced by the effective temperature. At relatively low temperatures ($\sim 3 \times 10^5$ K) the NLTE model spectra resemble the LTE model spectra, since the LTE spectra do not show the strong absorption edges that cut off the LTE emission spectrum at higher temperatures. Folding those low temperature models with the low energy resolution of ROSAT PSPC, the edges, either in absorption or emission, are smeared out and both models result in similar fits.

At increasing gravity the emission edges gradually disappear and may even turn into absorption edges. At the increased density the collisional ionizations start to dominate over photoionizations and thermal equilibrium is restored. Still the spectra differ significantly from LTE model atmospheres at the same temperature and gravity, so one should be careful assuming that LTE is a good approximation at high gravity. It may be worthwhile to investigate whether this assumption becomes valid at even higher gravities, e.g. in accretion discs around neutron stars which may have $\log g > 9$.

The abundances have an important impact upon the shape of the spectrum. At low metallicities emission edges show up, which is not the case at high metallicities. It is therefore important to have at least some knowledge about the abundances of the atmosphere that one tries to fit.

The first conclusion is that the white dwarf effective temperature, gravity and abundances cannot be determined uniquely, since the spectral continuum of a high gravity, high temperature atmosphere resembles a lower gravity atmosphere at a lower temperature. Therefore it is necessary to have additional information about the source, e.g. from observations in other wavelengths or on theoretical grounds, in order to constrain the free parameters and to be able to determine effective temperatures and luminosities unambiguously. The second conclusion that can be drawn is that the NLTE model spectrum is not cut off at a certain dominant edge as is the case for LTE models. This effect is important for soft X-ray sources which emit a considerable amount of their flux above this cutoff energy. These sources can be fitted with models at lower temperatures because NLTE models are more efficient X-ray emitters than LTE models above the cutoff energy. The lower temperature is compensated by a higher radius such that the NLTE and the LTE luminosities are comparable.

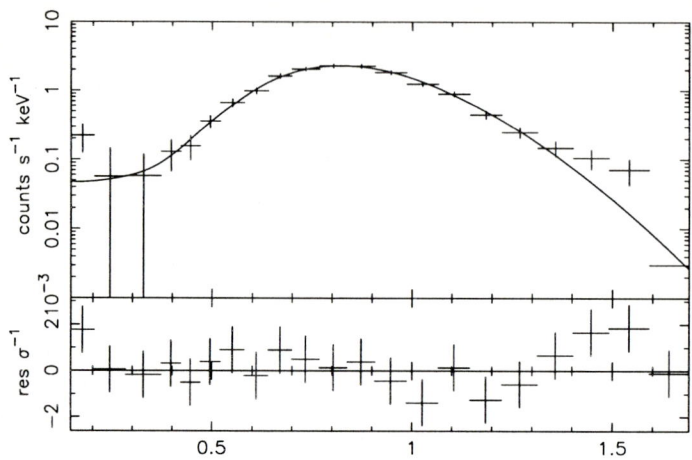

Fig. 4. NLTE models fitted to the galactic supersoft source RX J0925.7−4758 . For fit parameters see Tab. 1.

The discrepancy between the observed spectrum of RX J0925.7−4758 and the fitted model above ~ 1.3 keV can be explained by the fact that the unfolded model spectrum has a strong Ne IX edge at 1.2 keV that cuts off the spectrum. RX J0925.7−4758 must either be fitted to models with even higher effective temperature and gravity or to more sophisticated model spectra. The high gravity suggests that RX J0925.7−4758 is a massive white dwarf with probably high O, Ne and Mg abundances.

Acknowledgement: This work has been supported by funds of the Netherlands Organization for Scientific Research (NWO).

References

Allen C.W., 1973, Astrophysical quantities, The Athlone Press, University of London
Dennefeld M., 1989, in: Recent developments of Magellanic Cloud research, K.S. de Boer, F. Spite & G. Stasińska (eds)., Paris
Ebisawa, K., Asai, K., Dotani, T., Mukai, K. & Smale, A., 1995, in: Röntgenstrahlung from the Universe, eds. H.U. Zimmermann et al., MPE report 263, p. 133
Heise J., van Teeseling A., Kahabka P., 1994, A&A 288, L45
van den Heuvel E.P.J., Bhattacharya D., Nomoto K., Rappaport S.A., 1992, A&A 262, 97
Hubený I., 1988, Comput. Phys. Commun. 52, 103
Motch C., et al., 1991, A&A 246, L24
Motch C., Hasinger G., Pietsch W., 1994, A&A 284, 827
Rauch T., 1995, in: Röntgenstrahlung from the Universe, eds. H.U. Zimmermann et al., MPE report 263, p. 63
van Teeseling A., Heise J., Paerels F., 1994, A&A 281, 119
Verner D.A., Yakovlev D.G., 1995, A&AS 109, 125

Luminous Supersoft X-Ray Sources in Globular Clusters

R. Di Stefano[1], M.B. Davies[2]

[1] Harvard-Smithsonian Center for Astrophysics, Cambridge, MA 02138
[2] Institute of Astronomy, Cambridge, CB3 OHA

1 Observations

Globular clusters have been well-studied by ROSAT. Yet, the existence of just a single luminous supersoft X-ray source (SSS) has been reported.

1.1 The Lone Observed SSS in a Globular Cluster

The single detected source, 1E 1339.8+2837, located in M3 (NGC 5272), was observed both in the ROSAT all-sky survey (Verbunt et al. 1994, 1995), and in pointed observations using the HRI (Hertz, Grindlay & Bailyn 1993). Estimates of the temperature and bolometric luminosity based on the all-sky survey were $kT \sim 45$ eV, and $L \sim 10^{35}$ erg s^{-1}. The source is a transient. Note that the inferred luminosity (though uncertain) is smaller than typical for SSSs found in the Galaxy, the Magellanic Clouds, and M31.

1.2 Null Results

Most ROSAT studies have not discovered evidence for SSSs in globular clusters. For example, Verbunt et al. (1995) reported that, of the 17 clusters in which the ROSAT XRT sky survey could have discovered a source like 1E 1339.8+2837, there were 15 null results, and only one possible detection (in NGC 1851), in addition to the detection of 1E 1339.8+2837 itself.

2 Comparison with LMXBs

Globular clusters are known to have a per capita population of low-mass X-ray binaries (LMXBs) that is roughly 100 times larger than that of the Galactic disk. This has been thought to provide evidence that stellar interactions within clusters are responsible for the formation of many cluster LMXBs. If the same ratio held for SSSs, then the globular cluster population would contain on the order of 100 SSSs—roughly one per globular cluster. If, on the other hand, the per capita population of SSSs in globular clusters is the same as that inferred for the Galactic disk, then the entire globular cluster system should contain $\mathcal{O}(1)$ SSS.

Table 1. The results of seeding 9 Globular clusters with SSSs

Globular Cluster	E(B-V)	Distance (kpc)	t_{exp} (ksec)	Galactic SSSs ⟨Counts⟩	Fraction	1E 1339.8+2837 ⟨Counts⟩	Fraction
NGC 6341	0.02	7.5	6.7	3.E5	0.93	2.E3	1.0
NGC 6752	0.04	4.2	5.2	7.E5	0.95	2.E4	1.0
NGC 7099	0.06	7.4	6.2	3.E5	0.93	2.E3	1.0
ω Cen	0.15	4.9	12.	8.E4	0.83	3.E2	1.0
NGC 6397	0.18	2.2	2.4	4.E4	0.82	1.E2	1.0
NGC 6656	0.36	3.0	8.4	1.E4	0.63	0	0.0
NGC 6642	0.37	8.0	7.6	1.E3	0.46	0	0.0
NGC 6626	0.38	5.9	4.3	1.E3	0.44	0	0.0
NGC 6544	0.74	2.5	2.7	1.E3	0.33	0	0.0

The results of ROSAT PSPC observations of these clusters were reported by Johnston, Verbunt, and Hasinger (1994). The 5th and 6th columns show the results of seeding the clusters with a population of SSSs with properties similar to those of sources found in our Galaxy, the LMC and SMC, and M31. ⟨Counts⟩ is the lifetime-weighted average number of counts per source that would have been recorded during an exposure of duration t_{exp}, and "fraction" is the fraction of seeded sources that would have been detected (i.e., which contributed 10 or more counts). The 7th and 8th columns show the analogous quantities when the clusters are "seeded" with the single observed source, 1E 1339.8+2837; the count rates are lower because the source has a smaller L and somewhat lower value of T than the "galactic" sources. Note the dramatic dependence on N_H.

3 Are There SSSs in Globular Clusters?

We want to turn the observational results, which have mainly been null, into limits on the total population of SSSs in globular clusters. To this end we have "seeded" each of the globular clusters observed by Johnston et al. (1994), with a population of SSSs, and have determined the count rate associated with each seeded source (see Di Stefano, Becker & Fabbiano 1996 for a description of an analogous procedure).

Table 1 indicates that SSSs like those detected in our own Galaxy cannot be hidden in globular clusters. The majority of such sources would have been detected if they were on during ROSAT observations. Furthermore, most detected sources would have contributed large numbers of counts. A significant fraction of sources from the lower L and T edges of the distribution could have been missed only in clusters viewed behind significant columns of gas and dust. Thus, even sources such as the single observed source are not common.

4 Should There Be SSSs in Globular Clusters?

The two main types of SSS in the Galaxy and the Magellanic Clouds are symbiotics and candidates for the close-binary supersoft source model (CBSSs; see

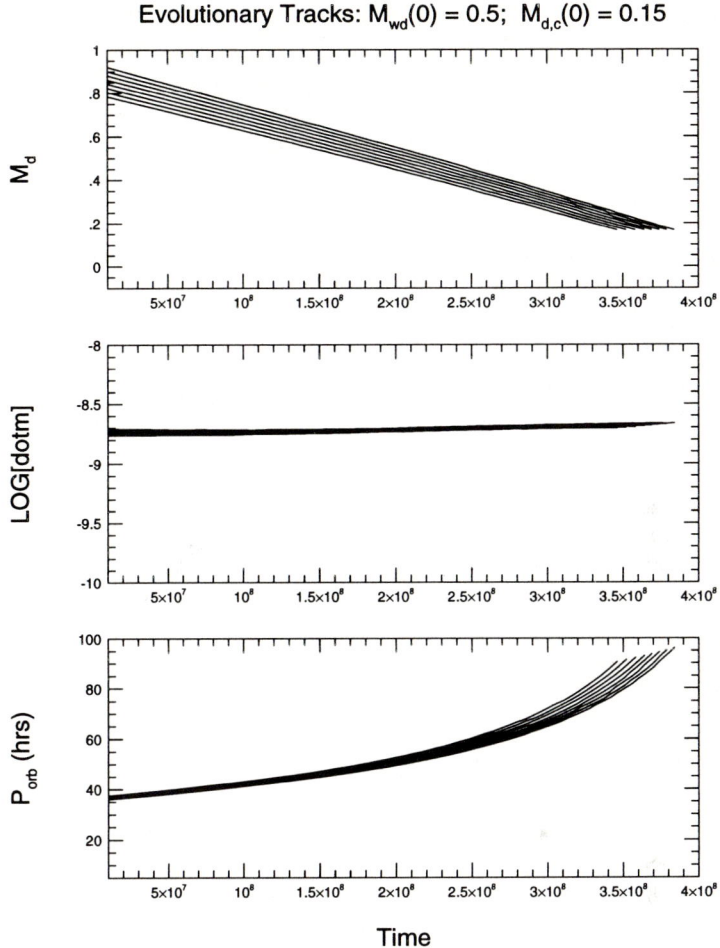

Fig. 1. The top, middle, and lower panels show the mass of the donor, its mass loss rate, and the orbital period, respectively, as functions of time.

Di Stefano & Nelson [DN; 1996]). Primordial binaries should not have yielded significant numbers of either system in globular clusters today. Stellar interactions, however, should create CBSSs in the dense central regions of globular clusters. The donor star would be a blue straggler, formed through the merging of two lower-mass stars. (In fact, M3 is rich in blue stragglers [Guhathakurta et al. 1994].) We have performed simulations for two clusters, 47 Tuc and ω Cen. These will be described in detail elsewhere (Di Stefano & Davies 1996). Briefly, we use a code that follows thousands of systems, each through multiple interactions (Davies and Benz 1995; Davies 1995). States in which a white dwarf is in a close orbit with a blue straggler are evolved with a binary evolution code (see, e.g., Di Stefano et al. and DN). Typical evolutions are shown in Figure 1; the systems formed tend to have values of $\dot{m}$ below the steady-burning region, and should therefore appear as SSS transients. The total numbers are small—fewer than 10 in each cluster during the past 2 Gyrs. Thus, given the short lifetime and low duty cycle of activity, we would not expect to find active SSSs in either cluster. The mismatch is not so large, however, as to preclude the existence of a few SSSs "on" at any given time in the globular cluster system.

5 Results and Prospects

Theory predicts that SSSs may be formed in clusters via interactions in or near the dense cluster cores. The rate of formation tends to be small ($< 10^{-8}$/yr), and the time duration of SSS activity is shorter by at least an order of magnitude than even the lowest estimates presently made for LMXB lifetimes. Thus, it may not be surprising that observations seem to rule out the possibility that SSSs are 100 times as numerous per capita in globular clusters as they are in the Galactic disk. The factor that makes it most difficult to place concrete observational limits on the SSS population in all Galactic globular clusters is that some are seen through a large column density (see, e.g., Rappaport et al 1994). We are working to refine our calculations. Preliminary indications are that the theoretical predictions will continue to be in harmony with the observations. Nevertheless, the process of carrying out a detailed test will lead to the derivation of useful ratios between observable systems, such as blue stragglers and CVs, and will help us to further classify the supersoft source and related occupants of the globular cluster binary zoo.

Acknowledgement: We thank A. Cool, J. Grindlay, P. Hertz, J. Mattei, and F. Verbunt for discussions. This work was supported by GER-9450087.

References

Davies M.B., 1995, MNRAS 276, 887
Davies M.B., Benz W., 1995, MNRAS 276, 876
Di Stefano R., Nelson L.A., 1996, this volume p. 3
Di Stefano R., Davies M.B., 1996 (in prep.)
Di Stefano R., Nelson L.A., Lee W., Wood T., Rappaport S., 1996, NATO ASI on Thermonuclear Supernovae, eds. Lapuente-Ruiz P., Canal R. (in press)
Di Stefano R., Becker C.M., Fabbiano G., 1996, this volume p. 37
Guhathakurta P., Yanny B., Bahcall J.N., Schneider D.P., 1994, ApJ 108, 1786
Hertz P., Grindlay J., Bailyn C., 1993, ApJ 410, L87
Johnston H., Verbunt F., Hasinger G., 1994, AA 289, 273
Rappaport S., Dewey D., Levine A., Macri L., 1994, ApJ 423, 633
Verbunt F., Johnston H., Hasinger G., Belloni T., Bunk W., 1994, MmSAI 65, 249
Verbunt F., Bunk W., Hasinger G., Johnston H., 1995, A&A 300, 732

The Integrated X-Ray Spectrum of Galactic Populations of Luminous Supersoft X-Ray Sources

R. Di Stefano[1], C.M. Becker[2], G. Fabbiano[1]

[1] Harvard-Smithsonian Center for Astrophysics, Cambridge, MA 02138
[2] Department of Physics and Center for Space Research, MIT, Cambridge, MA 02139

Abstract. We compute the composite X-ray spectrum of a population of unresolved SSSs in a spiral galaxy such as our own or M31. The sources are meant to represent the total underlying population corresponding to all sources which have bolometric luminosities in the range of $10^{37} - 10^{38}$ ergs s^{-1} and kT on the order of tens of eV. These include close-binary supersoft sources, symbiotic novae, and planetary nebulae, for example. In order to determine whether the associated X-ray signal would be detectable, we also "seed" the galaxy with other types of X-ray sources, specifically low-mass X-ray binaries (LMXBs) and high-mass X-ray binaries (HMXBs). We find that the total spectrum due to SSSs, LMXBs, and HMXBs exhibits a soft peak which owes its presence to the SSS population. Preliminary indications are that this soft peak may be observable.

1 Motivation

Computations of the fraction of luminous supersoft X-ray sources (SSSs) that could have been detected by ROSAT surveys indicate that SSSs form a significant population in the Local Group (Di Stefano & Rappaport 1994; DR). DR estimated that the numbers of presently active and not self-obscured SSSs in M31, the Milky Way, and the Magellanic Clouds are $800 - 5000$, $400 - 1000$, and $50 - 100$, respectively.

Observations of some spiral galaxies have revealed the presence of soft excesses in their X-ray spectra (see, e.g., Kim, Fabbiano, and Trinchieri 1992). These observations, in concert with the predictions of large galactic SSS populations for spiral galaxies, lead to the natural question: Are SSSs responsible for the observed soft excesses?

2 Method

We approach the answer in two steps. First we compute the composite X-ray spectrum of a population of unresolved SSSs in a "typical" spiral galaxy. Spirals with a significant hot ISM, or with an active nucleus would require a separate treatment. The galaxy is "observed" by folding the composite spectrum through the ROSAT PSPC response matrix (pspcb_gain2_256.rsp); the simulated pulse height spectrum is computed. Second we add the composite spectra due to other types of sources, to determine if any unique signature of the SSS population can be unambiguously identified. To test the sensitivity of the results to input assumptions, we carry out a number of different simulations. We vary (a) the input spectra of LMXB and HMXB "contaminants", and (b) the angle, θ, the disk of the external spiral galaxy makes with our line of sight to it.

Interstellar Gas In our simulations, the X-ray sources are embedded in the galaxy's gas distribution, which is modeled as an exponential disk with scale height z_{gas}. We take the column density along a line of sight perpendicular to the disk to be 10^{21} cm^{-2}. To this we add a column density chosen uniformly from the range $2 - 5 \times 10^{20}$ cm^{-2}, comparable to that associated with looking out of the plane of the Galaxy.

Luminous Supersoft X-Ray Sources Using the results of DR, we seed the galaxy with 1000 SSSs. The temperature and luminosity distribution is taken from the work of Rappaport, Di Stefano, & Smith (1994; RDS). Although RDS modeled close-binary systems, the distributions of properties compare well with known SSS of all types (e.g., old novae). DR and Motch, Hasinger, and Pietsch (1994) found that SSSs are likely to be part of a disk population. We have therefore chosen $z_{SSS}/z_{gas} = 0.4$.

Other X-Ray Sources Fabbiano (1989) indicates that X-ray emission from normal spiral galaxies is dominated by neutron star binaries. We therefore consider LMXBs and HMXBs as "contaminants" to the SSS contribution. Based on observations of the Milky Way and M31 (Fabbiano 1989), we seed our galaxy with 30 HMXBs (scale height: $z_{HMXB}/z_{gas} = 0.4$) and 100 LMXBs (scale height: $z_{LMXB}/z_{gas} = 1.5$).

In addition, other sources of X-radiation are present in typical galaxies; these include cataclysmic variables, stars with active coronae, supernovae, and black hole candidates. Since we are primarily interested in the sources as "contaminants" to the composite spectrum of SSSs, it is the low energy portion of the source X-ray spectra that is of most concern to our simulations. Unfortunately, the effects of interstellar absorption at these wavelengths lead to significant uncertainties. Given these uncertainties, we have attempted to simplify our simulations by (1) explicitly including only LMXBs and HMXBs as contaminants to the composite SSS spectrum, and (2) characterizing the associated spectra so as to systematically vary the soft component most relevant to our study. We carry out all of our simulations three times, first, with spectra for both LMXBs and HMXBs that we regard to be "pessimistic", in that the soft components of these sources are overemphasized by choosing only the softest observed spectra for LMXBs and HMXBs. In this "worst case" scenario, the copious soft radiation emitted by our too-soft population of LMXBs and HMXBs should also provide an upper limit to the soft radiation likely to be associated with the other types of X-ray source that we have not explicitly included in our simulations. In two separate sets of simulations we have used spectra that are in the mid-range (the "realistic" case) and upper end ("optimistic" case) of hardness observed for both LMXBs and HMXBs (see the caption to Figure 2 for details).

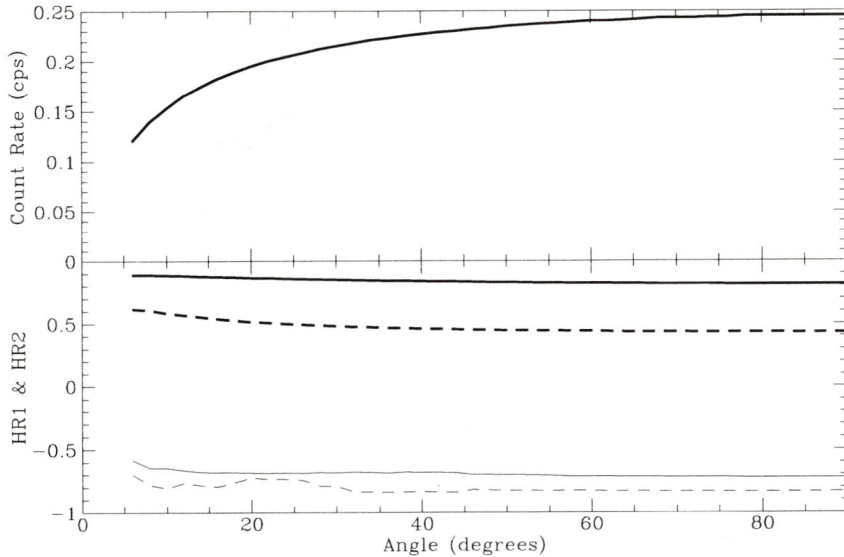

Fig. 1. The angular dependence of the (a) count rate, (b) HR1 = (C+B-A)/(C+B+A) (solid) and HR2 = (C-B)/(C+B) (dashed), where A = (0.1 – 0.4 keV), B = (0.4 – 0.9 keV), and C = (0.9 – 2.0 keV). The SSS component of the composite spectrum is represented by the thin line. Thick lines correspond to the composite spectrum due to all sources in the "realistic" case.

3 Results

Significant features of the composite SSS spectrum as observed by the ROSAT PSPC are shown in Figure 1. Our simulation results, which include the spectra of LMXBs (thermal bremsstrahlung) and HMXBs (power law), are shown in Figure 2. The SSS component is clearly identifiable in the realistic and optimistic cases, where the soft excess can not be explained through any linear combination of LMXB and HMXB components in the optimistic case. The more complete analysis required for the pessimistic case is underway.

4 Conclusions

If SSSs exist in typical spiral galaxies in the numbers inferred by Di Stefano & Rappaport (1994) for both our own Galaxy and M31, then they are likely to be associated with a soft X-ray excess. Further work will quantify the effect and explore whether the observed soft excesses in some spirals may in fact be due to the presence of a population of luminous supersoft X-ray sources. A parallel investigation for elliptical galaxies is also underway. The present work indicates the possibility of detecting evidence of SSSs populations in distant galaxies through the observation of soft excesses in galactic spectra is promising.

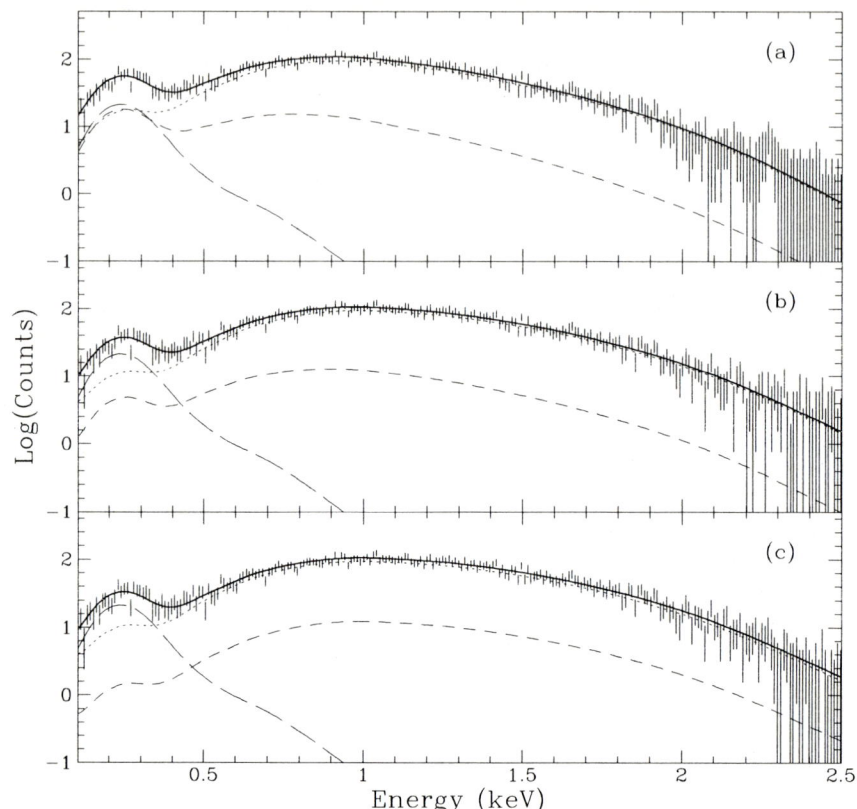

Fig. 2. Simulated 50 ks observation of a galaxy at 5 Mpc in the (a) pessimistic case: $kT = 1.3$, $\alpha_{ph} = 3.5$; (b) realistic case: $kT = 5.0$, $\alpha_{ph} = 2.5$; and (c) optimistic case: $kT = 10.0$; $\alpha_{ph} = 1.5$. The dashed traces (longest to shortest) are the contributions to the composite spectrum from the SSS, HMXB, and LMXB components. Note that all cases are somewhat "pessimistic", in that HMXBs are embedded with the same scale height used for SSSs.

Acknowledgements: Thanks to J. Greiner, J. Grindlay, J. McClintock, R. Remillard and S. Vrtilek for discussions. The work was supported by NSF: GER-9450087.

References

Di Stefano R., Rappaport S., 1994, ApJ 437, 733
Fabbiano G., 1989, ARA&A 27, 87
Kim D., Fabbiano G., Trinchieri G., 1992, ApJS 80, 645
Motch C., Hasinger G., Pietsch W., 1994, A&A 284, 827
Rappaport S., Di Stefano R., Smith J.D., 1994, ApJ 426, 692

Neutron Stars Can Do It Too if They Want to

N.D. Kylafis[1,2]

[1] University of Crete, Physics Department, 710 03 Heraklion, Crete, Greece
[2] Foundation for Research and Technology-Hellas, 711 10 Heraklion, Crete, Greece

Abstract. Neutron stars typically appear as hard X-ray sources. This is because the emission occurs near the neutron-star surface and the lower limit to the characteristic temperature of emission is of order 1 keV. In order to have characteristic temperature of 50 eV and near-Eddington luminosity, the photosphere must be at 1000 neutron-star radii. For sub-Eddington luminosities, a spherically symmetric accretion flow onto a neutron star is optically thin. For near-Eddington luminosities, the accretion flow can be either optically thin or optically thick. It all depends on how the spherically symmetric flow starts at large distances from the neutron star surface. If it starts subsonically, the flow remains subsonic, the velocity decreases with decreasing radius, the optical depth of the flow is huge and the photosphere is pushed out. Thus, a neutron star could, under special conditions, appear as a supersoft source. Such special conditions may occur in transient X-ray sources. Indeed, the source RX J0059.2-7138 is a transient X-ray pulsar in the Small Magellanic Cloud with 90% of its luminosity emitted as a supersoft X-ray spectrum.

1 Introduction

It is by now certain that most supersoft X-ray sources are white dwarfs and the accepted model for them is that proposed by van den Heuvel *et al.* (1992). The supersoft spectrum (i.e., blackbody temperature of the order of tens of eV) is explained as arising from steady nuclear burning of hydrogen accreting onto massive white dwarfs.

With this introduction it should be obvious that I am not pushing the neutron-star model. I only want to point out that, if a neutron star wanted to appear as a supersoft X-ray source, there is a simple way it could do it. I will present evidence below that one neutron star has done it already, however not many of them choose to do it.

An Eddington luminosity and a blackbody spectrum of temperature 50 eV for a supersoft X-ray source imply that the emitting surface is of order 10^{19} cm^2 or a sphere of radius 10^9 cm. This immediately suggests the identification of supersoft X-ray sources with white dwarfs.

When the photosphere of an accreting neutron star is at or near its surface, the emitted blackbody spectrum has temperature of order 1 keV for an Eddington luminosity. Thus, the only way for a neutron star to masquerade as a white dwarf is to push its photosphere at about 10^3 neutron star radii. In the next section I will present a simple way for this to happen.

2 Model

Consider near-Eddington spherical accretion onto a neutron star. If the flow is supersonic (as it is in hard X-ray sources), the flow velocity increases inwards (except possibly near the neutron-star surface), the optical depth to absorption in the accreting matter is negligible and the photosphere is at the neutron-star surface. A lower bound to the temperature at the photosphere is 1 keV. Thus, under such conditions, a neutron star cannot appear as a supersoft X-ray source.

However, if the spherical accretion flow is subsonic, the inflow velocity *decreases* inwards, the optical depth to absorption in the accreting matter is huge and remains larger than one throughout most of the flow. In such a case, the photosphere moves out to very large radii. If the photosphere is at about 10^9 cm, the neutron star masquerades as a white dwarf and the emitted spectrum is that of a supersoft X-ray source.

I want to stress here that only for mathematical simplicity a spherical accretion flow was discussed above. Any distribution of matter (e.g., a torus) that is optically thick to absorption and extends to about 10^9 cm will emit a supersoft spectrum if it is fed from the inside with a near-Eddington luminosity. If the extend r of the optically thick matter is different than 10^9 cm, the blackbody temperature will be $50(L/L_E)^{1/4}(r/10^9 \text{ cm})^{-1/2}$ eV, where L_E is the Eddington luminosity. Thus the peak of the emitted spectrum may shift outside the ROSAT window.

3 Calculations

The hydrodynamic equations (mass, momentum and energy conservation) have been solved numerically assuming LTE from a few thousand neutron star radii to the neutron star radius (Kylafis & Xilouris 1993). These calculations are similar to those of Vitello (1978) and as a check his results were reproduced.

Given the extend of the flow, the only remaining parameter is $\epsilon \equiv 1 - \dot{M}/\dot{M}_E$, where $\dot{M}$ is the radial accretion rate and $\dot{M}_E$ is the Eddington rate. Here ϵ is a measure of the relative importance of the radiation force on the radial flow because the luminosity at infinity is $L_\infty = (1-\epsilon)L_E$. For $\epsilon \lesssim 0.1$, one can always find radial subsonic flows which are optically thick to free-free absorption throughout the flow. In such a case, the emitted spectrum is a blackbody in the tens of eV range.

The character of the radial flow is completely insensitive to the initial velocity. If the initial velocity is subsonic, the flow is subsonic throughout. Similarly for a supersonic initial velocity. In a subsonic flow the velocity decreases as the radius decreases and for a typical example see Fig. 2 of Kylafis & Xilouris (1993).

The LTE temperature in a subsonic flow increases inwards and for a typical example see Figure 3 of Kylafis & Xilouris (1993). The temperature reaches high values as the matter approaches the neutron star and nuclear reactions will take place there. Nevertheless, the increase in the luminosity is small and no qualitative changes in the flow occur due to this increase.

The effective optical depth to free-free absorption from r to infinity $\tau_*(r) \equiv (3\tau_s\tau_a)^{1/2}$ is also increasing dramatically inwards (see Figure 4 of Kylafis & Xilouris 1993). Here τ_s is the corresponding electron scattering optical depth and τ_a is the corresponding free-free absorption one.

Despite the huge optical depth of the accreting matter in a subsonic flow, the amount of matter there is a tiny fraction of a solar mass.

4 The Unique Source RX J0059.2–7138

The source RX J0059.2-7138 is a truly unique X-ray source. It is a transient source in the Small Magellanic Cloud and it was discovered by ROSAT (Hughes 1994). It was also seen by ASCA (Hughes 1996) almost simultaneously.

Its spectrum can be described as consisting of three components: Below 0.5 keV, the spectrum is similar to that of any other supersoft X-ray source. It can be fitted by a blackbody curve with temperature ~ 35 eV, it is unpulsed and it contains 90% of the total luminosity.

Above about 3 keV, the spectrum is a power-law with index 0.7. It is pulsed and at 3 keV the modulation is $\sim 35\%$ (preliminary value). Such a spectrum is typical of X-ray pulsars.

Between 0.5 and 3 keV the spectrum is a power-law with index 2 and has a modulation which is decreasing with decreasing energy.

A reasonable model for this source may be the following: An X-ray pulsar is surrounded by a torus or a thick disk whose outer boundary is at a few thousand neutron-star radii. The observation direction is essentially along the axis of the torus and thus we are able to see the spectrum of the X-ray pulsar unaffected ($E \gtrsim 3$ keV). Most (90%) of the luminosity however is absorbed by the torus and re-emitted from its outer boundary as a supersoft spectrum ($E \lesssim 0.5$ keV). The spectrum between 0.5 and 3 keV is a pulsar spectrum probably contaminated by scattered photons. The lower the energy the more the contamination and the less the modulation.

5 Summary

Neutron stars can in principle appear as supersoft X-ray sources but in practice only one neutron star appears to have done it so far. It is quite likely that this peculiar source is the "Rapid Burster" of the supersoft X-ray sources.

An optical monitoring program of the source RX J0059.2–7138 could in principle tell us when this transient source will be bright again in X-rays. Then we could observe it again and try to understand its details better.

Acknowledgment: I would like to thank Jack Hughes for useful discussions regarding the source RX J0059.2–7138 and for sending me some preliminary data on this source.

References

Hughes J.P. 1994, ApJ 427, L25
Hughes J.P., 1996 (private comm.)
Kylafis N.D., Xilouris E. M., 1993, A&A 278, L43
van den Heuvel E.P.J., Bhattacharya D., Nomoto K., Rappaport S.A., 1992, A&A 262, 97
Vitello P.A.J., 1978, ApJ 225, 694

SSSs as Progenitors of the BHCs

W. Kundt

Institut für Astrophysik der Universität Bonn, 53121 Bonn, Germany

Abstract. In contrast to white dwarfs, neutron stars with near companions can get surrounded by massive accretion disks when their donor stars transfer matter at a greater-than-Eddington rate for sufficiently long. When the disk gets massive, the system will successively pass through the stages of super-Eddington (SES) and/or super-soft (SSS), and end up as a black-hole candidate (BHC).

1 The Known Supersoft Sources – What Are They?

At the Garching workshop, 34 bright supersoft X-ray point sources (SSSs) have been discussed: 6 of them in the LMC, 5 in the SMC, 8 in the Galaxy, and 15 in the Andromeda galaxy (M31). Whereas the class of known supersoft X-ray point sources in the sky consists of over 4000 members, and includes AGN, solitary neutron stars, and probably white dwarfs, the subclass of SSSs considered here is restricted to compact binaries with spectral peaks between 20 and 60 eV, and with peak X-ray powers near the Eddington luminosity of a solar mass, some 10^{38} erg/s. What do these sources consist of?

Work in the past has preferentially concentrated on modelling these sources with white-dwarf binaries, because the deep potential well of a neutron star (n-star) is expected to lead to a harder (X-ray) source than observed. The class of SSSs contains, however, the Galactic Be-star transient RX J0059.2–7138 and the super-Eddington source SMC X-1, both of which reveal their n-star primary through its (short) spin period of 2.7s and 0.71s , respectively. Neutron-star binaries are therefore contained in the class of SSSs. Other members, like the LMC-source CAL 83, suggest their neutron-star affiliation through the presence of a Wolf-Rayet star – as is shown by the Cyg X-3 system, and by Kepler's SNR (Bandiera 1987) as well as by the W.-R. nebulae S 308 and NGC 6888 whose morphologies reveal their being SN shells of tomorrow (Kundt 1996a).

At this point, one enters the controversy of whether or not W.-R. stars are the progenitors of SNe of type Ia : Whereas Oemler and Tinsley (1979) argued strongly against SNe of type Ia stemming from white dwarfs, the consensus of the last decade has changed. My own conviction – of the core-collapse of a massive star being the only viable SN mechanism – was explained in 1990 and has been independently considered by Colgate (1991). The similar energetics, cooling timescales, opacities, (effective) temperature evolutions, relativistic electron populations, and fine structures of all SNe can be understood in terms of similar progenitors which differ by no more than the extent and chemical composition of their envelope. At this meeting, I heard Mario Livio cast (statistical) doubts on the white-dwarf-explosion scenario; cf. Branch et al. (1995).

Which of the remaining SSSs do not contain a neutron star? There is no clear case. None of the presented models – involving a white dwarf – was without difficulties. Crampton et al. (1996) infer from their radial velocities that the compact components in the 3 LMC SSSs RX J0513.9–6951, CAL 83 and CAL 87, and in the Galactic SSS RX J0019.8+2156 all have higher than white-dwarf masses. Their orbital periods (in days) are short: 0.7628, 1.0475, 0.44, and 0.6604; and their velocity amplitudes (in km/s) are small: 11, 35, 40, and 67, whereby the 3rd system eclipses (!), so that 40 km/s is its unprojected velocity amplitude. All four systems therefore look like (becoming) black-hole candidates (BHCs).

Best studied is (the 'Schaeidt source') RX J0513.9–6951, in particular through its (optical, 3-year) MACHO lightcurve reported by Southwell et al. (1996). Its orbital separation is only $10^{11.5}$cm$(M/3M_\odot)^{1/3}$ whereas for a visual brightness of $M_V = -2$ mag, its equivalent visual blackbody radius is $10^{12.2}$cm$T_4^{-1/2}$ (for an effective temperature T in units of 10^4 K), some 5 times larger! This implies that the optical lightcurve cannot be modeled by either star(s) or disk screening each other, rather by an illuminated, extended windzone - like for the neutron-star binary SS 433 (Kundt 1991). The equivalent blackbody radius of the X-ray source, on the other hand, is $10^{9.5}$cm (Schaeidt et al. 1993), i.e. somewhat smaller than the expected size of the accretion disk in the system. How can the X-ray outbursts influence – reduce by typically one magnitude – the optical output? When the (almost pointlike) X-ray source is the lamp that illuminates the extended windzone, and when its increase by a factor of $\gtrsim 10^3$ reduces (rather than enhances) the optical output, I conclude that (not the illumination but) the windzone gets depleted during X-ray outbursts. How ? Kundt and Fischer (1989) propose a wind from the disk's inner edge, driven centrifugally by the star's corotating magnetosphere; see also Horn and Kundt (1989). Such a centrifugally driven wind may well be reduced during X-ray flares of the disk - most likely when the disk is massive (so that it can strongly confine the star's magnetosphere).

Are we dealing with a white-dwarf or a neutron-star system? Apart from the feeding requirements – which are some 10^3 times more modest for a neutron star – and the energy budget – a neutron star can store some 10^3 times the energy in its spin motion – there is the qualitative difference that an accretion disk around a white dwarf would require a feeding rate of more than $10^{-5} M_\odot/yr$ in order to fill up to a sizable weight whereas the corresponding (Eddington) limit for a neutron star is only $10^{-8} M_\odot/yr$. The situation can be compared to the filling of a funnel with a wide or narrow neck, respectively. Neutron stars are expected to surround themselves with self-gravitating disks whenever their companion donates more than the Eddington rate for sufficiently long; (cf. Krolik 1984).

In other words: I propose that both the super-Eddington X-ray sources (SESs), and the SSSs contain massive accretion disks, i.e. are progenitors of the BHCs. In the first case, only a massive disk is thought to be able to supply more than the Eddington mass rate down to the neutron star surface, in the form of thin, heavy 'blades' in the orbit plane (Kundt et al. 1987). (Less heavy feeding would lead to near-spherical scenarios for which the radiated power stays below the Eddington limit). Known SESs are: SMC X-1, the LMC transient A0535–668, LMC X-3, LMC X-4, and Cir X-1, with respective X-ray powers L_X given by $\log(L_X/\text{erg s}^{-1}) = 38.8, \lesssim 39.1, 38.5, 38.6$, and 38.9.

In the case of the SSSs, note that the gravitational potential of a massive disk is of order 2GM(disk)/R(disk), roughly 3 times shallower than that of a white dwarf, so that its formation power – limited by $L_{\text{Edd}}[M(disk)]$ – should be radiated at supersoft X-ray temperatures. X-ray outbursts result whenever new material is added to the disk and/or sinks deeper into the disk's (plus the n-star's) gravitational well. During such re-settlings ('disk quakes'), the windzone of the system is temporarily depleted because the neutron star's magnetosphere is more strongly confined by disk material at the inner edge. Quasi-periodic X-ray outbursts thus occur due to accretion instabilities, during which the wind-mass loss from the compact component is strongly reduced, leading to a systematic reduction of the optical output, by as much as one magnitude (in most systems).

As has already been mentioned, this interpretation (of an SSS) agrees with our understanding of the sources RX J0059.2–7138, CAL 83, SMC X-1, and SS 433, and yields an explanation of how the optical luminosity – with its distinctly larger radiating area than that of the binary orbit – can vary in anticorrelation with the (quasi-pointlike) X-ray source. It is consistent with the (large) mass estimates by Crampton et al. (1996) for the compact components, with the absence of nova explosions and planetary nebulae (except for CAL 83), and it is demanded – as their necessary progenitors – by the existence of the (class of) black-hole candidates which will be discussed in the next section. Moreover, a bipolar flow is indicated in (the Schaeidt source) RX J0513.9–6951 whose speed (of $\gtrsim 3800$ km/s: Pakull et al. 1993; Crampton et al. 1996) exceeds the record for outflows in PNe (of 1600 km/s, in the core of M 2-9) by more than twice: Balick (1989), Mellema (1996), i.e. asks for a more powerful (than white-dwarf) engine.

2   The Black-Hole Candidates

The BHCs apparently fill the (large) mass gap between the low-mass X-ray binaries (LMXBs) and HMXBS. They are commonly thought of as consisting of stellar-mass black holes surrounded by low-mass accretion disks, the latter filled by a near companion. High-mass disks tend to be excluded from consideration – with the notable exception of Krolik (1984) – because they are thought to be unstable. This despite the fact that all of us live happily in the (self-gravitating) Milky Way! What are the possible fates of a heavy disk? Note that both the clumping instability and the bar-mode instability of a gaseous disk are strongly reduced in the (relevant) case of a degenerate disk, so that a thin Maclaurin spheroid may not be a poor approximation to a massive disk in a compact binary (Kundt 1979).

How stable is the neutron star's location at the center of a massive disk? Heemskerk et al. (1992) assess the development of m = 1 instabilities in a self-gravitating gaseous disk around a central star with arbitrary mass ratio μ of disk and central star, and find dynamical instability for $\mu > 1$. There are at least three reasons, however, for why their interesting results may not conflict with the presence of massive disks in neutron star binaries. First, experience with pottery and pizzas tells that existing instabilities need not be excited: if the disk around a neutron star is filled up sufficiently slowly and smoothly, throughout some 10^7 yr, the location instability may never grow important. Second, all known neutron stars are believed to have surface magnetic fields above 10^{11} G with which they push against the inner edge of the disk, thereby reducing the location instability, (Kundt et al. 1987, Kundt 1990; note in particular that the msec pulsars are predicted to be γ-ray pulsars – instead of radio pulsars – if their surface fields were as weak as their dipole components). Third, even if the location instability were fully excited, and the neutron star would drift into the disk, both star and disk would still conserve their identities because of the Eddington limit (which forbids the star to swallow its disk rapidly).

Neutron stars surrounded by massive disks are therefore expected, and may well explain the BHCs, (Kundt and Fischer 1989; Lewin and Tanaka 1995; Kundt 1996a). They are particularly wanted for the following seven reasons: • (1) The X-ray lightcurves of the BHCs during outburst all peak near the Eddington luminosity of a neutron star, $L_{Edd}(1.4M_\odot)$, rather than near that of their (several times larger) mass. • (2) After recovery, during (decade-long) quiescence, their X-ray luminosities fall to extremely low values, between $10^{30.8}(erg/s)(d/kpc)^2$, in the case of A 0620–00 (McClintock et al. 1995), and $10^{33.9}(erg/s)(d/3.5\ kpc)^2$, in the case of V404 Cyg (Wagner et al. 1994). Implausibly low disk viscosities would have to be postulated for a low-mass disk in order to model this unexpected behaviour – whereas there are no difficulties for a rigidly rotating spheroid. • (3) The (stacked, noisy) optical lightcurves of A 0620–00 and Nova Muscae in quiescence are almost indistinguishable, and lack reflection symmetry about the two minima which would be implied in the absence of a luminous, asymmetric windzone, (Remillard et al. 1992). A BH–disk system would be unable to blow such a dense windzone: the corotating magnetosphere of a neutron star is required. • (4) An extended, variable, luminous windzone is likewise signalled by the strong Balmer emission lines, and by the noisy LOS line velocities. • (5) During high state, the spectral power of BHCs can peak above 1 MeV, see Fig. 1. Note that the equivalent blackbody area is as small as your hand, i.e. we deal with microscopic emitters. No stellar-mass BH can radiate such a spectrum; sparks from a solid surface are required. • (6) The intensity-fluctuation power spectra look like those of the LMXBs of high neutron star spinrate (Van der Klis 1994). • (7) The BHCs share the following further properties with established neutron star sources: (a) polarized optical emission, (b) high-low state bimodality, (c) superhumps, (d) Li absorption (observed in V404 Cyg and A 0620–00, like in the X-ray pulsator Cen X-4: Martin et al. 1992), (e) X-ray dipping, (f) formation of (radio) jets, i.e. of a relativistic electron population, (cf. Kundt 1996b). Any of these seven properties argues in favour of a neutron star as the core of the compact component.

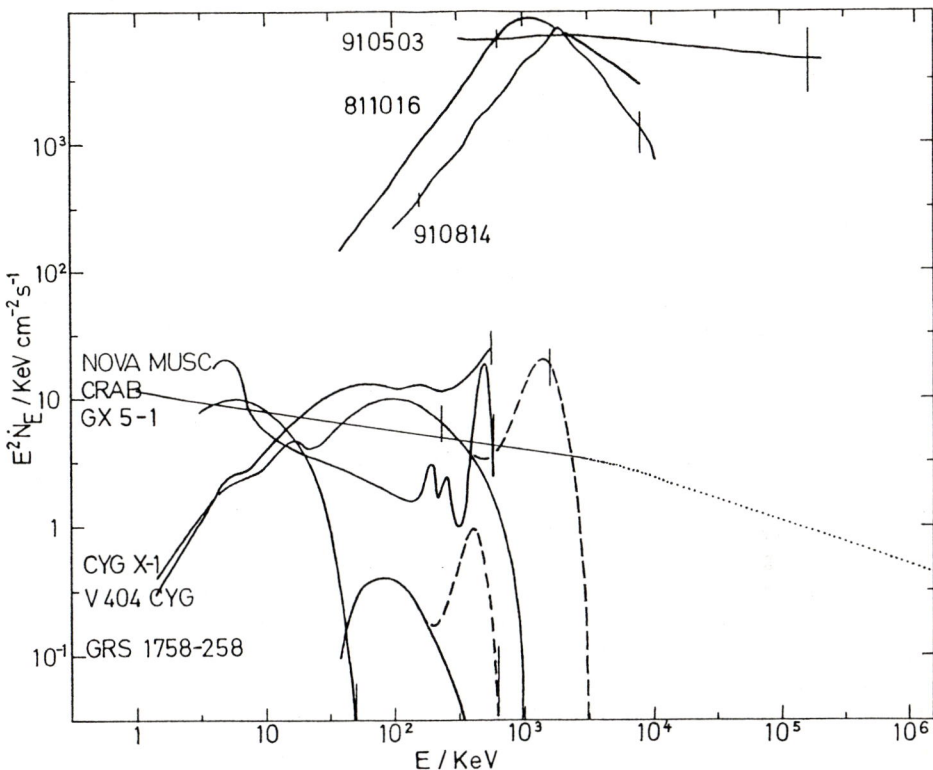

Fig. 1. A few (hard) X-ray and γ-ray spectra of binary X-ray sources, BHCs, γ-ray bursts, and the Crab pulsar, taken from Kundt and Chang (1993). Note that the BHCs in their high state (broken) can have spectra whose power (per logarithmic frequency) peaks above 1 MeV.

3 Summary

For no obvious reason, self-gravitating accretion disks have been widely ignored in the literature. They are likely to form in close neutron-star binaries, and to give rise in succession to the SESs, SSSs, and BHCs.

Acknowledgement: I have greatly enjoyed the open-minded atmosphere provided by the LOC, and reflected by all the participants. My thanks for the manuscript go to Duncan Lorimer.

References

Balick B., 1989, AJ 97, 476
Bandiera R., 1987, ApJ 319, 885

Branch D., Livio M., Yungelson L.R., Boffi F.R., Baron E. 1995, PASP 107, 1019
Colgate S.A., 1991, in Supernovae, S.E. Woosley (ed.), Springer, p. 585
Crampton D., Hutchings J.B., Cowley A.P., Schmidtke P.C., McGrath T.K., O'-Donoghue D., Harrop-Allin M.K., 1996, ApJ 456, 320
Heemskerk M.H.M., Papaloizou J.C., Savonije G.J. 1992, A&A 260, 161
Horn S., Kundt W., 1989 Ap&SS 158, 205
Krolik J., 1984, ApJ 282, 484
Kundt W., 1979, A&A 80, L7
Kundt W., 1990, in Neutron Stars and their Birth Events, W. Kundt (ed.), NATO ASI C 300, Kluwer, p. 40
Kundt W., 1991, Comments Astrophys. 15, 255
Kundt W., 1996a, in Multifrequency Behaviour of High-Energy Cosmic Sources, F. Giovannelli (ed.), Frascati Workshop 1995 (in press)
Kundt W., 1996b, in Jets from Stars and Galactic Nuclei, W. Kundt (ed.), Lecture Notes in Physics 471, Springer, p. 1
Kundt W., Chang H.-K., 1993, Ap&SS 200, 151
Kundt W., Fischer D., 1989, JA&A 10, 119
Kundt W., Özel M., Ercan E.N., 1987, A&A 177, 163
McClintock J.E., Horne K., Remillard R.A., 1995, ApJ 442, 358
Lewin W.H.G., Tanaka Y., 1995, in X-ray Binaries, W.H.G. Lewin, J. van Paradijs, E.P.G. van den Heuvel (eds.), Cambridge Astrophys. Series, Chapter 3
Martin E.L., Rebolo R., Casares J., Charles P.A., 1992, Nat. 358, 129
Mellema G., 1996, in Jets from Stars and Galactic Nuclei, W. Kundt (ed.), Lecture Notes in Physics 471, p. 149
Oemler A., Tinsley B.M., 1979, AJ 84, 985
Pakull M.W., Moch C., Bianchi L., Thomas H.-C., Guibert J., Beaulieu J.P., Grison P., Schaeidt S., 1993, A&A 278, L39
Remillard R.A., McClintock J.E., Bailyn C.D., 1992, ApJ 399, L145
Schaeidt S., Hasinger G., Trümper J., 1993, A&A 270, L9
Southwell K., Livio M., Charles P.A., et al., 1996, this volume p. 165
Van der Klis M., 1984, A&A 283, 469

Part II

Winds and Accretion Disks in SSS

Simulation of the Visual Light Curve of CAL 87

S. Schandl, E. Meyer-Hofmeister, F. Meyer

Max-Planck-Institut für Astrophysik, Postfach 1523, 85740 Garching, Germany

Abstract. We model the visual light curve of CAL 87 based on the assumption that an accreting steadily burning white dwarf irradiates the accretion disk and the secondary star, as suggested by van den Heuvel et al. (1992). We use constraints on the geometry derived from the known orbital period. As sources of visual light we include the secondary star and an accretion disk with an optically thick, cold, clumpy spray at its rim, presumably caused by an accretion stream of high mass flow rate impinging on the disk at the hot spot. This spray moving around the disk can account for the asymmetry in the light curve and the depth of the secondary minimum. It also might be the cause of the observed low X-ray luminosity if the white dwarf is permanently hidden by this disk.

1 Introduction

The ROSAT observations led to the discovery of a series of luminous X-ray sources with very soft spectra. For several of these supersoft sources optical counterparts were found, mainly blue objects. It became clear, that these objects might come from various different types of systems, including nuclei of planetary nebulae and novae. An interesting subclass are the binary supersoft sources, in which mass flows over from a secondary star on to the white dwarf primary. If the orbital period is known clear constraints for the geometry of the binary follow. The mass of the secondary, assumed to be on or near to the main sequence, is connected with the size of the Roche lobe. The size of the accretion disk is limited correspondingly. As suggested by van den Heuvel et al. (1992, later referred to as vdH) the mass transfer rate lies in a narrow range of the order of $10^{-7}\,M_\odot$/yr to allow stationary nuclear burning on the white dwarf and the mass ratio cannot be very different from $q = M_{\text{secondary}}/M_{\text{wd}} = 2$.

For several systems orbital periods are known: e.g. CAL 83 (25 hr, Smale et al. 1988), CAL 87 (10.6 hr, Callanan et al. 1989), RX J0019.8+2156 (15.8 hr, Beuermann et al. 1995), RX J0513.9-6951 (18.2 hr, Crampton et al. 1996) and 1E 0035.4-7230 (4.1 hr, Schmidtke et al. 1996). For low inclination systems where secondary star and disk do not eclipse each other one would expect the variation of the optical light with phase to be nearly sinusoidal, caused mainly by the varying aspect of the partially illuminated secondary. For higher inclinations the eclipses produce a signature of the contributing disk areas and of parts of the secondary, both illuminated by the hot white dwarf. Cal 87 is a unique test case because of its high inclination.

Table 1. The System Parameters

Observations:

orbital period [1]	P =	10.6 h
distance to LMC	d =	57.5 kpc

SSS-model dependent parameters [2]:

white dwarf	M_1 =	0.75 $M_\odot$
secondary star	M_2 =	1.5 $M_\odot$
mass accretion rate	$\dot{M}$ =	$0.8 \, 10^{-7}$ $M_\odot$/yr
=> separation	a =	$2.2 \, 10^{11}$ cm
=> luminosity of WD	L =	$2.2 \, 10^{37}$ erg/sec

Parameters used in the simulation:

temperature of the non-irr. sec. star [3]	$T_\star$ =	8000 K
efficiency of radiation reprocessing	η =	0.5
inclination	i =	77°

References: 1 Callanan et al. (1989), 2 van den Heuvel et al. (1992), 3 Schaller et al. (1992)

In an earlier attempt Callanan and Charles (1989) had already tried to fit the optical light curve, at that time interpreting the source as a LMXB and therefore using a mass ratio very different from what is expected now.

It is the aim of our present investigation to determine a theoretical light curve for CAL 87, based on the model of vdH for steady nuclear burning on accreting white dwarfs. We model the contributions from the accretion disk, hot spot spray and the secondary star, each of them illuminated by the central hot white dwarf (see also Schandl et al. 1996). Following the results of vdH we have a narrow range of possible mass accretion rates, white dwarf masses and secondary star masses. We obtained values listed in Tab. 1. Additionally the well observed optical light curve puts severe constraints on the theoretical model.

The success of our model led us to calculate the light curve of CAL 87 if it would be seen under different inclinations. We found solutions similar to the light curves of RX J0019.8+2156 and CAL 83 (Schandl et al. 1996).

2 Observational and Theoretical Constraints

Looking to the light curve of CAL 87 its asymmetry will be noticed immediately. In this paper we want to follow the idea that the interaction of the accretion stream with the disk rim could be the origin of this: Looking to the basic constituents of a binary system, the compact star, its companion, the disk and the mass flow, only the effects of the hot spot can add an asymmetric component to the light curve. The high accretion rate suggested in the vdH model on the order of $10^{-7}\,M_\odot$/yr could lead to rather dramatic effects when this gas impinges the disk rim. In other X-ray binaries such sprays have already been considered.

The asymmetry of the light curves of eclipsing nova-like cataclysmic variables is also understood as resulting from the hot spot (e.g. Smak 1971, Smak 1994). In these systems the companion does not significantly contribute to the optical light and one observes only the disk and the hot spot. Their maximal optical light is observed just before the eclipse of the disk when we look directly towards the hot spot which is then located in front of the white dwarf (see Fig. 1 for the geometry which is the same for all close binaries with mass transfer). – In contrast, CAL 87 shows more optical luminosity *after* eclipse than before. In the model this is a natural consequence of the hot spot and the irradiation effect: The geometrical thick hot spot region acts as a large screen collecting a lot of radiation flux of the white dwarf. The emitted reprocessed optical light can be well observed at orbital phases where the hot spot is behind the white dwarf, that is after the eclipse (Fig. 1). Before eclipse we look against the non-irradiated outer rim of this area which covers the illuminated inner disk regions (Callanan et al. 1989, Cowley et al. 1990).

The optical light curve of LMXB 4U 1822-37 shown by Mason et al. (1980) is similar to that of CAL 87. Also there the intensity drop at the phases before eclipse was connected to the interaction between accretion stream and disk. Hellier & Mason (1989) modelled the optical and X-ray light curves including a bulge around phase 0.8 or two bulges around phase 0.8 and 0.2 where they connect the 0.8 – bulge with the impact of the accretion stream on the disk (where their phase 0.8 corresponds to 72° in our notation, see e.g. Fig. 2 below). In their simulation they added normalization parameters for the emission of the disk rim and the inner disk and also took for the illuminated secondary an optional constant plus sinusoidal component.

Noting this similarity Callanan & Charles (1989) calculated a fit to the light curve of CAL 87 using a similar model to that used in determining the disk structure of LMXB 4U 1822-37 by Hellier & Mason (1989). The stellar masses

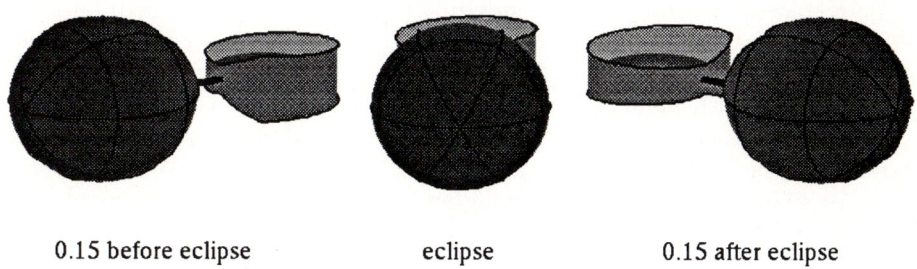

0.15 before eclipse   eclipse   0.15 after eclipse

Fig. 1. A model of CAL 87 around eclipse.

in their LMXB model for CAL 87 are presumably different from those expected now (mass ratio $q \simeq 0.2$ in Hellier & Mason (1989), $q = 2$ here). The phase dependence of the secondary's light described as a constant plus sinusoidal seems not to be adequate (compare the light curve in our Figure 3, model a and b).

For our analysis we first developed models for the hot spot region with a small vertical extension, without success. The total luminosity and the shape of the light curve (secondary minimum) did not nearly fit the observations. Instead of this, models with a thick structure at the disk rim allow to find a proper fit with small variations of the involved quantities (radius, height and temperature) in agreement with a similar result by Hellier and Mason (1980) for their simulations before.

This description is also supported by the low L_x/L_{opt} ratio suggesting that the central X-ray source is hidden from direct view similar to LMXBs, as supposed by Pakull et al. (1988). Additionally, there would be a deep eclipse in the X-ray light curve if the observed X-rays originated from the central white dwarf and not from scattering of surrounding gas as the observations suggest (Schmidtke et al. 1993, Kahabka et al. 1994). In order to cover the source the optically thick region has to have a vertical extension at each angle of azimuth of $z/r \geq (\pi/2 - inclination)$ which is about 0.25 in our calculations. The maximal simulated opening angle of the gas will be 0.45.

We compare this to the scale height of the disk $H_{\mathrm{disk}}/r = 0.04\, T_4^{1/2}$ and of an X-ray induced corona $H_{\mathrm{cor}}/r = 0.25\, T_{5.6}^{1/2}$ where the temperature $T = 10^{5.6}$ K corresponds to the X-ray flux obtained by the vdH model. Therefore, cold matter can not reach the required vertical height in form of a hydrostatic layering. Although the coronal scale height is of the right order it will not be optically thick to cover the central source continuously.

Therefore, the observed optical light has to be emitted by a cold and clumpy gas. This spray is embedded in the hot corona and both are in pressure equilibrium. All spray matter should then cover the central white dwarf all the time. Expelled in all directions the spray must move along free fall trajectories around the accretion disk. This can explain the large vertical height required to reproduce the observations mentioned above.

Although the hydrodynamic generation of such a two phase medium is not yet fully understood there are several conclusive arguments for their existence. Howarth & Wilson (1983) and Karitskaya et al. (1986) observed cold blobs ($T \simeq 10^{4.4}$ K) with number densities of $n \simeq 10^{13.3}$ cm^{-3} of the X-ray binary HZ Her/Her X-1 in optical and UV data. Bochkarev & Karitskaya (1989) suggested a physical model for these observed blobs. Schandl (1996) showed that the spray generated by a stream impinging onto a warped disk surface generates the preeclipse dips in the X-ray light curve of Her X-1. Frank et al. (1987) explained with such a two phase medium the optical light curves of dipping and coronal LMXBs. Also the observations of Cal 87 (Hutchings et al. 1995) showing gas in the temperature range between 25000 K and 29000 K around eclipse are consistent with such a spray.

Summarizing, the asymmetry in the optical light curve, the depth of the secondary minimum, the low L_x/L_{opt} ratio and the small decrease in X-rays during the deep optical eclipse may be understood by an optically thick spray of large vertical height which permanently covers the white dwarf.

3 The Simulation

3.1 The Secondary Star

We describe the surface of the secondary star with a spherical grid of $1°$ resolution in the polar and the azimuthal angles. Further on, the grid points lie on the equipotential surface of the Roche lobe of the star.

The observation of the secondary star is difficult because of a star which is in the close projected vicinity of CAL 87. After subtracting the light of this star the remaining spectrum shows apparently no signature of the secondary star (Pakull et al. 1988 and private communication 1995). Therefore we assume a temperature for the non-irradiated secondary of about $T_\star = 8000$ K as calculated for a main sequence star of 1.5 $M_\odot$ with a radius of $9.7 \, 10^{10}$ cm (= Roche radius) and 0.25 solar metallicity specific for the LMC (Schaller et al. 1992).

We describe the source of radiation, the white dwarf, as a point source. The irradiation of each surface element of the secondary depends on the angle of incidence θ and one obtains for the temperature of the irradiated star

$$\sigma T_{\star\mathrm{irr}}^4 = \sigma T_\star^4 + \eta \frac{L}{4\pi r^2} \cos\theta \qquad (1)$$

with σ Stefan-Boltzmann constant, r distance between the white dwarf and the surface element. The parameter η is the efficiency of the reprocessing of the illuminating radiation from the white dwarf to thermal radiation, comparable to $(1-a)$ where a is the albedo.

Depending on the geometry there is no irradiation of surface areas located behind the illumination horizon or in the shadow of the accretion disk (see Fig. 3 model a or b).

In some models below we include energy transport from irradiated to non-irradiated parts of the stellar surface. We took a simple description because the correct solution of this problem is complex and beyond the scope of this paper. We spread the luminosity $\sigma T_{\star\mathrm{irr}}^4 \, dq$ of each surface element with area dq weighted by a Gaussian kernel and integrate over the whole surface. The resulting luminosity of each surface element is

$$\sigma T_{\star\mathrm{irr},\mathrm{spread}}^4 \, dq = \mathcal{N} \, dq \int \sigma T_{\star,\mathrm{irr}}'^4 \exp\left(-\beta'^2/B^2\right) dq'. \qquad (2)$$

β is the angle between the normal vectors of the surface elements dq and dq' and the parameter B describes the angular width of the spreading. In our simulations we use $B = 45°$. The normalization constant $\mathcal{N}$ achieves the equality of the total luminosity of the star with and without energy transport:

$$\int \sigma T_{\star\mathrm{irr},\mathrm{spread}}^4 \, dq = \int \sigma T_{\star\mathrm{irr}}^4 \, dq. \qquad (3)$$

We take a black body spectrum for the emitted radiation of the secondary star where λ is the wavelength, h the Planck constant, c velocity of light and k the Boltzmann constant,

$$B_\lambda(T) = \frac{2\,h\,c^2}{\lambda^5} \frac{1}{\exp\left(\frac{hc}{kT\lambda}\right) - 1}. \tag{4}$$

This radiation flux is folded with an optical filter function $w(\lambda)$ (Allen 1973) to get the optical flux of each surface element

$$f_{\text{opt}} = \int_{\text{filter}} \pi\, B_\lambda(T)\, w(\lambda)\, d\lambda. \tag{5}$$

Finally the total observed optical flux is calculated using the angle ϑ between the normal vector of the surface element and the direction to earth:

$$F_{\star\text{total,opt}} = \frac{1}{\pi d^2} \int f_{\text{opt}} \cos\vartheta\, dq \tag{6}$$

3.2 The Accretion Disk

This section describes the disk without any interaction with the accretion stream. The modifications due to the spray are the topic of the next section.

The disk height z_0 (height of the photosphere above midplane) of the irradiated disk was determined from vertical structure computations for several values of $\dot{M}$, M_1 and α consistently. Approximation by power law gives

$$z_0/r = 10^{-0.95}\, r_{11}^{0.093} \left(\frac{M_1}{\rm M_\odot}\right)^{-0.38} \dot{M}_{-7}^{0.17}\, \alpha_{0.3}^{-0.12}. \tag{7}$$

The disk temperature is determined by frictional heating and irradiation, with θ again angle of incidence

$$\sigma T_{\text{d,irr}}^4 = \frac{3}{8\pi} \frac{GM_1 \dot{M}}{r^3} + \eta \frac{L}{4\pi\, r^2} \cos\theta \tag{8}$$

with G the constant of gravitation. The visual light of the black body radiation is calculated following Eqs. (4) - (6).

The disk size is taken as $r_{\text{d}} = 0.8\, r_{\text{L}}$ (r_{L} Roche lobe radius) according to the models of Paczyński (1977) and Papaloizou & Pringle (1977).

3.3 The Spray

We model the optically thick surface of the spray matter in form of an increase of the disk size, height and temperature. As observed by Hutchings et al. (1995) the temperature around eclipse is between 25000 K and 29000 K. Due to the illumination by the white dwarf, the corresponding surface temperature increases (see Eq. 8) and increases the observed mean temperature of all surface elements.

We assume the lower observed limit, 25000 K, for the temperature of the unilluminated spray.

The spray moves not only along the disk rim but also towards smaller disk radii ($r < 0.8\,r_\mathrm{L}$). But during our investigations we found that there is no real difference in the final light curve between solutions which also modify temperature and height at these inner disk regions. The reason is the high inclination of CAL 87 which makes the vertical extension much more relevant than the radial one. Therefore we restrict the modification of the disk shape to radii $r \geq 0.8\,r_\mathrm{L}$.

In Fig. 2 the radial and vertical modifications are shown in dependence on the azimuth ϕ starting at the phase of impact. The radius increases up to $r = 1.2\,r_\mathrm{L}$ at about $\phi = 45°$ and decreases at later phases to its original value. This model follows calculations of free fall trajectories of matter expelled at the hot spot. The maximal vertical extension of the disk of about $z/r = 0.45$ and the maximal radial extension are placed at about $45°$ after the impact.

The irradiation of the spray is calculated according to Eq. (1), the spectrum is taken as black body and the final optical flux calculation is performed in the same way as for the secondary star.

Numerically we describe the disk modified by the spray in polar coordinates with a resolution of $1°$ in azimuth using 200 concentric rings. Because the disk size varies with the azimuth some outer surface elements do not contribute.

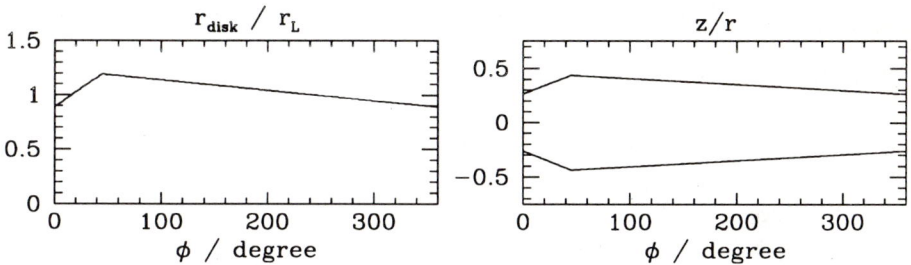

Fig. 2. The left panel shows the disk radius dependent on the azimuth where $\phi = 0$ is the phase of impact. The maximal disk height z/r is represented in the right diagram.

4 The Effect of Spray and Energy Transport on the Light Curve

Figure 3 shows four different models with and without the effects of the spray and the energy transport (see table 2). For comparison we set the inclination of all models equal to $77°$ as this is the best value for our final fit model d. The efficiency parameter $\eta = 0.5$ is changed only for model b to $\eta = 1$ to reproduce the total luminosity.

4.1 Model a

Model a consists of a disk not modified by the spray and a star without energy transport. This solution shows that the contribution of the disk itself is too small to reproduce the high luminosity around eclipse. At these phases ($\phi = 0.9 - 1.1$) the non-illuminated surface of the star shows only a constant signal. In contrast to this the width of the simulated eclipse fits the data very well. In connection with the V-shaped eclipse, this indicates an extension of the eclipsed body comparable to the disk size. a hot spot located only in the vicinity of the impact would produce a narrower U-shaped eclipse like that of a point source. Thus the effect of the stream–disk interaction has to modify the whole disk rim as the spray might do.

Because of the thin disk, the shadow on the secondary is small. The resulting large irradiated area yields a high luminosity around $\phi = 0.5$ where we look directly to the bright front side. Due to the small projected area of a thin disk also the depth of the secondary minimum around $\phi = 0.5$ is weak. This disk barely covers the secondary.

4.2 Model b

In contrast, the disk together with the spray produces a deeper secondary minimum shown in model b (see Fig. 4 for illustration of the different orbital phases). Its irradiated area is smaller than in model a and the overall contribution of the star is less although we set the efficiency parameter $\eta = 1$.

This model of a spray at the disk rim reproduces the observed data much better compared to model a. Only the width of the eclipse is too small although most optical light comes from the spray at the outer disk regions. This indicates that the disk is even larger or that the secondary contributes to the luminosity at these phases ($\phi = 0.0 \pm 0.1$). For the latter possibility we now investigate the effect of a transport of energy on the secondary.

Table 2. The differences between the models

model	spray	energy transport	η
a	no	no	0.5
b	yes	no	1.0
c	no	yes	0.5
d	yes	yes	0.5

4.3 Model c

We set the energy transport width $B = 45°$ in Eq. (2). This value also fits the averaged blue light curve of the X-ray binary Her X-1 which is a binary with an irradiated 2.2 $M_\odot$ companion suggesting a comparable stellar structure as in CAL 87. The advantage of that system is that there the contribution of the accretion disk to the optical light curve is small because of its special geometrical structure and thus nearly the whole light can be assigned to the secondary.

In general we find, the larger the transport width B, the larger the range of suitable fits. Model c corresponds to model a, but energy transport of width $45°$ is included. Even at phase $\phi = 0.0$ the star is brighter due to energy transported to the back side of the star. Due to this the width of the minimum of the stellar contribution is smaller when compared to models a and b. The clear increase of the optical stellar light between phases $\phi = 0.0$ and $\phi = 0.1$ reduces the demands on the disk luminosity.

The stellar luminosity is larger than in model a even around $\phi = 0.5$ where we look directly to the irradiated stellar surface. The directly illuminated parts contribute less than in model a, but the heated regions in the shadow and behind the illumination horizon contribute more than before, especially because of the lower bolometric corrections for these lower temperatures (Eq. (5)).

This shift of energy into the wavelength range of the optical filter is even more evident in the comparison of the stars of model b and d, although η of model b is twice that of model d.

4.4 Model d

Model d includes energy transport on the secondary and the spray described in section 3.3. It combines two striking features of the light curve: (1) the strong depth of the secondary minimum is obtained by the spray because its large projected area covers the secondary well, (2) a slightly better fit around eclipse compared to model b results from the energy transport on the secondary, which contributes to the light curve at these phases, reducing the strong demand on the disk. Additionally, because of the more realistic efficiency parameter $\eta = 0.5$ we prefer model d to model b.

The solution agrees with the predictions of Cowley et al. (1990) who suggest that there is no significant heating of the secondary. vdH described the heated side of the star to be at least three times brighter than the non-heated side. This is also reproduced in model d.

5 Conclusions

Following the model of vdH we find a reasonable fit for the visual light curve of CAL 87 including the secondary, the disk and an extended spray region. This extended region presumably created by the material impinging on the disk acts like a screen for the radiation from the hot white dwarf and provides the high amount of optical light, which cannot originate in a thin disk (Meyer & Meyer-Hofmeister 1995, and model a). This results in a permanently hidden white dwarf, covered by the spray. This is supported by the observed low ratio of X-ray to optical luminosity on the order of 1–10 and the missing deep and wide X-ray eclipse of the white dwarf.

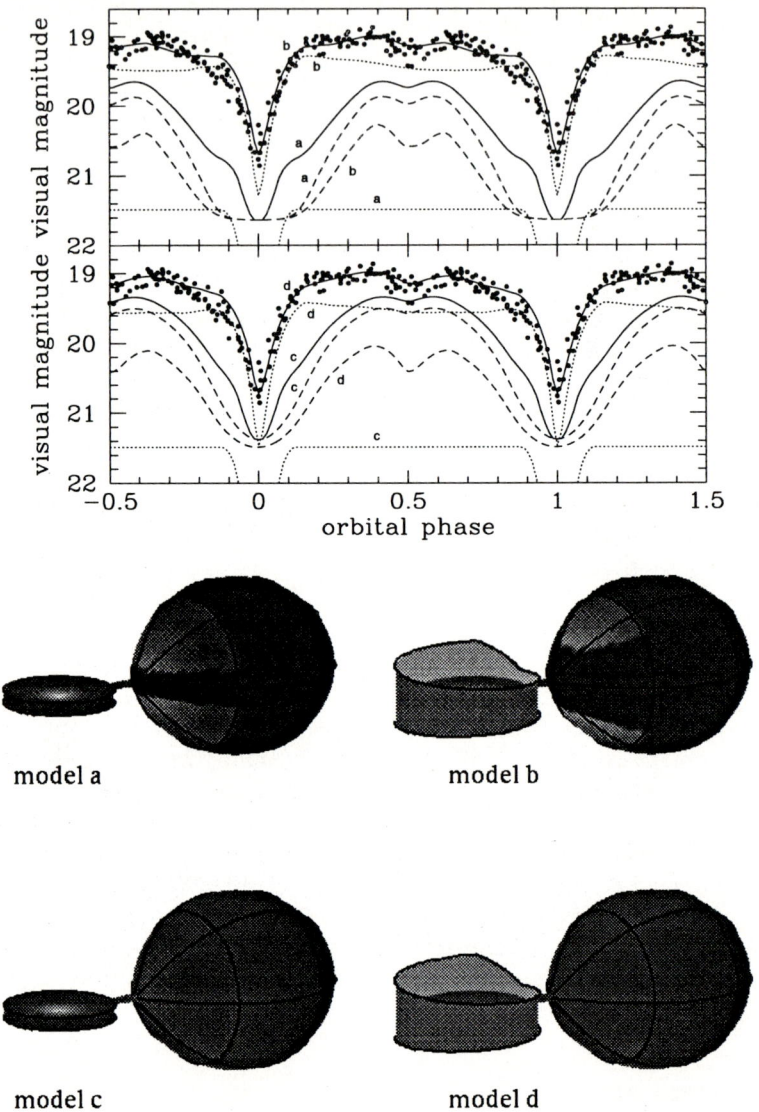

Fig. 3. Shown are light curves of four simulations corresponding to model a and b (upper panel), c and d (lower panel). For a detailed description see section 4. For each model the simulated optical light curve (solid lines) and the contribution of the star (dashed lines) and the disk (dotted lines) are drawn. The dots show the composite V light curve of CAL 87. The photometry is done between 1985 November and 1992 December (Schmidtke et al. 1993). A view to the systems at $\phi_{\rm orbit} = 0.35$ is shown below (1/4 resolution of the calculations).

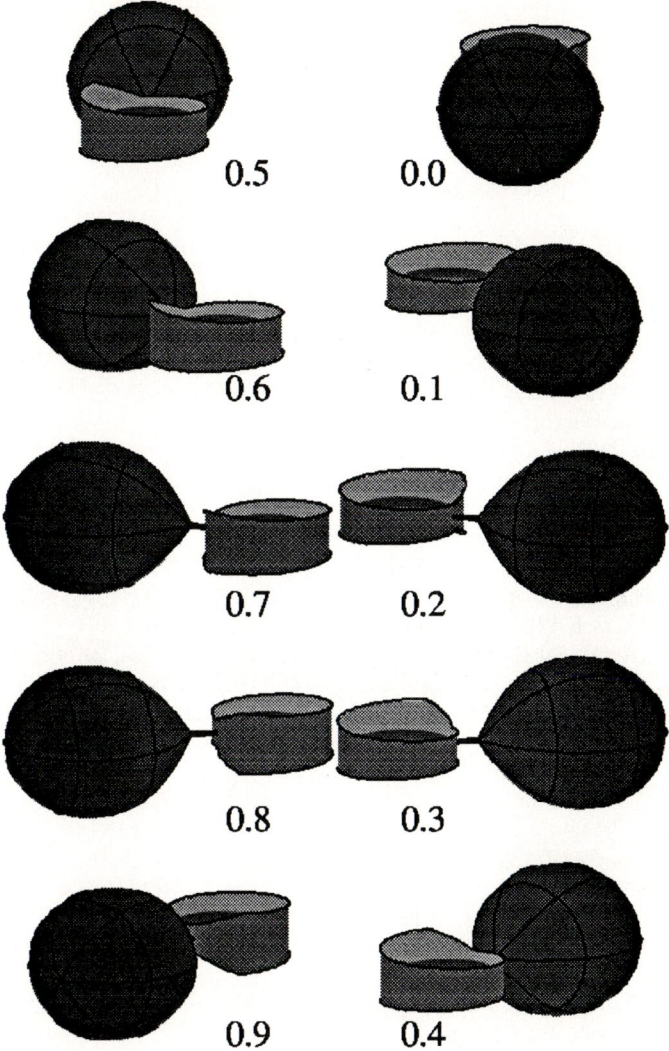

Fig. 4. This figure illustrates an orbit of the system (model d).

Including energy transport of the irradiated flux to non-illuminated parts on the secondary surface yields slightly better results around primary eclipse. It increases the visual light from the secondary because of lower bolometric corrections. Therefore, one obtains a high stellar contribution to the light curve using a reasonable efficiency parameter of radiation reprocessing, even if the irradiated surface area is small because of the large disk shadow.

Acknowledgements: We are grateful to Paul Schmidtke who kindly provided original data of CAL 87.

References

Allen C.W., 1973, Astrophysical Quantities, 3rd edition, Univ. London, The Athlone Press, p. 205
Beuermann K., Reinsch K., Barwig H. et al., 1995, A&A 294, L1
Bochkarev N.G., Karitskaya E.A., 1989, Ap&SS 154, 189
Callanan P.J., Charles P.A., 1989, ESA SP-296, 23rd ESLAB symposium, p. 139
Callanan P.J., Machin G., Naylor T., Charles P.A., 1989, MNRAS 241, 37
Cowley A.P., Schmidtke P.C., Crampton D., Hutchings J.B., 1990, ApJ 350, 288
Crampton D., Hutchings J.B., Cowley A.P., Schmidtke P.C., McGrath T.K., O'-Donoghue D., Harrop-Allin M.K., 1996, ApJ 456, 320
Frank J., King A.R., Lasota J.-P., 1987, A&A 178, 137
Hasinger G., 1994, Reviews in Modern Astronomy 7, ed G. Klare, Astron. Gesellschaft, p. 129
Hellier C., Mason O., 1989, MNRAS 239, 715
Howarth, I.D., Wilson, B., 1983, MNRAS 204, 1091
Hutchings J.B., Cowley A.P., Schmidtke P.C., Crampton D., 1995, AJ 110, 2394
Kahabka P., Pietsch W., Hasinger G., 1994, A&A 288, 538
Karitskaya E.A., Bochkarev N.G., Gnedin Y.N., 1986, Astron. Zh. 63, 1001
Mason K.O., Middleditch J., Nelson J.E., et al., 1980, ApJ 242, L109
Meyer F., Meyer-Hofmeister E., 1995, Cataclysmic Variables, A. Bianchini et al. (eds.), Kluwer Academic Publishers, p. 463
Paczyński B.,1977, ApJ 216, 822
Pakull M.W., Beuermann K., van der Klis M., von Paradijs J., 1988, A&A 203, L27
Papaloizou J., Pringle J.E., 1977, MNRAS 181, 441
Schaller G., Schaerer D., Meynet G., Maeder A., 1992, A&AS 96, 269
Schandl S., 1996, A&A (in press)
Schandl S., Meyer-Hofmeister E., Meyer F., 1996, A&A (subm.)
Schmidtke P.C., Cowley A.P., McGrath T.K., Hutchings J.B., Crampton D., 1996, AJ 111, 788
Schmidtke P.C., McGrath T.K., Cowley A.P., Frattare L.M., 1993, PASP 105, 863
Smak J., 1971, Acta Astr. 21, 15
Smak J., 1994, Acta Astr. 44, 59
Smale A.P., Corbet R.H.D., Charles P.A. et al., 1988, MNRAS 233, 51
van den Heuvel E.P.J., Bhattacharya D., Nomoto K., Rappaport S.A., 1992, A&A 262, 97 (vdH)

Accretion Disks in Supersoft X-Ray Sources

R. Popham, R. Di Stefano

Harvard-Smithsonian Center for Astrophysics, MS 51, 60 Garden St., Cambridge, MA 02138 USA

Abstract. We examine the role of the accretion disk in the steady-burning white dwarf model for supersoft sources. The accretion luminosity of the disk is quite small compared to the nuclear burning luminosity of the central source. Thus, in contrast to standard accretion disks, the main role of the disk is to reprocess the radiation from the white dwarf. We calculate models of accretion disks around luminous white dwarfs and compare the resulting disk fluxes to optical and UV observations of the LMC supersoft sources CAL 83, CAL 87, and RX J0513.9–6951. We find that if the white dwarf luminosity is near the upper end of the steady-burning region, and the flaring of the disk is included, then reprocessing by the disk can account for the UV fluxes and a substantial fraction of the optical fluxes of these systems. Reprocessing by the companion star can provide additional optical flux, and here too the disk plays an important role: since the disk is fairly thick, it shadows a significant fraction of the companion's surface.

1 Introduction

Supersoft sources have been modeled as white dwarfs accreting matter from a companion at a high enough rate to produce steady nuclear burning (van den Heuvel et al. 1992). In this model, accretion takes place through a disk; however, the total luminosity of the white dwarf due to nuclear burning is much greater than the total accretion luminosity of the disk. Thus it might appear that the disk is of little importance in supersoft systems - that it simply "adds fuel to the fire".

There is an important difference, however, between the disks in supersoft sources and ordinary accretion disks. In most accreting systems, the gravitational potential energy of the accreting material provides the main source of energy. In supersoft sources, the system's primary energy source is nuclear burning of the accreted material. A large fraction of the energy produced by nuclear burning irradiates the surface of the disk, and this exceeds the accretion energy dissipated in the disk by a large factor. Thus, in supersoft sources, the primary role played by the disk is to reprocess the copious radiation produced by the central source. The predominant role of reprocessing is a natural consequence of the steady-burning white dwarf model for these systems. It is a unique feature of supersoft sources which provides an opportunity to test both the model and our understanding of reprocessing in disks.

The plan of this paper is as follows: in §2 we show that nuclear-burning white dwarfs alone cannot provide the optical and UV fluxes observed in supersoft sources. We briefly describe our disk model in §3, and we show that simple accretion disks which do not include reprocessing also fall well short of the observations. We make our model progressively more realistic by including reprocessing of the white dwarf radiation by the disk, the effects of disk flaring, and other sources of reprocessed radiation in §4.1, 4.2, and 4.3, respectively.

Throughout the paper, we compare our model spectra to the observed optical and UV fluxes for three LMC supersoft sources: CAL 83, CAL 87, and RX J0513.9–6951. Sources in the LMC have the advantages of having a known distance and little extinction. These fluxes are approximate, and are shown in the figures as line segments. They are taken from the following sources: CAL 83 (UV) Bianchi & Pakull 1988; CAL 83 (optical) Smale et al. 1988; CAL 87 (UV) Hutchings et al. 1995; CAL 87 (optical) Pakull et al. 1988; RX J0513.9–6951 (UV and optical) Pakull et al. 1993. It is important to remember that all of these systems are variable, and that the optical and UV data for each system are not simultaneous. Of the three systems, CAL 83 probably provides the best comparison with models. RX J0513.9–6951 is a transient X-ray source which also shows substantial variability in the optical (Reinsch et al. 1996, Schaeidt 1996, and Southwell 1996). CAL 87 is an eclipsing system, and this presumably decreases its brightness and changes the shape of its spectrum (see contribution by Schandl et al. 1996a,b). Nonetheless, the available measurements give us a good idea of the fluxes from these systems.

2 Nuclear-Burning White Dwarfs

It is easy to see that a hot white dwarf alone cannot account for the very large optical and ultraviolet fluxes of supersoft sources. This is illustrated in Fig. 1, where we show blackbody spectra for 5 steady-burning white dwarfs which sit at the low-luminosity end of the steady-burning tracks calculated by Iben (1982). Their properties are listed in Tab. 1. We have scaled the spectra to a distance of 50 kpc. The resulting optical and ultraviolet fluxes are far smaller than those observed from CAL 83, CAL 87, and RX J0513.9–6951.

3 Simple Disk Models and Spectra

Can the disk provide the observed optical and UV flux? In order to answer this question, we model the disk in the following way. The radial structure of the disk is calculated using the "slim disk" equations. These were developed by Paczyński and collaborators to model disks around black holes (Paczyński & Bisnovatyi-Kogan 1981; Muchotrzeb & Paczyński 1982). They are a more sophisticated version of the standard thin disk equations in that they dispense with some of the simplifying assumptions made by the thin disk model, and allow a more accurate treatment of the disk, particularly in the boundary layer region near the accreting star. The radial disk equations are described in detail by Popham & Narayan (1995), who used them to model boundary layers in cataclysmic variables with high values of $\dot{M}$.

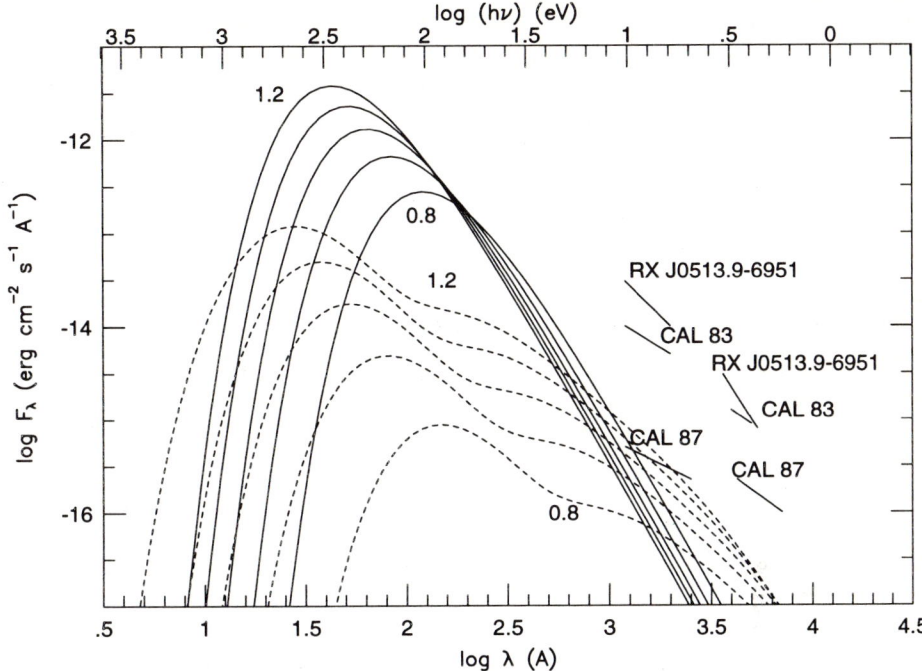

Fig. 1. Blackbody spectra of white dwarfs undergoing steady nuclear burning (solid lines), and of the corresponding accretion disks (dashed lines). The spectra are labeled by the white dwarf mass, and the white dwarfs sit at the lower limit of the steady-burning region, with parameters listed in Tab. 1. Observed optical and UV spectra of CAL 83, CAL 87, and RX J0513.9–6951(references in §1) are shown as line segments.

After calculating the radial structure of the disk, we calculate the vertical structure for each disk annulus. This involves solving the equations of hydrostatic equilibrium and simple two-stream radiative transfer equations, for values of the disk surface density, effective temperature and gravity given by the radial solution. We then use this vertical solution to calculate the continuum spectrum for that disk annulus. By adding the spectra of all of the annuli, we obtain the total disk spectrum.

We calculate the observed flux by simply dividing the luminosity by $4\pi d^2$, with $d = 50$ kpc. Note that the projected area of a flat disk varies with inclination as $\cos i$. The observed flux is often taken to be $F = L\cos i/2\pi d^2$, so that the flux integrated over a sphere of radius d is L. Using this expression, our choice of $F = L/4\pi d^2$ corresponds to implicitly assuming that $i = 60°$. Note that the flux from a face-on disk is twice as large as this, $F(i = 0°) = L/2\pi d^2$.

How will the inclusion of the disk flux change the overall spectrum? The dashed lines in Fig. 1 show simple disk spectra for the same values of $\dot{M}$, $M_{\rm WD}$,

Table 1. Parameters for steady-burning white dwarfs and accretion disks at the low-luminosity edge of the steady-burning region (first five lines), and at a higher luminosity (last line).

$M_{\rm WD}$ $M_\odot$	$\dot{M}$ $10^{-7}\,M_\odot\,{\rm yr}^{-1}$	$L_{\rm WD}$ $10^{37}\,{\rm ergs\,s}^{-1}$	$R_{\rm WD}$ $10^8\,{\rm cm}$	$T_{\rm WD}$ $10^5\,{\rm K}$	L_{acc} $10^{35}\,{\rm ergs\,s}^{-1}$
0.8	0.51	1.53	25.3	2.41	1.36
0.9	0.85	2.53	15.7	3.47	4.08
1.0	1.27	3.82	11.3	4.53	9.47
1.1	1.82	5.44	8.78	5.61	19.1
1.2	2.50	7.48	7.14	6.74	35.3
1.2	4.98	14.9	18.4	4.99	27.2

and $R_{\rm WD}$ as the nuclear burning white dwarfs. Note that two components are visible in the disk spectra: a hot component due to the boundary layer, and a cooler component due to the disk. Despite being less luminous overall than the white dwarfs, the disk spectra nevertheless dominate in the optical, and also in the UV for the higher values of $M_{\rm WD}$. Nonetheless, they still fall well short of the observed fluxes from CAL 83 and RX J0513.9–6951.

4 Reprocessing Disks

The major role played by the disk in the steady-burning white dwarf model of supersoft sources is to reprocess the radiation from the white dwarf. This is a result of the huge luminosity of the white dwarf, which is $\sim 20-100$ times larger than the accretion luminosity for the parameters listed in Tab. 1. Adams & Shu (1986) calculated that 1/4 of the luminosity of the central star will hit the disk, assuming a flat, infinitely thin disk. Thus, the reprocessing luminosity should exceed the accretion luminosity by a large factor. We include the effects of this reprocessed luminosity in our disk spectra by changing the boundary condition at the disk surface in our vertical structure model to include the incident flux.

Another form of reprocessing is direct radiative energy flux across the star–disk boundary. It is less certain what fraction of the white dwarf luminosity will enter the disk in this way, but as a first estimate we simply take the fraction of the surface area of the star which is covered by the disk, which is given approximately by $H(R_{\rm WD})/R_{\rm WD}$, where H is the disk height. This direct heating requires a change to the inner boundary condition on the radial radiative flux in our radial disk model.

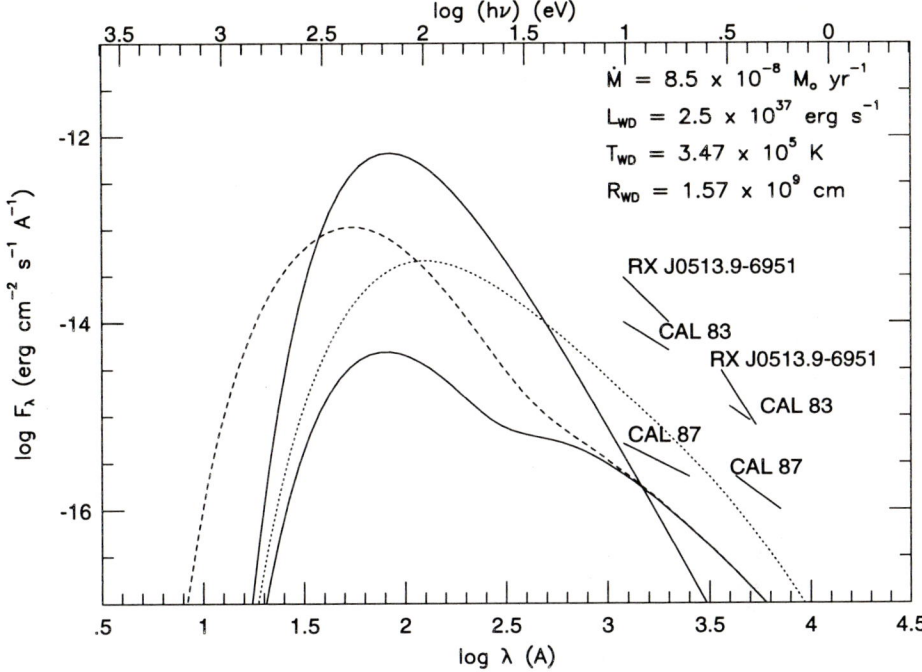

Fig. 2. Spectra of a disk with the $M_{WD} = 0.9\,M_\odot$ parameters listed in Tab. 1: spectra of the white dwarf and the disk with no external heating (solid lines), the disk spectrum including radiative heating through the disk–star boundary (dashed line), and the disk spectrum including the effects of reprocessing (dotted line).

4.1 Disk Spectra Including Reprocessing

Fig. 2 shows disk spectra using the parameters for $M_{WD} = 0.9\,M_\odot$ listed in Tab. 1. When direct heating through the star–disk boundary is included, the inner disk is much hotter, and the high-energy end of the spectrum is much brighter. However, the disk spectrum hardly changes in the optical and UV, since direct heating only affects the inner part of the disk.

Reprocessing, on the other hand, heats the entire disk surface, and produces a substantial increase of a factor of 3–5 in the optical and UV flux from the disk. Nonetheless, the fluxes still fall short of the observed ones by a factor of ~ 10. Thus, it seems clear that a substantially larger white dwarf luminosity will be required to explain the data. Accordingly, we take a white dwarf solution with $L_{WD} \simeq 1.5 \times 10^{38}$ ergs s^{-1}. The parameters for this solution are shown in the last line of Tab. 1. When reprocessing is included for this solution, we obtain the spectrum shown in Fig. 3. This comes much closer to the optical and UV data for CAL 83 and RX J0513.9–6951, but it still falls short by a factor of a few in the UV and a factor of 5–10 in the optical.

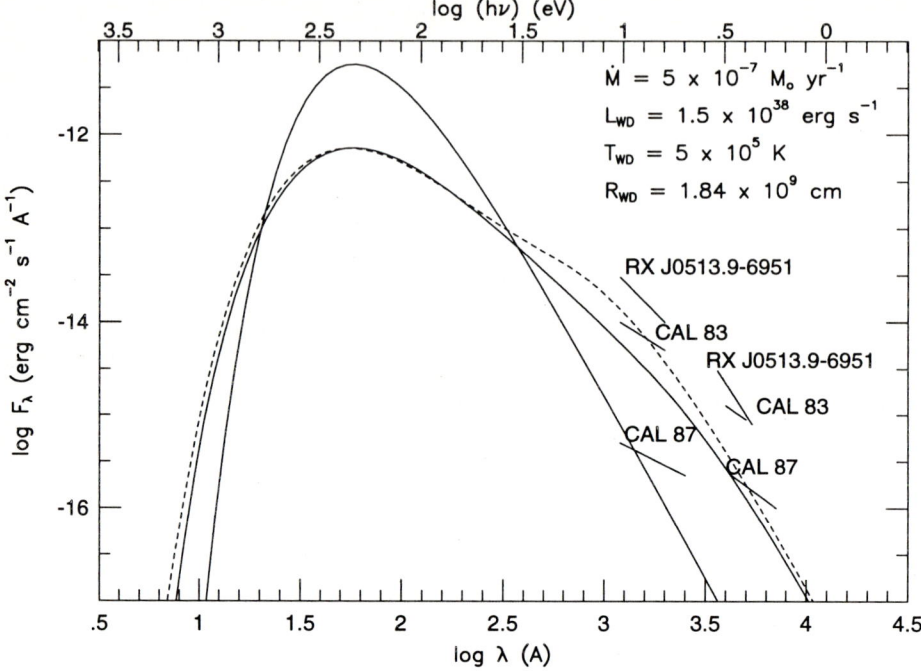

Fig. 3. Spectra of the steady-burning white dwarf and the disk including reprocessing (solid lines), and of the disk including reprocessing with disk flaring (dashed line), for a higher-luminosity solution with the parameters listed in the last line of Tab. 1.

4.2 Disk Flaring

Thus far we have calculated the flux incident on the disk surface by assuming a flat, infinitely thin disk. However, we can use our vertical structure code to compute the disk height $H(R)$. We find that the disk has a substantial thickness, with $H/R \gtrsim 0.25$. We also find that the disk surface is not flat; $d\ln H/d\ln R$, which is 1 for a flat disk with constant H/R, is $\simeq 1.1 - 1.15$ for the disk solution shown in Fig. 3. This increases the incident flux on the disk by a large factor, particularly in the outer parts of the disk.

The spectrum of the disk including the additional reprocessing flux due to flaring is shown by a dashed line in Fig. 3; the UV flux now is approximately the same as that of CAL 83, although it has a steeper slope. The optical flux still falls short of the observations. Note that the optical flux is constrained largely by the size of the disk; we have set the outer edge of the disk at 1.35×10^{11} cm $\simeq 75 R_{\rm WD}$. This is slightly smaller than the Roche lobe radius of a $1.2\,M_\odot$ star with a $2\,M_\odot$ companion and a 1-day orbital period.

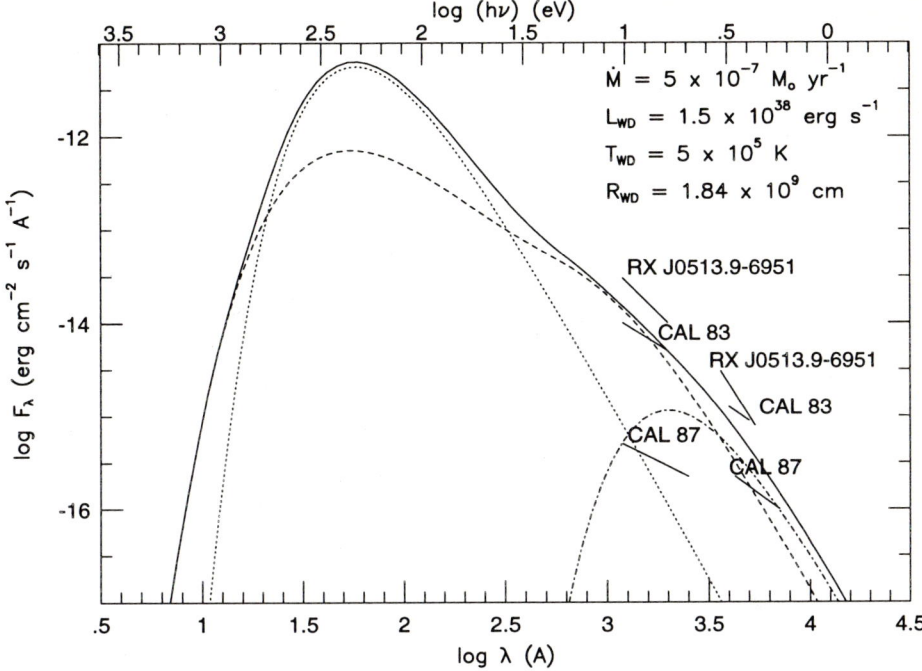

Fig. 4. Spectra for the same solution as in Fig. 3; the spectra of the white dwarf (dotted line), the disk with reprocessing and flaring (dashed line), and the donor star (dash-dotted line), and the combined spectrum of the three (solid line).

4.3 Other Sources of Reprocessed Radiation

Radiation from the white dwarf is also reprocessed by the donor star. Since the donor fills its Roche lobe, its radius expressed as a fraction of the binary separation a depends only on the mass ratio. For a $2\,M_\odot$ donor and a $1.2\,M_\odot$ white dwarf, we find $R_{donor}/a \simeq 0.42$. We have done numerical integrations to determine the fraction f of the white dwarf luminosity that reaches the donor surface. If no disk is present, f is just the fraction of the total 4π solid angle which is subtended by the donor, so $f = 0.047$. However, a flat, infinitely thin disk prevents radiation emitted from the upper half of the white dwarf from reaching the lower half of the donor, and vice versa, reducing f to 0.028. If the disk has a finite thickness, then some fraction of the donor surface is completely shadowed by the disk; for a disk with $H/R = 0.25$, $f = 0.0085$.

In the solution shown in Fig. 3, the disk height at the outer edge of the disk, taken from our vertical structure solution, corresponds to $H/R \simeq 0.275$, so that $f = 0.0071$. Despite this, reprocessing by the donor is still a significant source of flux from the system, particularly in the optical. Fig. 4 shows the spectrum of the white dwarf, the disk, and the donor star, assuming that the absorbed white dwarf luminosity is radiated evenly from the donor surface, and the combined spectrum of the three. The combined spectrum matches the observed UV and optical fluxes of CAL 83 fairly well. It falls short of the RX J0513.9-6951 fluxes by about a factor of 2, but reproduces the slopes quite well; however, as noted above, our model spectra would predict twice as much flux if viewed face-on. Our spectrum exceeds the CAL 87 fluxes, as expected since this system has a high inclination.

Additional optical flux may originate from the outer edge of the disk or from a region of increased disk height resulting from the impact of the accretion stream on the disk, as modeled in CAL 87 by Schandl et al. (1996).

5 Summary

We have examined the role of the accretion disk in supersoft X-ray sources, assuming that the steady-burning white dwarf model applies. Reprocessing of white dwarf radiation by the disk plays a uniquely important role in supersoft sources, and appears to be able to explain their large optical and UV fluxes. Our disk models also provide a foundation for more phenomenological models like the one by Schandl et al. (1996a,b), which agrees quite well with the lightcurve of CAL 87. In the near future we plan to use disk models in conjunction with evolutionary calculations (Di Stefano & Nelson 1996) to attempt to constrain M_{WD} and $\dot{M}$ in a number of close binary supersoft sources.

Acknowledgement: We thank Ramesh Narayan and Jochen Greiner for helpful discussions. This work has been supported in part by grants NASA NAG5-2837, NSF AST-9423209, and NSF GER-9450087.

References

Adams F.C., Shu F.H., 1986, ApJ 308, 836
Bianchi L., Pakull M., 1988, in A Decade of UV Astronomy with the IUE Satellite, E.J. Rolfe (ed.), ESA SP-281, Vol. 1, p. 145
Di Stefano R., Nelson L., 1996, this volume p. 3
Hutchings J.B., Cowley A.P., Schmidtke P.C., Crampton D., 1995, AJ 110, 2394
Iben I., 1982, ApJ 259, 244
Muchotrzeb B., Paczyński B., 1982, Acta Astronomica 32, 1
Paczyński B., Bisnovatyi-Kogan G., 1981, Acta, Astronomica 31, 283
Pakull M.W. et al. , 1993, A&A 278, L39
Pakull M.W., Beuermann K., van der Klis M., van Paradijs J., 1988, A&A 203, L27
Popham R., Narayan R., 1995, ApJ 442, 337
Reinsch K., van Tesseling A., Beuermann K., Abbott T.M.C., 1996, A&A (in press)
Schaeidt S., 1996, this volume p. 159
Schandl S., Meyer-Hofmeister E., Meyer F., 1996a, A&A (subm.)
Schandl S., Meyer-Hofmeister E., Meyer F., 1996b, this volume p. 53
Southwell K., et al. , 1996, this volume p. 165
Smale A.P. et al. , 1988, MNRAS 233, 51
van den Heuvel E.P.J., Bhattacharya D., Nomoto K., Rappaport S., 1992, A&A 262, 97

Part III

X-Ray and Optical Observations of SSS

Supersoft X-Ray Sources in M 31

J. Greiner[1], R. Supper[1], E.A. Magnier[2,3]

[1] Max-Planck-Institute for Extraterrestrial Physics, 85740 Garching, Germany
[2] University of Amsterdam, Astronomy Dept, 1098 SJ Amsterdam, The Netherlands
[3] University of Washington, Dept of Astronomy FM-20, Seattle, WA 98195, USA

1 Introduction

As the most massive galaxy in the Local Group the early-type spiral M31 is close enough to be studied in detail at various wavelengths. A ROSAT PSPC mosaic of 6 contiguous pointings with an exposure time of 25 ksec each was performed in July 1991 (first M31 survey). A raster pointing of 80 observations with 2.5 ksec exposure time each covering the whole M31 disk was made in July/August 1992 and January 1993 (second M31 survey). A multi-step detection algorithm involving a maximum likelihood technique as final step was applied to the individual pointings of the first survey with a likelihood threshold of 10. A total of 396 X-ray sources were detected within the field of view of about 6.3 deg^2 (Supper et al. 1996). For each source several quantites are determined such as the position, the total number of background subtracted counts, the corresponding countrate and two hardness ratios for a crude spectral characterisation.

CCD photometry has been performed in four passbands (*BVRI*) of the entire optical disk of M31 in September 1990 and September 1991 using the McGraw-Hill 1.3m telescope of the Michigan-Dartmouth-MIT Observatory at Kitt Peak (Magnier et al. 1992, Haiman et al. 1993). The observations have typical completeness limits of (22.3,22.2,22.2,20.9). Different selection criteria on these data sets allow the extraction of different classes of objects. Of particular interest are likely Galactic foreground stars which have been identified with the criteria V$\leq$18 and B–V$>$0.4 (predominantly main sequence and giant stars).

2 Selection Criteria and Source Sample

Two selection criteria were applied to the total sample of detected X-ray sources:
- the hardness ratio has to satisfy HR1 + $\sigma_{HR1} \leq -0.80$. The hardness ratio HR1 is defined as the normalized count difference $(N_{50-200} - N_{10-40})/(N_{10-40} + N_{50-200})$, where N_{a-b} denotes the number of counts in the PSPC between channels a and b. Similarly, HR2 is defined as $(N_{91-200} - N_{50-90})/N_{50-200}$. We note that HR1 is sensitive to the absorbing column, and that HR2 is basically zero (within the errors) for supersoft sources. The error in the hardness ratio ($\sigma_{HR1/2}$) was calculated by applying Gaussian error propagation.
- no correlation with any optical object in the error box which has been classified as a foreground object.

Table 1. Summary of the supersoft X-ray sources in M31 according to our selection criteria (see text for details) plus the source proposed by White et al. (1995) as supersoft transient (last row). Given for each source are the name (column 1), the best fit X-ray position (2), the 3σ location error (3), the total number of counts collected in the corresponding pointing (exception: for the White et al. source all pointings are used; 4), the PSPC countrate in the 0.1-2.4 keV band (5), the hardness ratio HR1 with error (6), the hardness ratio HR2 with error (7), and the maximum blackbody temperature (8, see text).

Name	Coordinate (2000.0)	Error (″)	N_{cts}	countrate (cts/ksec)	HR1	HR2	T_{bb}^{max} (eV)
RX J0037.4+4015	$00^h37^m25\overset{s}{.}3$ $+40°15'16''$	18	10	0.31±0.31	−0.93±0.07	0.02±0.71	43
RX J0038.5+4014	$00^h38^m32\overset{s}{.}1$ $+40°14'39''$	33	33	0.80±0.28	−0.92±0.08	−0.49±0.53	45
RX J0038.6+4020	$00^h38^m40\overset{s}{.}9$ $+40°20'00''$	15	74	1.73±0.29	−0.93±0.06	0.32±0.66	43
RX J0039.6+4054	$00^h39^m38\overset{s}{.}5$ $+40°54'09''$	21	22	0.44±0.44	−0.92±0.07	−0.04±0.71	45
RX J0040.4+4009	$00^h40^m26\overset{s}{.}3$ $+40°09'01''$	27	31	0.85±0.32	−0.94±0.06	−0.90±0.10	42
RX J0040.7+4015	$00^h40^m43\overset{s}{.}2$ $+40°15'18''$	21	50	1.26±0.32	−0.94±0.06	−0.31±0.64	42
RX J0041.5+4040	$00^h41^m30\overset{s}{.}2$ $+40°40'04''$	18	16	0.32±0.18	−0.95±0.05	−0.62±0.44	40
RX J0041.8+4059	$00^h41^m49\overset{s}{.}9$ $+40°59'21''$	27	23	0.49±0.24	−0.93±0.07	−0.63±0.43	43
RX J0042.4+4044	$00^h42^m27\overset{s}{.}6$ $+40°44'32''$	15	72	1.69±0.32	−0.93±0.07	−0.07±0.70	43
RX J0043.5+4207	$00^h43^m35\overset{s}{.}9$ $+42°07'30''$	15	57	2.15±0.55	−0.92±0.08	−0.27±0.66	45
RX J0044.0+4118	$00^h44^m04\overset{s}{.}8$ $+41°18'20''$	15	69	2.46±0.42	−0.94±0.06	0.11±0.81	42
RX J0045.5+4206	$00^h45^m32\overset{s}{.}3$ $+42°06'59''$	24	86	3.14±0.34	−0.89±0.07	−0.29±0.65	42
RX J0046.2+4144	$00^h46^m15\overset{s}{.}6$ $+41°44'36''$	15	63	2.15±0.39	−0.93±0.07	0.62±0.40	38
RX J0046.2+4138	$00^h46^m17\overset{s}{.}8$ $+41°38'48''$	27	30	1.12±0.40	−0.91±0.09	−0.27±0.65	40
RX J0047.6+4205	$00^h47^m38\overset{s}{.}5$ $+42°05'07''$	30	29	1.05±0.36	−0.92±0.07	0.06±0.70	39
RX J0045.4+4154	$00^h45^m29\overset{s}{.}0$ $+41°54'08''$	18	1040	29.63±0.98	+0.78±0.03	−0.59±0.03	128

Among all 396 detected sources a total of 15 fulfill these two requirements. These sources are listed in Tab. 1 and are identical to the sources #3, #12, #18, #39, #78, #88, #114, #128, #171, #245, #268, #309, #335, #341, and #376 in Supper et al. (1996). Fig. 1 shows the positions of these sources overplotted on an optical image of M31.

3 X-Ray Characteristics

We have applied a spectral fit only to the brightest source of our sample, namely RX J0045.5+4206. A blackbody model fit gives an absorbing column of 1.2×10^{21} cm^{-2}, a factor of two higher than the Galactic value (0.6×10^{21} cm^{-2}, Dickey and Lockman 1990). But the uncertainty is large, and the line of sight Galactic absorption is well within the 3σ contour (see Fig. 2). The resulting blackbody temperature is 30^{+20}_{-10} eV at the Galactic absorbing column, similar to the known supersoft sources (Greiner et al. 1991, Kahabka et al. 1994). The bolometric luminosity for this model blackbody is about 7×10^{37} erg/s. A comparison of blackbody and white dwarf atmosphere models has shown that while the temperature estimates did not differ much, the bolometric luminosity can be a factor 10–100 lower when using a white dwarf atmosphere model (Heise et al. 1994). We there-

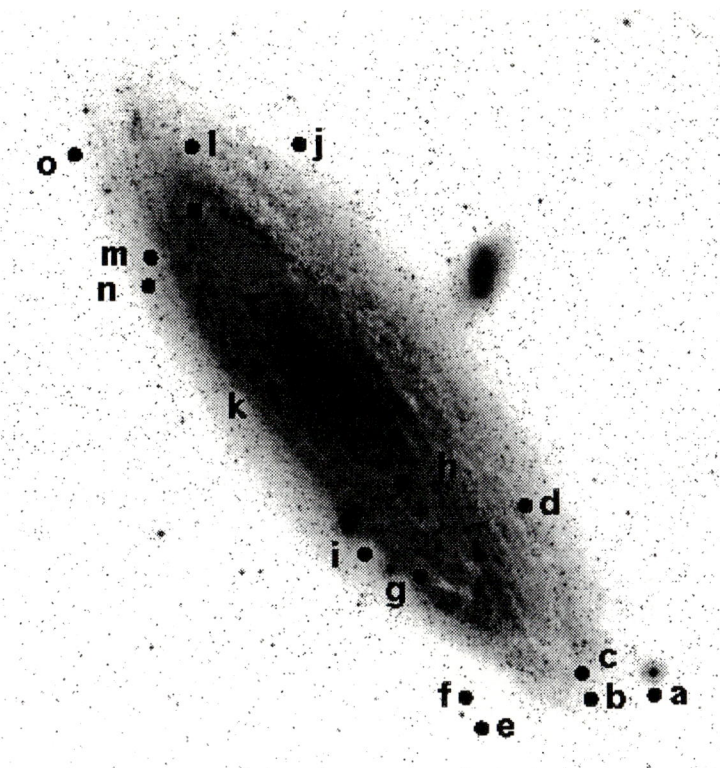

Fig. 1. The location of the 15 supersoft sources plotted over an optical image (POSS O plate) of M31. Letters (a through o) denote the X-ray sources in Tab. 1 from top to bottom. The source RX J0045.4+4154 (White et al. 1995) is shown as a square. Sources h, k and l have objects brighter than 20th mag in their X-ray error box (see text for possible counterpart types). Sources e and m lie at the edge of the *BVRI* survey field, while a, j and n are outside of this survey.

fore think that the above given bolometric luminosity derived at the Galactic absorbing column should be a better representation than the best-fit value.

In order to get some crude spectral information also for the other, fainter X-ray sources we have adopted the following scheme: We calculate the hardness ratio HR1 for blackbody models with different temperatures and absorbing columns (Fig. 3). We then take the measured HR1 value, and determine the model blackbody temperature at the Galactic absorbing column. Since for a given hardness ratio the temperature decreases with increasing N_H, these temperatures are upper limits. As could be expected already from the rather similar values of HR1, these maximum blackbody temperatures lie in a quite narrow range between 40–50 eV (last column in Tab. 1).

Recently, White et al. (1995) have reported on a recurrent X-ray transient which they claim to be supersoft. We have determined the maximum temper-

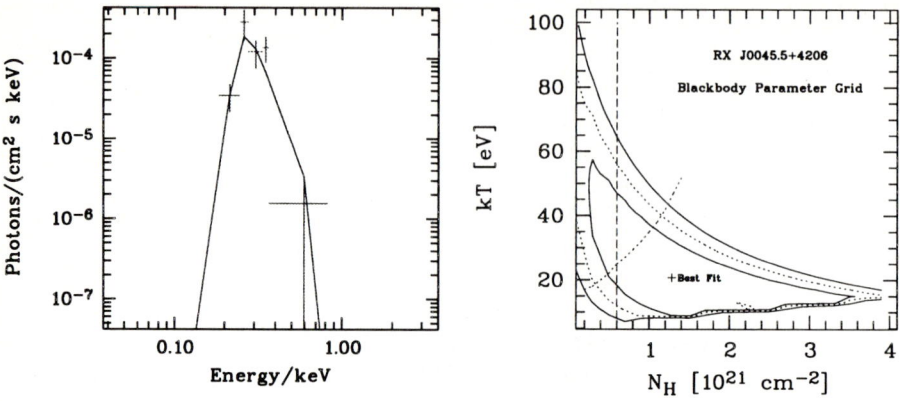

Fig. 2. Best fit blackbody model of the brightest of our supersoft X-ray sources RX J0045.5+4206 (left) and the $1, 2, 3\sigma$ error contours in the $kT - N_H$ plane (right panel). Only photons from one single 25 ksec pointing are used. Some photons had to be discarded because of another bright source at only 1.8 distance making the flux somewhat uncertain. Due to small photon numbers, the parameter range is not too well constrained. and thus introduces an additional uncertainty in the luminosity. In the right panel the dashed vertical line corresponds to the mean total Galactic absorption, and the dotted line marks the Eddington limit of the blackbody model.

ature with the same method as our source sample, using the merged pointings of the second PSPC survey of M31. This temperature and the hardness ratio of this source (Tab. 1) clearly demonstrate that it is considerably harder than the sources we have extracted here. From the spectral similarity we speculate that it is rather a Her X-1 like object than a supersoft X-ray source.

All 15 X-ray sources have been checked for temporal variability against the August 1992 observations of the second M31 survey. None of the sources fulfilled the criterion for variability (flux difference in the two observations divided by the quadratic sum of the flux errors has to be larger than 3σ).

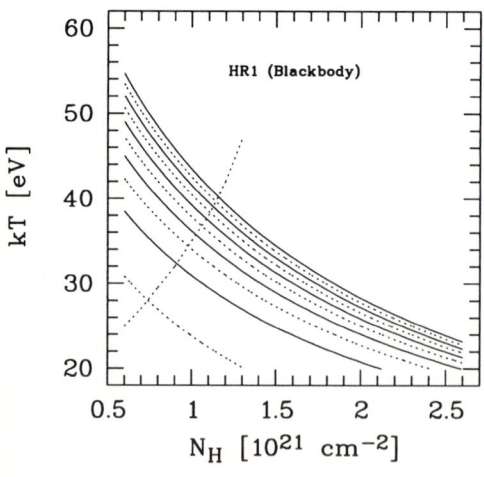

Fig. 3. The hardness ratio HR1 as a function of blackbody temperature kT and absorption column density N_H. HR1 increases from the upper line with HR1$= -0.80$ to the bottom line with HR1$= -0.98$ in steps of 0.02. For a given hardness ratio the blackbody temperature decreases with increasing column density. Therefore, the kT value at $N_H = 0.6 \times 10^{21}$ cm^{-2} gives the maximum blackbody temperature. The dotted line marks the Eddington limit of the blackbody model.

4 Source Positions and Identifications

The 396 X-ray source positions as they result from the maximum likelihood detection algorithm have varying errors depending on the off-axis angle in the field of view of the PSPC. Since for nearly a quarter of X-ray sources optical counterparts could be identified due to the positional coincidence plus the colour information from the *BVRI* survey, we have used a subset of these optically identified X-ray sources (namely the 29 globular clusters) to improve the systematic errors in the X-ray source determination. We have determined the matrix coefficients for a unique rotation plus translation which gives the smallest residual position differences for the optical and X-ray positions of the optically identified X-ray sources. The same matrix has then been applied to all X-ray sources.

There are several source positions (Fig. 1) with optical objects inside the error box. The optical selection criterion is effective mainly against red objects of some of our supersoft sources. Single, i.e. non-interacting white dwarfs (WDs) and many magnetic cataclysmic variables (CVs) also have supersoft X-ray spectra, and such objects are blue. From the known ratio of optical to X-ray luminosity and the observed X-ray fluxes we might expect Galactic, single WDs at typical V magnitudes of 18–20 mag and CVs at even fainter magnitudes. These type of objects thus constitute a possible contamination of our sample. In particular, objects RX J0041.8+4059, RX J0044.0+4118 and RX J0045.5+4206 have blue objects of 19–20th mag in their respective error box, and thus might be Galactic WDs or CVs. On the other hand, blue stars in M31 are not distributed randomly, but in the spiral arms. The above mentioned objects are in regions with many blue stars, i.e. in associations. Thus, the probability of having a coincident CV or WD in these areas is not the sky density of CVs or WDs times the area of the survey, but rather only times the total area of the associations, which is substantially smaller. Spectroscopic observations are clearly necessary to distinguish between these alternatives.

There is recent evidence that also some active galactic nuclei (AGN) have supersoft X-ray spectra (Greiner et al. 1996). At least one AGN (WPVS007, Grupe et al. 1995) has HR1<−0.8, and there are a number of further secure identifications with HR1<−0.6. We therefore caution that one or the other source in our list which is located in the outskirts of M31 might be an AGN shining through M31. RX J0040.4+4009 and RX J0040.7+4015 are plausible candidates.

In addition to the *BVRI* survey, images were also aquired in the Hα line. Only five of the sources turned out to lie in the Hα fields: RX J0040.7+4015, RX J0041.5+4040, RX J0041.8+4059, RX J0042.4+4044 and RX J0044.0+4118. None of these reveals any Hα emission. The upper limits in Hα luminosity are 10^{35} erg/s, i.e. a CAL 83 like nebula (Pakull & Motch 1989, Remillard et al. 1995) would have been detected. This result is not surprising in view of the fact that in a search for nebulae around known supersoft sources in the Galaxy and the Magellanic Clouds only one out of 10 sources has revealed a nebula (Remillard et al. 1995).

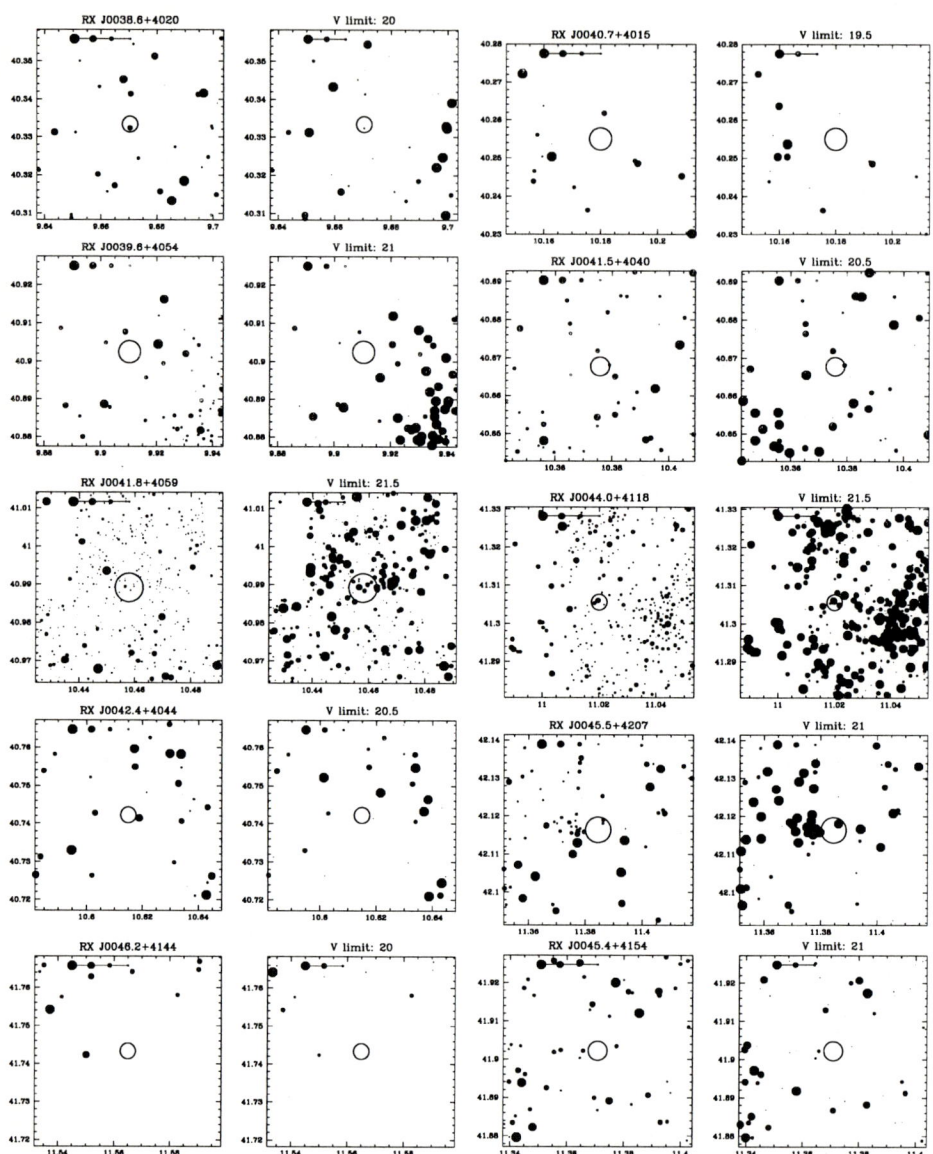

Fig. 4. Finding charts for those supersoft sources covered by the *BVRI* survey which have V limiting magnitudes fainter than 20 mag. For each source two panels are given: The right (with the ROSAT name on the top) is the V image with the dot size scaled according to the brightness with the scaling of V=15,17,19,21 shown in the upper left. In the right panel (with the limiting V magnitude on top) the size is scaled by the color (B-V=0.0,0.4,0.8 from left to right in the upper left scaling).

5 Conclusions

Our selection criterion according to the X-ray hardness ratio is very restrictive. An identical selection criterion has been applied to all sources in the *ROSAT* all-sky survey, and only about 30% of the known supersoft sources in the LMC and SMC were found (Greiner 1996). For instance, even a M31 source identical to CAL 83 would probably been missed with the present selection criterion. This is due to the fact that at lower intensities the error in the hardness ratio increases readily above 0.1–0.15 and removes the sources from the sample. Also, the selection criterion is more sensitive to low-temperature objects, i.e. sources with emission above 0.4 keV are already excluded. Therefore, a more relaxed hardness ratio selection will certainly increase the number of further sources which could qualify as supersoft sources.

The optically brightest supersoft X-ray source in the Large Magellanic Cloud has V=16.2 mag. A similar source in M31 is expected to have V=21.7 mag, still within the range of the *BVRI* survey. Those objects which have optical counterparts brighter than this might be expected not to be burning WDs in close binaries, but possibly Galactic single WDs or CVs or bright blue objects in M31. Also, a supersoft AGN is not excluded a priori as counterpart for X-ray sources in the outskirts of M31. We therefore conclude that not all of the sources described here in more detail may qualify as luminous close binaries like CAL 83 after spectroscopic identification has been succeeded.

The location distribution of our supersoft source sample across M31, especially the lack of sources in the inner bulge, clearly suggests that they belong to the disk population. Therefore, the detection probability is seriously affected by the absorbing column between the source and the observer.

Acknowledgement: JG is supported by the Deutsche Agentur für Raumfahrtangelegenheiten (DARA) GmbH under contract FKZ 50 OR 9201. The *ROSAT* project is supported by the German Bundesministerium für Bildung, Forschung, Wissenschaft und Technologie (BMBW/DARA) and the Max-Planck-Society.

References

Dickey J.M., Lockman F.J., 1990, Ann. Rev. Astron. Astrophys. 28, 215
Greiner J., Hasinger G., Kahabka P. 1991, A&A 246, L17
Greiner J., 1996, this volume p. 285
Greiner J., Danner R., Bade N., Richter G.A., Kroll P., Komossa S., 1996, A&A (in press, MPE prepr. 342)
Grupe D., Beuermann K., Mannheim K., Thomas H.-C., Fink H.H., de Martino D., 1995, A&A 300, L21
Haiman Z., Magnier E.A., Lewin W.H.G., Lester R.R., van Paradijs J., Hasinger G., Pietsch W., Supper R., Trümper J., 1993, A&A 286, 725
Heise J., van Teeseling A., Kahabka P., 1994, A&A 288, L45
Kahabka P., Pietsch W., Hasinger G., 1994, A&A 288, 538

Magnier E.A., Lewin W.H.G., van Paradijs J., Hasinger G., Jain A., Pietsch W., Trümper J., 1992, A&A Supp 96, 379
Magnier E.A., Prins S., van Paradijs J., Lewin W.H.G., Supper R., Hasinger G., Pietsch W., Trümper J., 1995, A&AS 114, 215
Pakull M.W., Motch C., 1989, in Extranuclear Activity in Galaxies, ed. E.J.A. Meurs, R.A.E. Fosbury, (Garching, ESO), p. 285
Remillard R.A., Rappaport S., Macri L.M., 1995, ApJ 439, 646
Supper R., Hasinger G., Pietsch W., Trümper J., Jain A., Magnier E.A., Lewin W.H.G., van Paradijs J., 1996, A&A (in press)
White N.E., Giommi P., Heise J., Angelini L., Fantasia S., 1995, ApJ 445, L125

X-Ray and Optical Observations of RX J0925.7−4758: Constraints on the Binary Structure

C. Motch

CNRS, Observatoire de Strasbourg, 11 rue de l'Université, F-67000 Strasbourg

Abstract. We report on preliminary results from an X-ray and optical observing campaign of the galactic supersoft X-ray source RX J0925.7–4758. The combined analysis of all available V band photometry indicates an orbital period of 3.79 ± 0.24 d. Medium resolution spectra collected during 5 consecutive nights show that the radial velocities of Hα and He II λ4686 vary together with the ≈ 3.8 day cycle and probably exclude short 1 day alias periods. This rather long period and the absence of marked stellar signatures in the optical spectrum indicate that the mass donor is probably a giant or subgiant K4 to A0 type star. A long ROSAT HRI observation reveals that the source is variable from 0.35 to 0.5 cts s^{-1} on a time scale of a few days suggesting that the X-ray flux of RX J0925.7–4758 could be modulated with orbital motion. The phasing of the photometric V band and radial velocity curves is consistent with a model in which the photometric modulation is caused by the heated hemisphere of the companion star and emission lines arise from a region dynamically linked to the X-ray source. If this interpretation is correct, then the companion star may indeed be significantly more massive than the X-ray source as expected in current models of steady nuclear burning supersoft sources.

1 Introduction

So far, only a handful of supersoft sources are known in the Galaxy and most of them have soft X-ray luminosities significantly smaller than those found in the Magellanic Clouds for instance. In particular, soft X-ray galactic surveys, including the ROSAT all-sky survey, failed to detect galactic equivalents of sources like Cal 83 which would yield PSPC count rates as high as 1,000 cts/s if they were located at typical galactic distances. A natural explanation for this is that interstellar absorption in the Galaxy heavily attenuates the soft X-ray emission from these sources and that consequently, their observed X-ray spectra should look quite different from those of their Magellanic counterparts. Using this simple idea led to the discovery of RX J0925.7–4758 in the low galactic latitude part of the ROSAT all-sky survey (Motch Hasinger & Pietsch 1994, hereafter MHP94).

The X-ray spectrum of RX J0925.7–4758 is highly unusual for supersoft sources as it displays peaked emission close to 1 keV with very little flux detected below 0.4 keV and above 1.5 keV. This peculiar energy distribution may be explained by intrinsically soft emission cut off at low energies by photoelectric absorption from a large column density $N_H \approx 10^{22}$ cm^{-2}. Recent ASCA SIS observations by

Ebisawa et al. (1996) reveal the probable presence of several absorption edges such as O VIII (0.9 keV) and Ne X (1.38 keV). The fact that only the very high energy tail of the emitted spectrum can be analyzed and the absence of strong constraints on the distance to the source allow a large range of possible bolometric luminosities ($L_{Bol} = 10^{34.8}$ to $\geq 10^{37.7}$ erg s^{-1}, Ebisawa et al. 1996).

RX J0925.7–4758 is identified with a V $\approx$ 17 highly reddened object (E(B-V) = 2.1) suggesting that most, if not all, of the large photoelectric absorption seen in X-rays is of interstellar origin. The optical spectrum is characterized by H$_\alpha$ and high excitation emission lines such as He II λ4686 and the N III - C III Bowen complex. V band photometry shows the presence of a modulation with a probable period in the range of 3.3 - 4.0 d (MHP94).

In this paper, we report on preliminary results from more recent optical photometric and spectroscopic observations and from a 4.1 day long ROSAT HRI observation.

2 Optical Observations

Observations collected in April and December 1992 were already described in MHP94. In February 1994, we obtained five consecutive nights of observations with the ESO 1.52 m telescope and the Boller & Chivens spectrograph. With grating # 23 (129 Å/mm) and CCD #24 FA2K (1.93 Å pixel^{-1}), the FWHM resolution was $\approx$ 4 Å covering the wavelength range between 4000 and 7300 Å. The typical exposure time was 30 min. On occasions, the entrance slit was wide opened to 8.3″ and aligned on RX J0925.7–4758 and comparison star 2 in MHP94. From these observations, it was possible to reconstruct a photometric V light curve simultaneous with spectroscopy.

Quasi-simultaneously with the ROSAT observations in June 1994, Dr. S. Mottola kindly obtained V band photometry using the ESO-Dutch 0.9 m telescope at ESO, La Silla and the TEK #29 CCD chip.

2.1 Timing Analysis of the 1992-1994 Photometric V Data

A modulation on a time scale of several days is present during all of the four observing runs. Merging the 1994 February and June photometry with the previously reported 1992 data basically yields the same periodicities as found in MHP94. On the long period side, there is a series of alias peaks in the range P = 3.55 - 4.03 d. Among this set of alias periods, the least square power spectrum (Lomb 1976, Scargle 1982) and the minimum string length periodogram (Dworetsky 1983) both give a 'best' period at P = 3.7790 ± 0.0020 d with maximum light at JD 2,448,700.88 ± 0.11 d. This 'best' period is slightly longer, but still consistent with that found in MHP94 (P = 3.55 d). However, since intrinsic variability is likely to be present in the light curve, an accurate determination

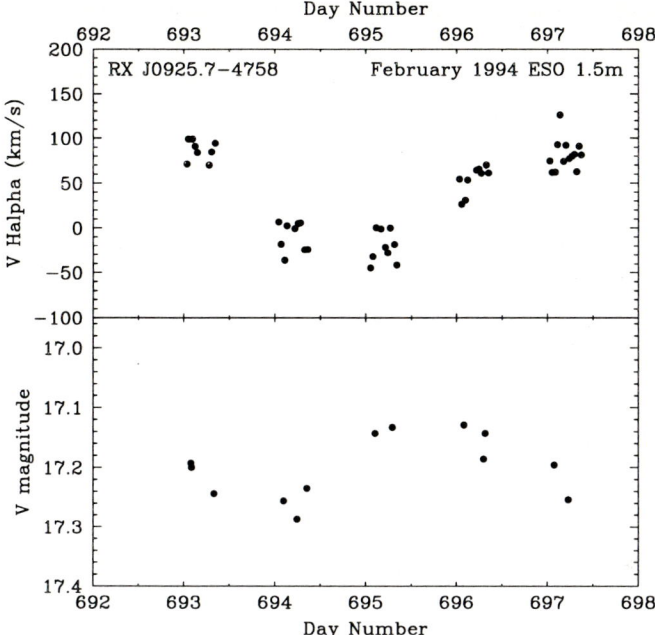

Fig. 1. Simultaneous H_α radial velocity and V magnitude light curves observed in February 1994. Day number is JD - 2,448,700.5.

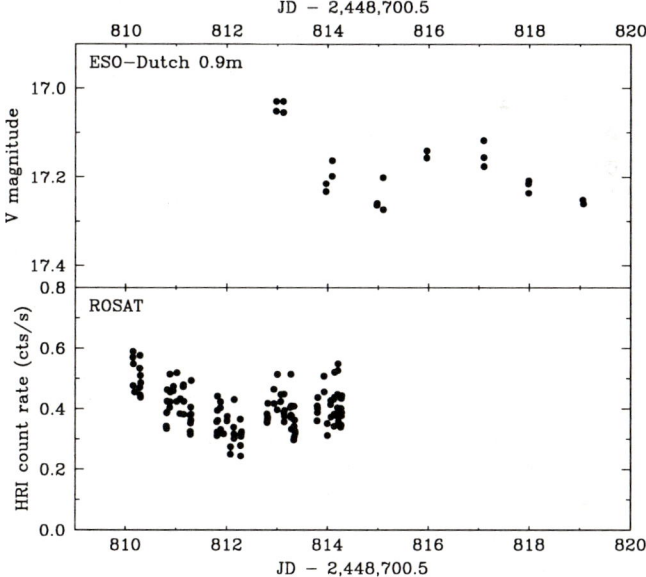

Fig. 2. ROSAT HRI and V photometric light curves obtained in June 1994. X-ray data are averaged into 300 s bins

of the photometric period would require more extended monitoring. For the moment we prefer to retain P = 3.79 ± 0.24 d as a conservative value. An one day alias series of slightly smaller significance occurs at P = 1.37 ± 0.03 d.

2.2 The Radial Velocity Curve

In Fig. 1 we show the H_α radial velocity curve together with the reconstructed V magnitudes. A smooth modulation appears both in the radial velocities and in the photometric light curves. The time scale of variability is compatible with the P = 3.55 - 4.03 d range of photometric periods. From the lack of significant radial velocity changes within a single night ($\approx$ 0.35 days) we conclude that short 1 day alias periods (P $\approx$ 1.37 d and P $\approx$ 0.774 d) are probably excluded.

H_α and He II λ4686 radial velocities vary together in phase. Assuming P = 3.779 d gives K_{H_α} = 63 ± 4 km s^{-1} and K_{HeII} = 84 ± 30 km s^{-1}. The optical photometric maximum occurs at inferior conjunction of the line emitting region, whatever is the exact value of the orbital period.

3 ROSAT HRI Monitoring

In June 1994 ROSAT monitored the source during 4.1 days with an effective exposure time of 55 ksec. We show in Fig. 2 the ROSAT HRI data together with the V band photometry obtained by Dr. S. Mottola with the ESO-Dutch 0.9 m telescope. The scatter in the HRI data is mostly due to photon counting statistics. The only significant variability is the $\approx$ 40% change in the X-ray flux from 0.35 to 0.5 cts/s on a time scale of a few days. Although the duration of the observation is not long enough to exclude the possibility that we simply witnessed a slow random variation, the time scale of variation is nevertheless comparable with the photometric period. Similar orbital phase dependent X-ray modulations are seen in some other supersoft sources (Kahabka 1996), and we may have detected such an effect in RX J0925.7–4758 as well. The maximum of the optical light curve, close to day 813, occurs $\approx$ 0.1-0.2 cycles after X-ray minimum.

4 Discussion

One of the straightforward result of this study is the apparent absence of large long term variations of the mean optical and X-ray flux of the source. From 1992 April till 1994 June the season averaged V magnitude of the optical counterpart did not change by more than a few percents. The source was always detected by ROSAT or ASCA and no X-ray flux variations larger than $\approx$ 50% are present from the survey discovery in 1991 till the most recent 1994 June observations. In the framework of the accreting white dwarf model (van den Heuvel et al. 1992), this behaviour is compatible with a genuine steady nuclear burning source or with a source decaying or rising on a time scale $\geq$ 3 yr.

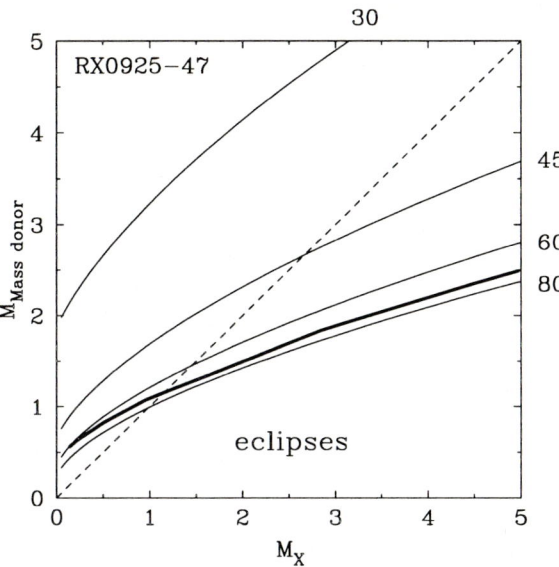

Fig. 3. Mass of the X-ray source plotted against the mass of the companion star for various orbital inclinations. We assume here $K_X = K_{HeII} = 84$ km s^{-1}. The dashed line represent equal masses. Solution below the thick line would produce X-ray eclipses.

The orbital period of RX J0925.7−4758 is most probably long, P = 3.79 ± 0.24 d, so far the longest known for supersoft sources, if one ignores the symbiotic systems. This readily indicates that the mass donor star is evolved. Roche lobe geometry implies a mean density of $\rho \approx 5\,10^{-3}$ - $10^{-2}\,\rho_\odot$ for the companion star corresponding to a giant or subgiant configuration. The absence of marked stellar features in the optical spectrum (MHP94) implies that as in other supersoft sources, the optical emission is dominated by the accretion disk, by the wind, and/or by the X-ray heated hemisphere of the companion star. If the visual absolute magnitude of RX J0925.7−4758 is similar to those of the Magellanic Cloud sources (i.e. up to $M_V = -2.1$ for RX J0513.9−6951, Pakull et al. 1993) the mass donor star can not be brighter than $M_V \approx 0.0$, indicating an A0 to K4 giant or subgiant spectral type.

The phasing of the He II λ4686 and H$_\alpha$ emission lines with respect to the optical photometric light curve is such that maximum light occurs when the line emitting region is closest to the observer. A similar behaviour is observed in Her X-1 and Sco X-1 and is generally interpreted by a model in which the high excitation lines are produced close to the X-ray source, probably in the inner region of the accretion disc, while the optical photometric modulation is essentially caused by the varying aspect of the X-ray heated hemisphere of the Roche lobe filling star. A similar phasing is found in the supersoft X-ray source Cal 87 (Cowley et al. 1990) and RX J0019.8+2156 (Beuermann et al. 1995). It has been argued, however, that He II $\lambda 4686$ lines may not exactly reflect the motion of the X-ray source since a sizeable fraction of the line emission may be produced in a wind emanating from the disc or from the companion star. Accordingly, the resulting emission line velocity may be smaller than that of the X-ray source and may even vary in antiphase if the companion star dominates (van den Heuvel et al. 1992). One should note, in addition, that the shape of the photometric light curve of RX J0925.7–4758 may not be compatible with that expected from a single hot stellar hemisphere. The presently available photometric data are still too scarce to allow to distinguish a truly periodic fine structure such as a narrow eclipse or dip from random cycle to cycle variations. However, at the 'best' period of P = 3.779 d, minimizing the scatter in phase, the folded light curve displays a rather peaked maximum with some evidence for a complex narrow absorption feature occuring $\approx$ 0.1 cycles before photometric maximum and apparently in phase with the X-ray minimum. A bulge on the edge of the accretion disc or a tilted disc such as proposed in Her X-1 (Gerend & Boynton 1976) could possibly explain the detailed features of the photometric light curve. The surprising occurrence of the X-ray minimum close to inferior conjunction of the X-ray source may reflect the fact that because of the large intervening interstellar absorption, ROSAT is only sensitive to the very high energy part of the X-ray spectrum. Subtle changes in the shape of the X-ray energy distribution may overwhelm any other bulk bolometric variations. Therefore, as long as the photometric light curve and its colour behaviour is not known with accuracy, the 'Her X-1' model for RX J0925.7–4758 should only be considered as a working hypothesis. If this interpretation is correct, then $K_X \geq K_{HeII} \approx 84$ kms^{-1} and the long orbital period point at a rather massive mass donor star. We show in Fig. 3 the possible relationships between the masses of the two components of RX J0925.7–4758. Very high inclinations would produce X-ray eclipses which are not observed. However, if the X-ray source is not compact or the X-ray eclipse narrow enough to have been missed in our data, this latter constraint may be somewhat relaxed. For $M_X = 1$ $M_\odot$, all non eclipsing solutions imply a companion more massive than the X-ray source as would be expected if thermally unstable Roche lobe overflow were the mass transfer mechanism driving the binary (van den Heuvel et al. 1992). Assuming $i \geq 30°$, the companion star may have a mass in the range of 1 to ≥ 3.5 $M_\odot$ and a radius in the range of 5 to 8 $R_\odot$. This range of values is fully consistent with the giant or subgiant A0 to K4 star inferred from the absence of visible stellar signatures in the optical spectrum. Non eclipsing solutions with companion stars less massive than the X-ray source are allowed if one uses the H_α velocities instead. However, February 1994 and April 1992 data (MHP94) both suggest that the amplitude of the He II $\lambda 4686$ radial velocities is indeed larger than that of H_α.

It is well known that the velocity dispersion of local disk populations of stars increases with age (Wielen 1977). Although the scattering mechanism is still controversial, it results in an increasing scale height with stellar age. If the mass of the companion star in RX J0925.7–4758 is larger than 2 $M_\odot$, then the system cannot be much older than 1.2 10^9 yr (e.g. Schaller et al. 1992). Such relatively young populations have a rather small scale height of $\approx 200\,\mathrm{pc}$ (e.g. Robin & Créze 1986). Objects like RX J0925.7–4758 may thus exhibit a strong concentration towards the galactic plane which would explain the scarcity of their detections so far.

Finally, we note that although the high mass transfer episode of RX J0925.7–4758 may have started in the giant-subgiant state, the evolved phase is still long enough ($\approx 10^8\,\mathrm{yr}$) to substantially increase the mass of the accreting white and drive it over the Chandrasekhar limit.

5 Summary

New spectroscopic and photometric observations show that the orbital period of RX J0925.7–4758 is in the range of 3.55 - 4.03 d . This is the longest orbital period known for supersoft sources if one excludes the symbiotic systems. In order to fill its Roche lobe, the mass donor must be slightly evolved, and the absence of detectable stellar absorption features in the optical spectrum argues in favour of an A0 to K4 giant or subgiant star. A 4.1 day long ROSAT HRI monitoring suggests that the X-ray flux of RX J0925.7–4758 varies with orbital phase. If the He II $\lambda4686$ radial velocity reflects the motion of the X-ray source then the mass donor star may be more massive than the accreting white dwarf. However, this conclusion depends on the validity of the 'Her X-1' model for RX J0925.7–4758 which cannot yet be firmly established from the available observations.

Acknowledgement: This work is based on observations obtained at the European Southern Observatory, La Silla, Chile with the ESO-Dutch 0.9 m and the ESO 1.5 m telescopes. We are especially grateful to Dr. S. Mottola who kindly carried out the optical photometric observations in June 1994. We also thank M. Pakull for useful discussions and for critical reading of this manuscript. The ROSAT project is supported by the Bundesministerium für Bildung, Wissenschaft, Forschung und Technologie (BMBF/DARA) and the Max-Planck-Gesellschaft. C.M. acknowledges support from a CNRS-MPG cooperation contract and thanks Prof. J. Trümper and the ROSAT group for their hospitality and fruitful discussions.

References

Beuermann K., Reinsch K., Barwig H., et al., 1995, A&A, 294, L1
Cowley A.P., Schmidtke P.C., Crampton D., Hutchings J.B., 1990, ApJ 350, 288
Dworetsky M.M., 1983, MNRAS 203, 917
Ebisawa K., Asai K., Dotani T., Mukai K., Smale A., 1996, in Röntgenstrahlung from the Universe, Eds. H.U. Zimmermann et al., MPE Report 263, p. 133
Gerend D., Boynton P.E., 1976, ApJ 209, 562
Kahabka P., 1996, A&A 306, 795
Lomb N.R., 1976, Ap&SS 39, 447
Motch C., Hasinger G., Pietsch W., 1994, A&A 284, 827 (MHP94)

Pakull M.W., Motch C., Bianchi L., Thomas H.-C., Guibert J., Ferlet R., Schaeidt S., 1993, A&A 278, L39
Robin A., Crézé M., 1986, A&A 157, 71
Scargle J.D., 1982, ApJ 263, 835
Schaller G., Schaerer D., Meynet G., Maeder A., 1992, A&ASS 96, 269.
van den Heuvel E.P.J., Bhattacharya D., Nomoto K., Rappaport S.A., 1992, A&A 262, 97
Wielen R., 1977, A&A 60, 263

The X-Ray Energy Spectrum of RX J0925.7–4758 with ASCA

K. Ebisawa[1], K. Asai[2], K. Mukai[1], A. Smale[1], T. Dotani[2], H.W. Hartmann[3], J. Heise[3]

[1] NASA/GSFC, Greenbelt, MD 20771 USA, and Universities Space Research Association
[2] Institute of Space and Astronautical Science, Yoshinodai, Sagamihara, Kanagawa, 229 Japan
[3] SRON Laboratory for Space Research, NL-3584 CA Utrecht, The Netherlands

Abstract. The super-soft X-ray source RX J0925.7–4758 was observed with the ASCA SIS with a superior energy resolution ($\Delta E/E \approx 10\%$ at 0.5 keV) for the first time. A simple blackbody model does not fit the data, and several absorption edges of highly ionized heavy elements are required. Introducing absorption edges significantly changes the best-fit blackbody parameters, such that the temperature is increased and the radius and bolometric luminosity are decreased. A massive white dwarf close to the Chandrasekar limit at a distance of $\gtrsim$ 10 kpc will be able to produce such an energy spectrum.

1 Introduction

Dozens of Super Soft Sources (SSS) have been discovered with Einstein and ROSAT (for a review, e.g., Hasinger 1994; Rappaport and Di Stefano 1996; Cowley et al. 1996; Kahabka and Trümper 1996). Steady nuclear burning on the surfaces of accreting white dwarves is a likely model for SSS (van den Heuvel et al. 1992; Heise et al. 1994). Observational information of SSS in the X-ray band is limited by spectral resolutions of the current instruments. While the ROSAT PSPC has a very suitable energy band (0.1 – 2 keV) for the study of SSS, its energy resolution ($\Delta E/E \sim 60\%$ at 0.5 keV) does not allow detection of local spectral structures such as emission lines or absorption edges.

Having a superior energy resolution ($\Delta E/E \sim 10\%$ at 0.5 keV), the ASCA Solid State Spectrometer (SIS) is potentially a powerful instrument for spectroscopic study of SSS, though its energy band (0.4 – 10 keV) is higher than the typical SSS energy range. RX J0925.7–4758 is an unusual SSS having most X-ray emission above 0.5 keV (Motch et al. 1993), and thus the most suitable target for the ASCA SIS.

The ROSAT spectrum of RX J0925.7–4758 is fit with the blackbody having $kT = 40 - 55$ eV with an inter-stellar absorption with $N_{\rm H} = (1.0 - 1.9) \times 10^{22}$ cm^{-2} (Motch et al. 1994). The large column density, which is consistent with the estimate from optical redenning of the continuum and strength of the interstellar absorption lines, suggests the source is located within or behind the Vela Sheet molecular cloud ($d = 425$ pc) (Motch et al. 1994).

2 Observation and Data Analysis

RX J0925.7–4758 was observed in the ASCA AO3 phase, on December 22 1994, for 20 ksec.

2.1 Simple Blackbody Model Fit

Fig. 1 is the result of a blackbody plus inter-stellar absorption model fit. We used the interstellar absorption model by Morrison and McCammon (1983). In Tab. 1, we show best-fit parameters, although these parameters have little sense, since the fit is not acceptable at all (reduced $\chi^2 \sim 10$). There are two noticeable features. First, absorption edge-like features are seen at around 0.7, 0.9 and 1.4 keV. Second, although a significant amount of the hydrogen column density is required to account for the low energy absorption, the oxygen K-edge feature at 0.54 keV is much weaker than is expected for the interstellar medium having the cosmic abundance.

Table 1. Spectral parameters for the blackbody fit

Temperature	46 eV
Blackbody Radius	1.6×10^5 km $(d/1\ \text{kpc})$
N_H	2.1×10^{22} cm^{-2}
Bolometric Luminosity	1.5×10^{40} $(d/1\ \text{kpc})^2$ ergs/s
χ^2/dof (dof)	10.0 (51)

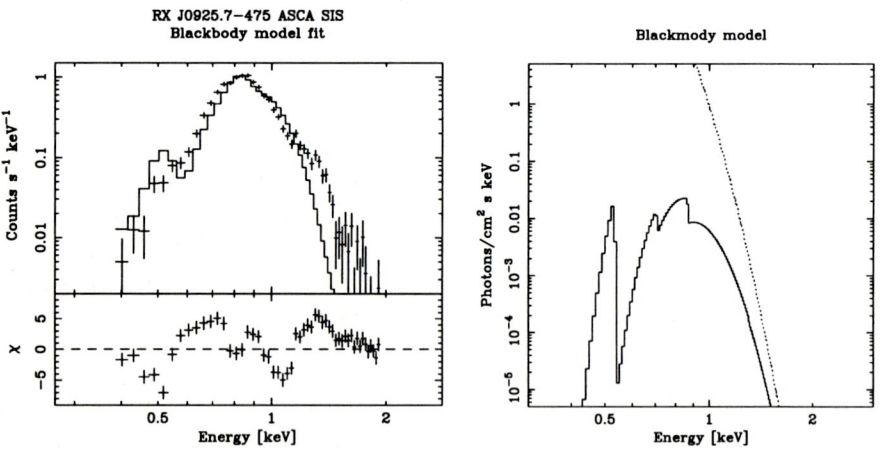

Fig. 1. Left: Blackbody model fit result. Right: the best-fit model with (solid-line) and without (dotted-line) the interstellar absorption.

2.2 Blackbody Model with Absorption Edges

Next we consider a simple model to fit the spectrum by introducing absorption edges to the blackbody model and changing elemental abundances of the intervening interstellar medium. We discuss origin of the absorption edges in section 3.1. At least three absorption edges are required at 0.94 keV, 1.04 keV and 1.43 keV (Tab. 2). In addition, in order to explain the small oxygen edge feature, we needed to reduce the oxygen abundance of the interstellar medium. Furthermore, we found reducing iron abundance significantly improves the fit at around the 0.71 keV Fe-L edge. Note that the blackbody temperature significantly increases, and the bolometric luminosity and radius decreases, compared to the fit without absorption edges (Tab. 1).

Table 2. Fit with a blackbody plus absorption edges

Temperature	96 ± 11 eV
Blackbody Radius	140^{+180}_{-70} km $(d/1\text{ kpc})$
$N_{\rm H}$	$1.0^{+0.08}_{-0.06} \times 10^{22}$ cm^{-2}
Bolometric Luminosity	$2.1^{+5.3}_{-1.3} \times 10^{35}$ $(d/1\text{ kpc})^2$ ergs/s
Edge Energies and Optical depths	0.94 ± 0.03 keV $\quad$ 0.65 ± 0.24
	1.04 ± 0.02 keV $\quad$ 1.19 ± 0.23
	1.43 ± 0.04 keV $\quad$ 1.2 ± 0.6
Oxygen abundance (cosmic = 1)	$0.38^{+0.11}_{-0.14}$
Iron abundance (cosmic = 1)	$0.2^{+0.5}_{-0.2}$
χ^2/dof (dof)	1.02 (42)

* Oxygen and iron abundances are relative to the cosmic abundance by Anders and Ebihara (1982), which gives 7.41×10^{-4} and 3.31×10^{-5}, for O/H and Fe/H respectively.

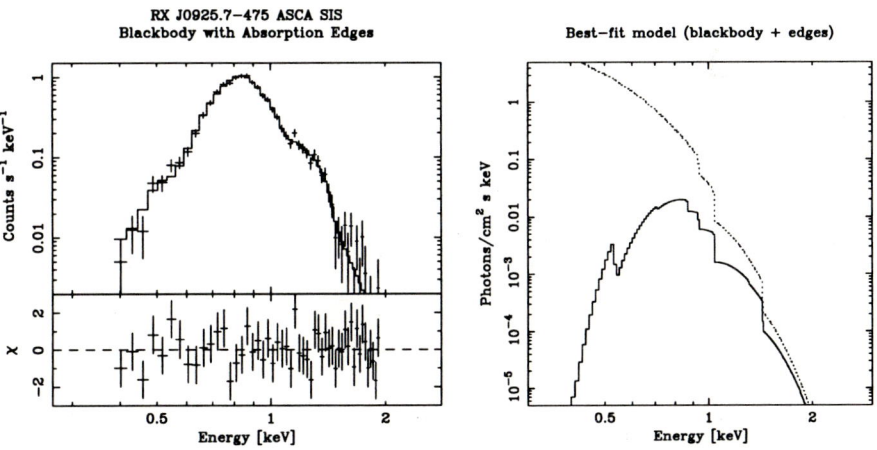

Fig. 2. Left: Fitting with a blackbody with absorption edges. Right: the best-fit model with (solid-line) and without (dotted-line) the interstellar absorption.

2.3 Application of a NLTE Spectral Model

We also tried a Non-Local Thermodynamical Equilibrium (NLTE) model by Hartmann and Heise (1996). They calculated the NLTE model spectral for two surface gravities, $\log g = 8.5$ and $\log g = 9.0$ (in the CGS unit). Effective temperatures are in the ranges of 5×10^5 to 8×10^5 K and 5×10^5 K to 10^6 K, respectively. In Fig. 3 and Tab. 3, we show results of the fitting with the $\log g = 9.0$ model. Oxygen and iron abundances for the interstellar medium were fixed to the values derived in the previous section.

Although the model fits the data well below the Ne IX K-edge (1.19 keV), the model predicts a much smaller flux above this energy. The Ne IX K-edge depth predicted in the model is much deeper than in the data, and the data rather indicates a higher energy edge feature which seems to correspond to the Ne X K-edge (1.36 keV). This indicates a higher ionization state is required than predicted by the present model.

Table 3. Spectral parameters with the NLTE model ($\log g = 9.0$) fit

Temperature	74 eV
Blackbody Radius	160 km $(d/1 \text{ kpc})$
$N_{\rm H}$	1.0×10^{22} cm^{-2} (fixed)
Bolometric Luminosity	9.7×10^{34} $(d/1 \text{ kpc})^2$ ergs/s
Oxygen abundance (cosmic = 1)	0.38 (fixed)
Iron abundance (cosmic = 1)	0.2 (fixed)
χ^2/dof (dof)	7.2 (51)

* The hydrogen column-density, oxygen and iron abundances are fixed to the values obtained in the previous section.

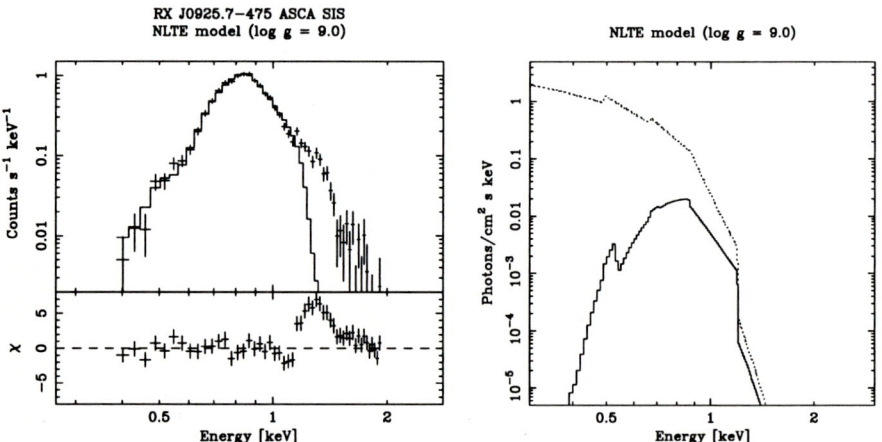

Fig. 3. Left: Fitting with the NLTE model by Hartmann and Heise (1996) for $\log g = 9.0$. Right: the best-fit model with (solid-line) and without (dotted-line) the interstellar absorption.

2.4 Comparison with the ROSAT Spectrum

ROSAT PSPC observation was carried out on November 20. 1992 (Motch et al. 1994). We obtained the PSPC data and responses of the same observation through two routes; one from Wouter Hartmann, which is based on the European ROSAT data analysis system, and the other from HEASARC at NASA/GSFC. For the latter, the off-axis mirror response was created with the program "pcarf". We found the normalizations of the two spectra differ by $\sim$ 30 % (HEASARC one is higher), although the spectral shapes are completely consistent. Origin of the normalization difference is under investigation.

In Fig. 4, we show the result of fitting the PSPC spectrum (taken from HEASARC) with the NLTE model by Hartmann and Heise (1996). Parameters are fixed to the best-fit values obtained with the ASCA SIS (table 3) besides the normalization. The normalization of the PSPC spectrum is 1.04 times that of ASCA, and the reduced χ^2 is 1.48 (dof=23).

Although energy resolution of the PSPC spectrum is not good enough to ascertain individual spectral edge features, systematic residuals are seen above $\sim$ 1.1 keV, suggesting the same high energy excess feature observed in the ASCA SIS spectrum. We conclude that the spectral shape has not changed systematically between the ROSAT PSPC and ASCA SIS observations which are 4 year apart. If the normalization of the HEASARC PSPC spectrum is correct, the luminosity does not show systematic variations either.

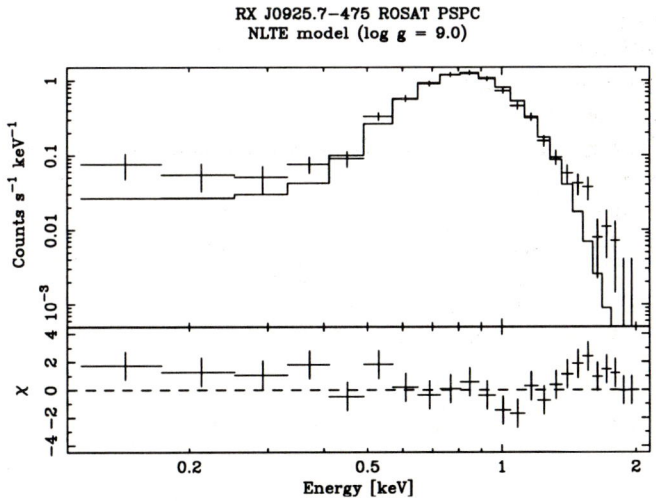

Fig. 4. Fitting the ROSAT PSPC spectrum (taken from HEASARC) with the NLTE model by Hartmann and Heise (1996) for $\log g = 9.0$. Parameters besides the normalization are fixed to those obtained with the ASCA SIS fit.

3 Discussion

3.1 Origin of the Absorption Edges

Let's consider origin of the observed absorption edges based on a simple model. If we assume the LTE condition, ionization balance is determined by the temperature, T, and the electron density, N_e, through the Saha relation. The electron density is of the order of the particle number density, and related with the temperature and the gas pressure as $P_g \approx N_e kT$. The gas pressure can be estimated from the hydrostatic equilibrium condition as $P_g \approx g/\kappa_R$, where g is the surface gravity and κ_R is the Rosseland mean opacity, ~ 0.4 cm^2 g^{-1}. Surface gravity of a white dwarf is typically 10^8 to 10^9 cm s^{-2}. Using the temperature obtained from the blackbody plus absorption edge model, ~ 96 eV, the typical electron number density in the atmosphere is estimated as $N_e \sim 10^{18}$ to 10^{19} cm^{-3}.

In Fig. 5, we show ion ratios for several abundant elements for $N_e = 10^{18}$ and 10^{19} cm^{-3}, as functions of the temperature. From this figure, considering the temperature is 90–100 eV, absorption edges which locate in the energy range 0.4–2.0 keV and may account for the observed spectral features will be the following: K-edges of O VIII (0.88 keV), Ne IX (1.19 keV), Ne X (1.36 keV) and Mg XI (1.75 keV), and L-edges of Fe XVII to $\sim$ Fe XXII (1.27 to $\sim$ 1.77 keV). Note that carbon and nitrogen are considered almost completely ionized.

The edge energies obtained through the blackbody plus absorption edge model fit (Tab. 2) do not precisely coincide these real edge energies. This is probably because (1) the continuum is not a blackbody shape and (2) the spectral features at the edges cannot be represented with the standard photoionization whose cross-section is proportional to E^{-3} above the edge. As a matter of fact, the model atmosphere spectrum (Fig. 3) indicates the continuum is represented with a combination of power-law spectra having discontinuities at edges. With these uncertainties, we may consider that the two apparent edges at 0.94 ± 0.03 keV and 1.04 ± 0.02 keV are due to the presence of O VIII (0.88 keV) and/or Ne IX (1.19 keV) edges. Note that the presence of Ne I edge at 0.86 keV due to the heavy interstellar absorption makes it difficult to ascertain the O VIII edge. The observed edge at 1.43 ± 0.04 keV is considered to correspond to the Ne X edge at 1.36 keV, although effects of the iron L-edges will not be negligible.

We tried a NLTE model by Hartmann and Heise (1996). Although the model could fit the data below the Ne IX edge at 1.19 keV, the strong Ne IX edge predicted by the model was not observed and significant residuals remain above this energy (Fig. 3). The data rather indicates presence of the Ne X edge, suggesting a higher ionization state. If the temperature is increased in this model, however, strong "emission edges" emerge in lower energies (see C VI and N VII edges at 0.49 keV and 0.67 keV respectively in Fig. 3 right), which do not fit the data. Such emission edges are caused by the outer temperature inversion layer, and it is known that temperature structure in the NLTE atmosphere is very sensitive to even small amounts of light metals (Li – Ca) (Rauch 1996). There will be more things to study in the NLTE model, and precise modeling of the ASCA spectrum is under progress.

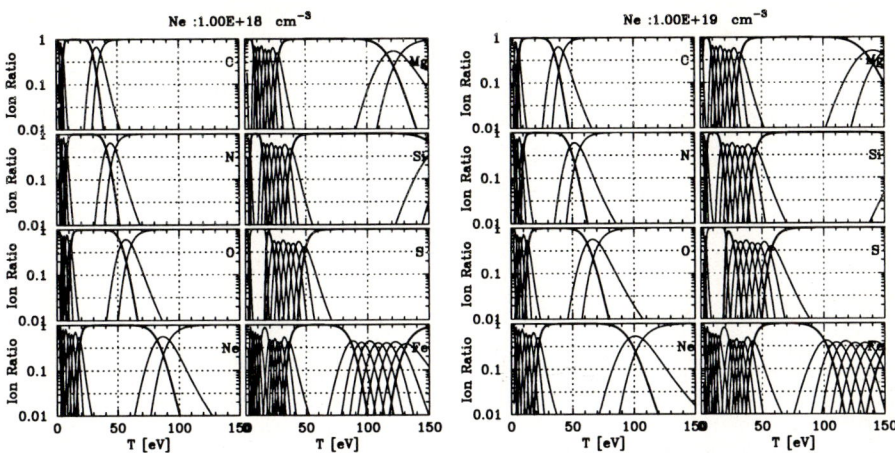

Fig. 5. Ion fractions for several elements determined from the Saha relation as functions of temperature. Shown for two electron densities, 10^{18} cm^{-3} (left) and 10^{19} cm^{-3} (right). Helium-like and Ne-like ions are shown with thick lines for each element.

3.2 A Model of RX J0925.7−4758

We found that the X-ray energy spectrum of RX J0925.7−4758 has several absorption edges of highly ionized heavy elements. Taking account of these absorption edges, the radius and the luminosity obtained from the blackbody model fit have become much smaller than those without assuming edges. Assuming the source is a white-dwarf, the minimum radius will be $\sim$ 3000 km at the Chandrasekar mass limit ($\sim 1.4 M_\odot$) (Fig. 6). For the radius obtained from the model fit (Tab. 2) to be consistent with the minimum white-dwarf radius, the distance

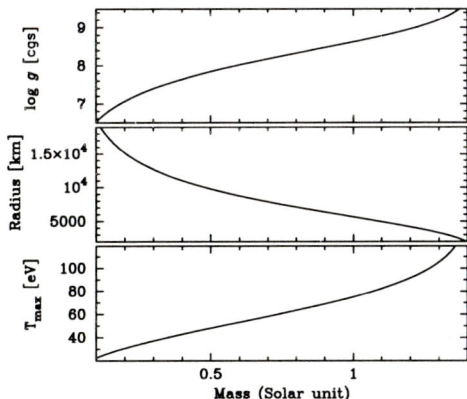

Fig. 6. Typical radius, surface gravity and the maximum temperature of a white dwarf as a function of the mass. The radius is derived from the formula by Pringle and Webbink (1975), and the surface gravity and the blackbody temperature corresponding to the Eddington limit are calculated.

will have to be ≳ 10 kpc. Such a large distance will be consistent with a heavy interstellar absorption derived from the present fitting and optical observations (Motch et al. 1994). A high temperature (∼ 96 eV) is required to account for the X-ray spectrum, and this is also consistent with a massive white-dwarf which can have such a high temperature (Tab. 2). Therefore, our conclusion is that RX J0925.7-4758 is a massive white-dwarf close to the Chandrasekhar limit at a distance of ≳ 10 kpc.

References

Anders E., Ebihara M., 1982, Geochimica et Cosmochimica Acta 46, 2363
Cowley A.P., Schmidtke P.C., Crampton D., Hutchings J.B., 1996, in "Compact Stars in Binaries", ed. E.P.J. van den Heuvel & J. van Paradijs (Dordrecht: Kluwer) p. 439
Hasinger G., 1994, in "The Evolution of X-ray Binaries", ed. S.S. Holt and C.S.R. Day, AIP Conference Proceedings 308, p. 611
Hartmann H.W., Heise J., 1996, this volume p. 6
Heise J., van Teeseling A., Kahabka P., 1994, A&A 288, L45
Kahabka P., Trümper J., 1996, in "Compact Stars in Binaries", ed. E.P.J. van den Heuvel & J. van Paradijs (Dordrecht: Kluwer) p. 425
Morrison R., McCammon D., 1983, ApJ 270, 119
Motch C., Hasinger G., Pietsch W., 1994, A&A 284, 827
Pringle J.E., Webbink R.F., 1975, MNRAS 172, 493
Rappaport S., Di Stefano R., 1996, in "Compact Stars in Binaries", ed. E.P.J. van den Heuvel & J. van Paradijs (Dordrecht: Kluwer) p. 415
Rauch T., 1996, this volume p. 3
van den Heuvel E.P.J., Bhattacharya D., Nomoto K., Rappaport S.A., 1992 A&A 262, 97

Optical Photometry of RX J0019.8+2156 over the Last Three Years

T. Will, H. Barwig

Universitäts-Sternwarte München, Scheinerstr. 1, 81679 München, FRG

Abstract. We present time resolved UBVRI photometry of RX J0019.8+2156 which we performed with a multi-channel-photometer from 1992 to 1995. The light curves we received show regular and irregular brightness variations. Prominent cyclic depressions of 0.5 mag which cover a phase interval of 0.4 are interpreted as eclipses of a white dwarf by a Roche lobe filling secondary. Furthermore, a shallow minimum at phase 0.5 suggests a secondary eclipse. The orbital period derived from our light curves shows a little, but systematic difference to the most precise period value given in literature. The varying depth of the primary minimum as another phenomenon may be interpreted either as a cycle to cycle variation or as a quasi-periodic long term modulation.

1 Introduction

RX J0019.8+2156 (henceforth RX J0019) was discovered in the course of a search for soft X-ray sources (SSS) from the ROSAT All-Sky-Survey. It was the first candidate of SSS that has been found in our galaxy (Reinsch et al. 1993). Due to the relatively high apparent brightness (12.5 mag) of its optical counterpart this system provides the rare opportunity of studying these binary systems in further details. The first photometric and spectroscopic observations are described by Beuermann et al. (1995). Preliminary system parameters were derived from the optical data. Two models were suggested to describe the observational results. The first one assumes a pre-cataclysmic, detached binary system with two components containing one solar mass each. But this model explains neither the observed flickering nor the X-ray emission sufficiently. The second model transferring mass from a Roche-lobe filling secondary onto a White Dwarf primary at such a high rate that nuclear burning on its surface seems possible, is more likely. Due to the relatively long orbital period of 16 hours, which is commensurable with two days, observations have to cover extended time intervals in order to extract a mean photometric light curve. In the following chapters, the results from the observations over the last three years are presented.

2 Observations

The observations were performed by using the high speed multi-channel-photometer MCCP (Barwig et al. 1987) at the 80-cm telescope of the Wendelstein Observatory. This photometer allows us to monitor a program star, a comparison star and the sky background simultaneously in UBVRI. By using this differential method, atmospheric transparency variations in particular during non-photometric nights can be taken into account. Our photometric data of RX J0019 covers 131 hours of observation within 25 nights which are spread over three years. The individual observing runs are listed in Tab. 1 (observations covering phase 0.0 are marked bold face).

Table 1. Journal of observations

Date of Observation	Start of measurement (HJD–2440000)	cycle (E)	duration (h)
21/09/1992	8887.296443	**0**	8.93
03/02/1993	9022.219093	**204**	3.28
07/02/1993	9026.240891	210	2.44
09/02/1993	9028.221953	213	2.91
11/02/1993	9030.233919	216	2.41
17/11/1993	9309.292191	638	4.78
06/12/1993	9328.245187	667	4.20
12/12/1993	9334.245408	676	2.01
16/12/1993	9338.201296	682	2.11
18/12/1993	9340.207739	685	6.45
28/12/1993	9350.196342	700	5.21
29/12/1993	9351.195015	702	4.42
09/01/1994	9362.237197	**719**	4.53
10/01/1994	9363.201174	720	5.36
11/10/1994	9637.339892	1135	8.22
12/10/1994	9638.299952	1136	3.46
13/10/1994	9639.252475	1138	7.16
14/10/1994	9640.412506	**1140**	6.23
15/10/1994	9641.301485	1141	3.21
16/10/1994	9642.299855	1142	3.65
18/10/1994	9644.266437	1145	8.44
20/10/1994	9646.283440	1148	6.77
21/10/1994	9647.284614	1150	5.52
22/10/1995	10013.224758	**1705**	10.16
23/10/1995	10014.249290	**1706**	9.31

3 Data Reduction

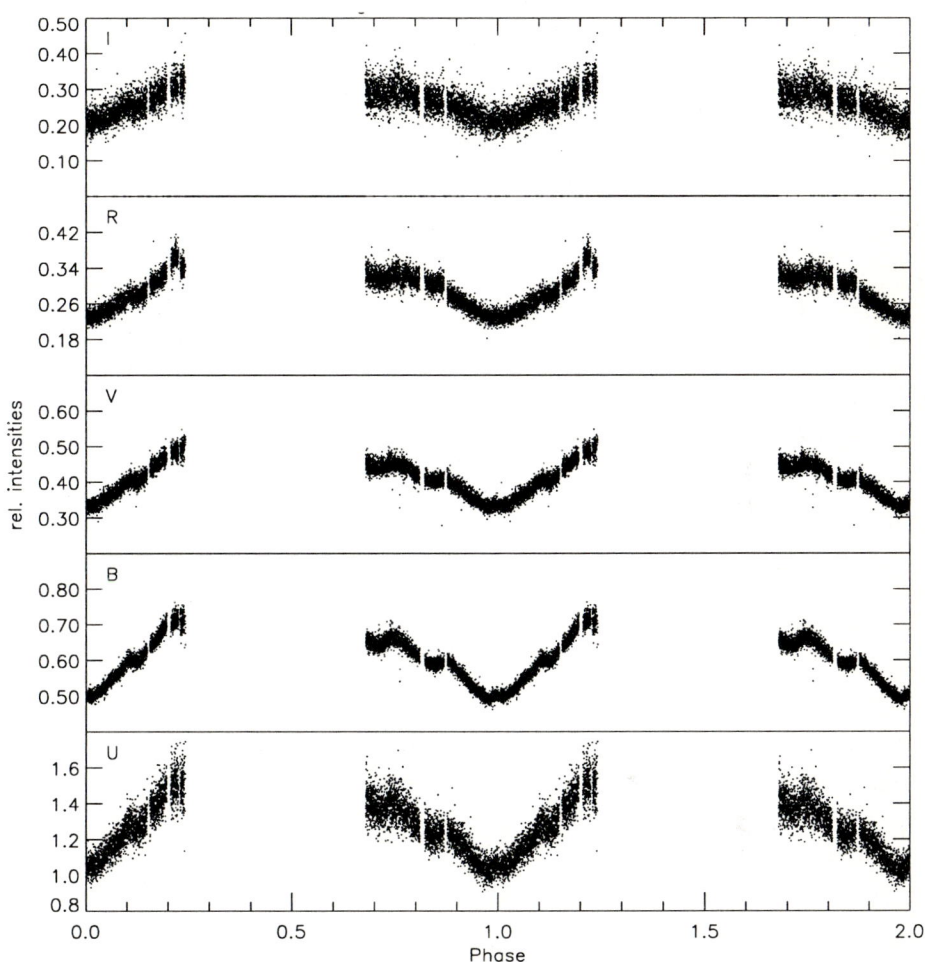

Fig. 1. Reduced light curves in UBVRI of September 21, 1992.

Usually, the original data of each channel is influenced by extinction and sky-background radiation. As the instrument measures simultaneously the brightness variations of a nearby comparison star as well as the sky straylight, these contributions can be eliminated by applying the so-called standard reduction procedure (see Barwig et al, 1987). In order to normalize the individual channels, calibration measurements were performed during photometric conditions by using the comparison star itself. The obvious constancy of the calibration coefficients during our observing campaign is indicative of a fairly high stability of all detectors. For this reason the observed non-periodic night-to-night variations of the light curves are very likely caused by instrinsic stellar variability of the X-ray binary system. The first reduced UBVRI light curves of RX J0019, which were also involved in the paper of Beuermann et al. (1995), are plotted in Fig. 1. They show a deep eclipse-like feature whose depth is dependent on the colour. The best S/N value was obtained in the B-filter, due to a combination of the instrumental quantum efficiency and the colour temperature of the target. Therefore, only B-light curves are displayed in the following figures.

4 Orbital Period

By examining our complete dataset, we could find only five light curves addition in which covered the phase of primary eclipse. The prominent deep eclipse features (see Fig. 2) were used to determine the orbital period of RX J0019. For this purpose, we fitted the light curves with a spline approximation assuming that the calculated zero points of their derivatives indicated the mid-time of the corresponding eclipses. As we used different spline parameters, we estimated the eclipse timing errors (see Tab. 2).

Table 2. Data used for period determination

cycle (E)	minima HJD–2440000 [d]	mean error [d]	O–C
0	8887.505	0.00071	0.00434
204	9022.250	0.00068	-0.00409
719	9362.382	0.00080	0.00683
1140	9640.457	0.00029	-0.00951
1705	10013.602	0.00067	0.01196
1706	10014.273	0.00085	0.00139

Applying a linear least-square-fit we received the ephemeris of the mid-eclipse time T_0:

$$T_0 = \text{HJD } 2448887.509(2)\text{d} + E \times 0.6604721(72)\text{d}$$

Our period value within the error bars coincides with the photometric and spectroscopic periods published by Beuermann et al. (1995). But obviously there is a systematic difference to the photometric period of Greiner and Wenzel (1995), which had been derived from photographic plates taken over a time span of 40 years (while investigating plates over more than 100 years) (see Tab. 3). Further studies are needed to decide whether a period change occured in the meantime or whether the definition of minima was incorrect, because only parts of the eclipse phase were covered by observation.

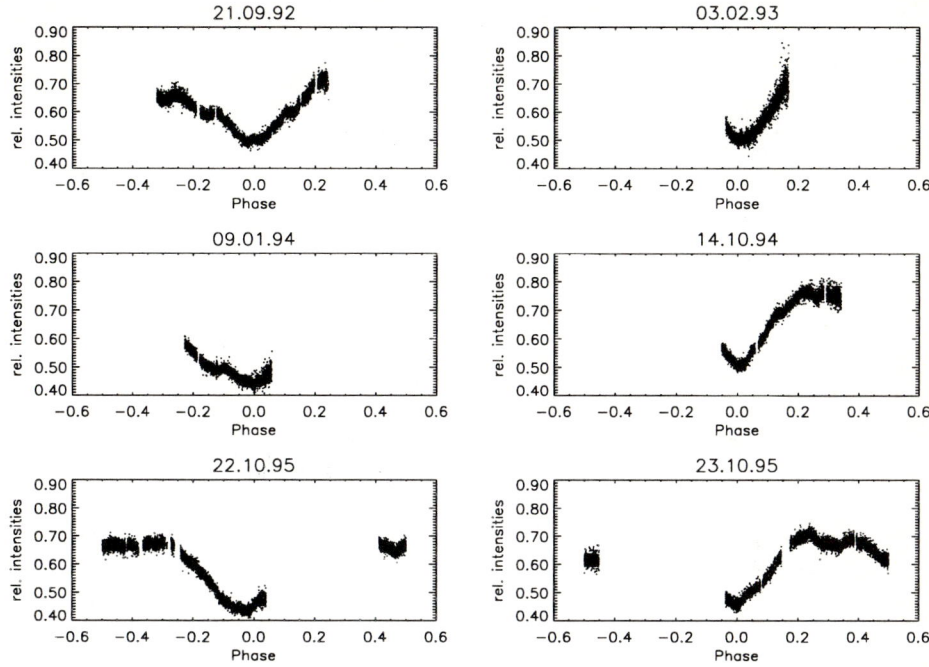

Fig. 2. Phase diagrams of all minima used for period determination (the phase was calculated from the period of Beuermann et al. 1995).

Table 3. Published orbital periods of RX J0019

Method	Period(error) [d]	Observations	Reference
photometry	0.66041(12)	Sep-Oct 92	Beuermann et al. (1995)
spectroscopy	0.66047(06)	Aug 92/93	Beuermann et al. (1995)
photometry	0.6604565(15)	1955–1993	Greiner et al. (1995)
photometry	0.6604721(72)	09/92–10/95	this paper

5 Combined Light Curves

In order to get a complete phase diagram of RX J0019, data of different nights had been combined. For this purpose only those measurements were selected which had been obtained under photometric conditions. During these nights the channel transformation coefficients could be determined at a high accuracy. Influences of instrumental instabilities on the light curves can be excluded applying the differential measuring method. In Fig. 3 (upper panel) two light curves obtained in 1992 and 1994, respectively, were combined with only a small

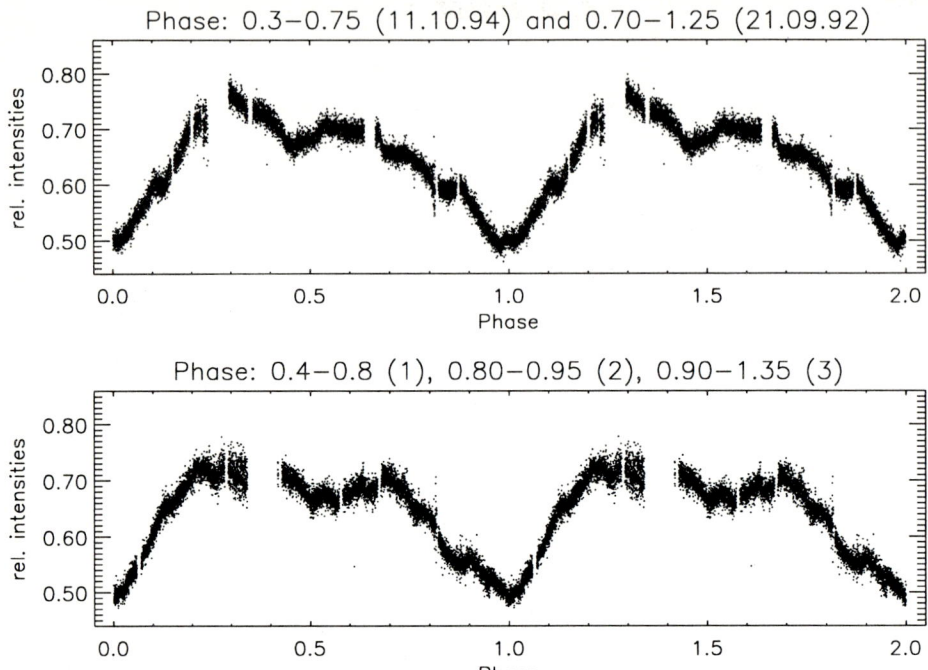

Fig. 3. Combination of light curves of different nights covering one complete orbital phase; in the lower plot the numbers in the brackets refer to the nights of December 18, 93, January 09, 94 and October 14, 94, respectively.

phase overlap. Their shape seems to be distorted by several step-like structures. Besides the most prominent primary eclipse feature, there is a significant depression at phase 0.5 suggesting a secondary eclipse. Gaps within the individual light curves are due to calibration measurements. In the lower plot of Fig. 3 another three light curves covering disjunctive phase intervals are plotted. The lightcurve also exhibits short-term variations which do not coincide with those of the corresponding phases of the upper plot while the phase of the secondary minimum has not changed significantly. In order to study the general photometric behaviour of RX J0019, all observational data within the time interval of three years were superimposed to a single phase diagram (Fig. 4). The individual short term variations obviously do not coincide with the orbital period and therefore disappear during superposition. Thus, only two eclipse features remain prominent. The appearance of a secondary minimum is of obvious discrepancy to the favoured model of RX J0019 for which an orbital inclination of 20 degrees is assumed. An interesting phenomenon hidden in the averaged phase diagram are the varying depths of the primary minima, plotted in Fig. 5. This variation may be approximated by a sine fit with a maximum period of 1.8 years. Another aspect is conspicious in this plot: minima of even cycles are brighter than those of odd cycles.

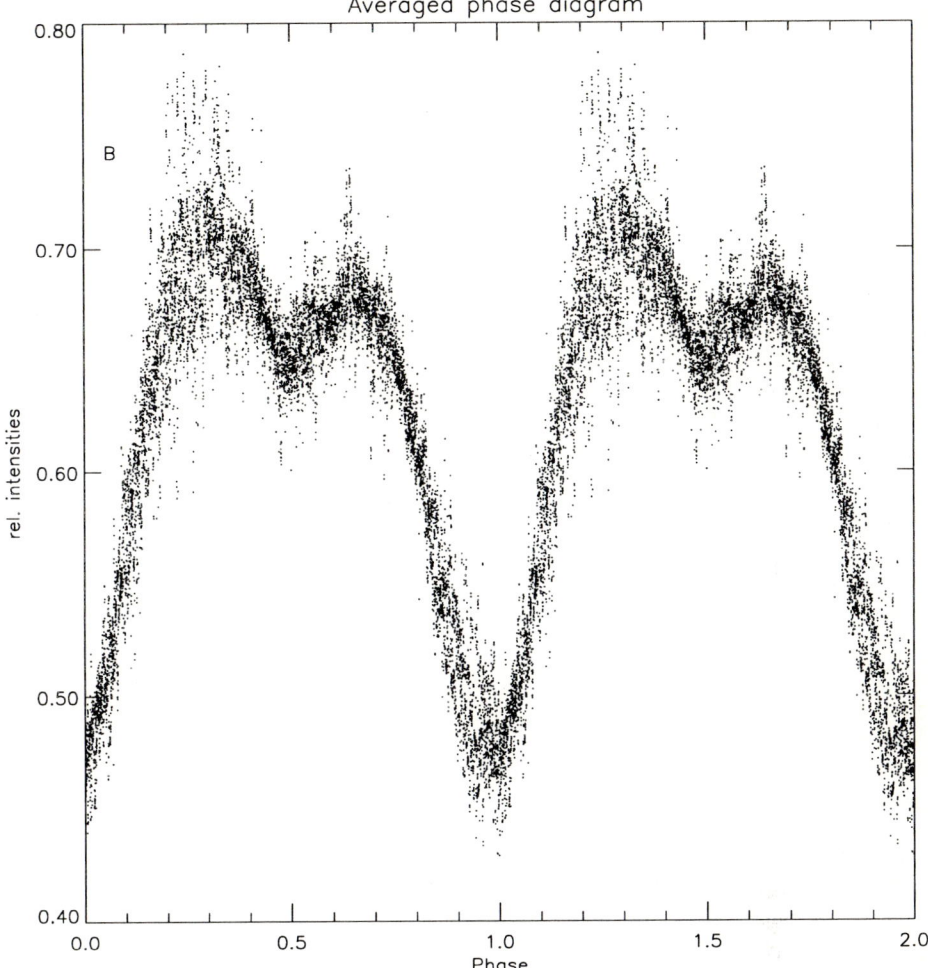

Fig. 4. Superposition of all light curves; the orbital phase is plotted twice.

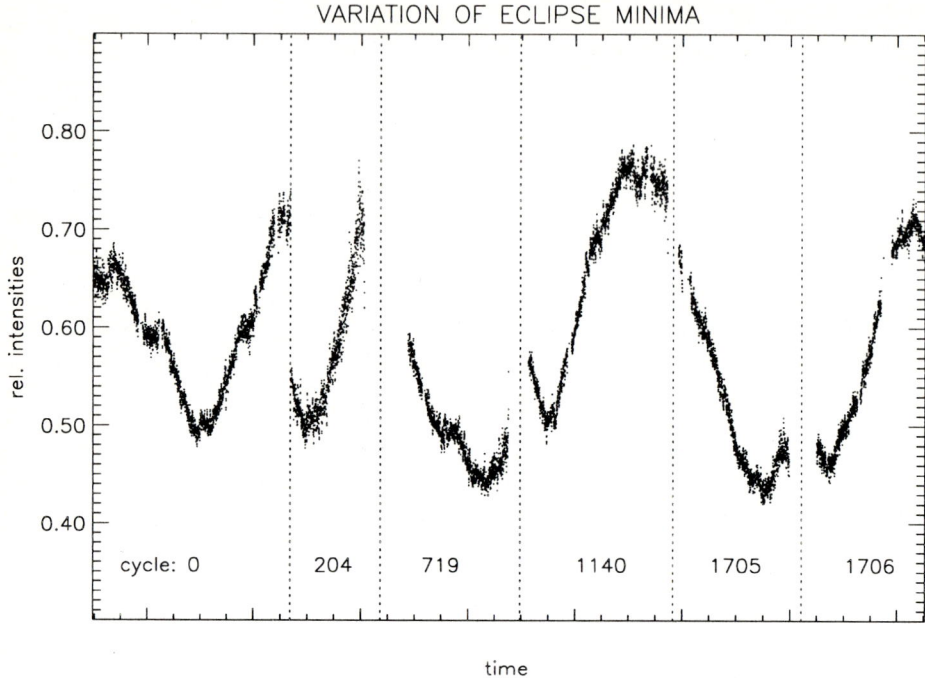

Fig. 5. Comparison of the depths of the primary minima. Each one is plotted in a seperate frame with its corresponding cycle number. Time interval between two tickmarks: 0.05 d

Acknowledgement: The authors like to thank the staff of the observatory for their permanent efforts not to waste a single photon. They are also very grateful to S. Wolf for graphical processing of the plots.

References

Barwig H., Schoembs R., Buckenmayer C., 1987, A&A 175, 327
Beuermann K., Reinsch K., Barwig H., Burwitz V., deMartino D., Mantel K.H., Pakull M., Robinson E.L., Schwope A., Thomas H.-C., Trümper J., Zhang E., 1995, A&A 294, L1
Greiner J., Wenzel W., 1995, A&A 294, L5
Reinsch K., Beuermann K., Thomas H.-C., 1993, Astron. Ges. Abstr. Ser. 9, 41

Phase Resolved UV Spectroscopy of RX J0019.8+2156

B.T. Gänsicke[1], K. Beuermann[1], D. de Martino[2]

[1] Universitäts-Sternwarte, Geismarlandstr. 11, 37083 Göttingen, Germany
[2] IUE Observatory VILSPA ESA, PO Box 50727, 28080 Madrid, Spain

Abstract. We present orbital phase-resolved UV spectroscopy of the supersoft X-ray source RX J0019.8+2156 obtained with IUE in 1992-93. The UV and optical light curves show a quasi-sinusoidal orbital modulation with wavelength dependent amplitude and a deep secondary minimum at $\Phi \sim 0.6$. The UV spectrum is of lower excitation than in other SSS. The various metal absorption lines are consistent with the source being located outside the gas layer of the galactic disc, though there is some evidence for intrinsic absorption. The UV spectrum of RX J0019 differs from those of the LMC sources CAL 83 and RX J0513.9–6951. A simple model where the UV light originates from the hot primary, the accretion disk and the heated secondary fails to reproduce the observed UV continuum.

1 Introduction

RX J0019.8+2156, henceforth RX J0019, is the only galactic supersoft source discovered from the ROSAT All Sky Survey at high galactic latitudes $|b| > 20°$ (Beuermann et al. 1995). In this paper, we will refer to supersoft sources (SSS) in the sense of close binaries containing an accreting, steadily burning white dwarf (van den Heuvel et al. 1992). Compared to the LMC SSS CAL 83 and RX J0513.9–6951 and to the only other galactic SSS known so far, RX J0925.7–4758, RX J0019 is optically brighter by ~ 5 mag and less absorbed by 1-2 orders of magnitude, underlining the potential of this object in understanding the physics of the SSS. RX J0019 shows variations of the optical, UV and X-ray flux as well as regular optical radial-velocity variations at a 15.85 hour period, interpreted as the binary orbital period (Beuermann et al. 1995). The moderately hot, irradiated accretion disc and the heated secondary star are expected to be the dominant sources of light at optical and UV wavelengths (Beuermann et

Table 1. Log of IUE observations.

Date	SWP spectra	LWP spectra	Orbital phase coverage
13/14 Jul. 1992	5	4	0.77-0.15
11 Jan. 1993	5	3	0.61-0.96
26-28 Aug. 1993	15	14	0.00-1.00

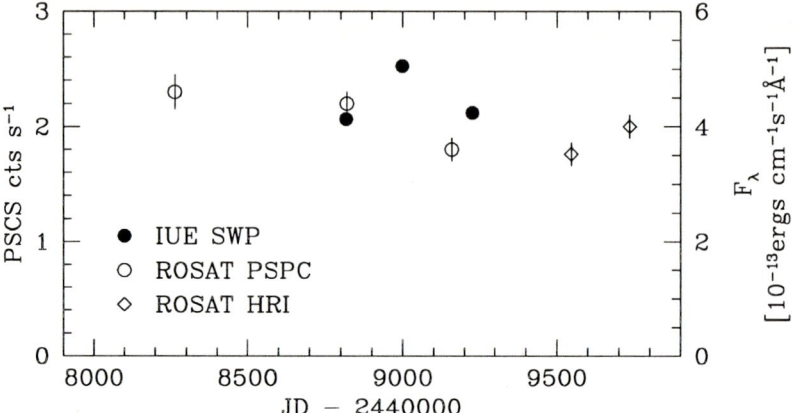

Fig. 1. UV and X-ray long term variability of RX J0019. The UV fluxes are averages over the SWP range (1250-1900Å), with statistical errors smaller than the point size. The ROSAT count rates are adopted from Reinsch et al. (1996a).

al. 1995). In addition to the orbital variability, the system shows erratic optical fluctuations of ~ 0.5 mag on timescales of weeks to months as well as a quasi-periodic long term variability of $\Delta m_V \sim 1$ mag with a cycle lengths of ~ 40 years, interpreted as oscillations caused by intermittent hydrogen burning on the white dwarf. The current model (Beuermann et al. 1995, Greiner & Wenzel 1995) describes RX J0019 as an SSS with an accretion rate of $\sim 10^{-7}\,M_\odot/\mathrm{yr}$, just a bit too low for stable hydrogen burning ($\sim 3 \times 10^{-7} M_\odot/\mathrm{yr}$ for a $1\,M_\odot$ white dwarf, Fujimoto 1982). The low X-ray luminosity of $4 \times 10^{36}\,\mathrm{erg\,s^{-1}}$ deduced from the ROSAT observations together with the optical brightness indicate that the system is momentarily in a low state.

2 Observations and Results

2.1 IUE Observations

UV observations of RX J0019 where carried out in July 1992, January 1993 and August 1993. A total of 25 SWP and 21 LWP exposures where acquired; full orbital coverage was achieved only in August 1993. Exposure lengths were typically 40 mins for the short wavelength range and 20 mins for the long wavelength range (Table 1). All spectra were obtained in the low resolution mode (~ 6 Å) and through the large aperture. All IUE images have been inspected for spurious features which have been removed.

The long-term variability of RX J0019 at UV and soft X-rays is shown in Fig. 1. The UV flux level of the January 1993 observations is $\sim 20\%$ higher than in July 1992 and August 1993, whereas the X-ray observations show a variability of $\sim 35\%$. From the sparse data, it seems that the irregular fluctuations on timescales of weeks to months discovered by Greiner & Wenzel (1995) in the

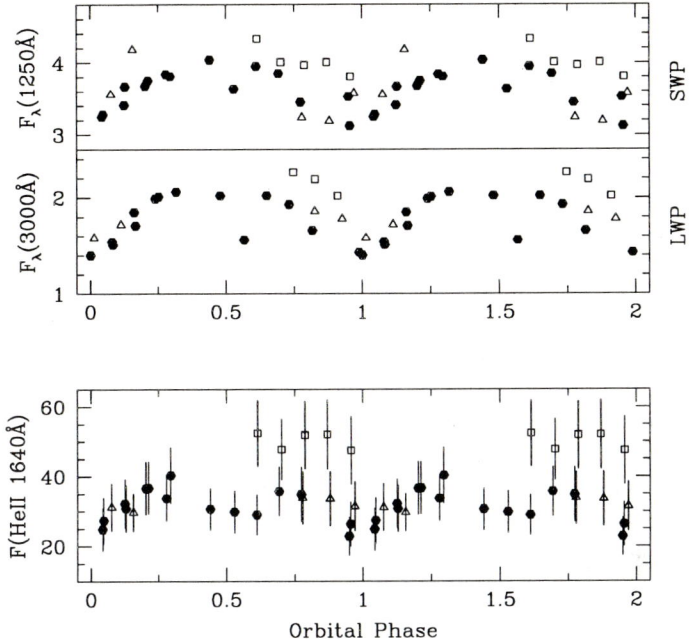

Fig. 2. UV light curves of RX J0019. Upper two panels: modulation of the continuum flux $[10^{-13}\,{\rm erg\,cm^{-2}s^{-1}\mathring{A}^{-1}}]$ at far (1225–1275Å) and near (2950–3000Å) UV wavelengths. Statistical flux errors are smaller than symbol size. Bottom panel: modulation of the HeII λ 1640 line $[10^{-13}\,{\rm erg\,cm^{-2}s^{-1}}]$. Errors are estimates from the uncertainty in defining the continuum. The individual IUE runs are coded as follows: $\triangle$ July 1992, $\square$ January 1993, $\bullet$ August 1993. Data are shown twice for clarity.

optical are present also at other wavelengths. It is interesting to note that the optical magnitude measured with the FES of IUE was $V \sim 12.8$, both for the July 1992 and the January 1993 observations. Unfortunately no measure of the optical level with the FES could be performed in August 1993.

2.2 Orbital Variability

We derived light curves for the near (LWP) and far (SWP) UV continuum as well as for the HeII λ 1640 emission line. All spectra were dereddened with $E(B-V) = 0.1$ (sect. 2.3) and orbital phases were calculated according to the ephemeris of Greiner & Wenzel (1995). The line fluxes of HeII were measured by fitting a straight line plus a Gaussian emission line to the observed spectra in the region around 1640 Å. The definition of the continuum is prone to uncertainties, resulting in relatively large errors in the derived line fluxes. The orbital light curve shows a quasi-sinusoidal modulation with a broad minimum at phase 1.0

(Fig. 2). The depth of the modulation is declining towards shorter wavelengths (LWP: 32%, SWP: 22%), but reaches a maximum in HeII (43%). The three IUE runs show that in addition to the flux level, also the form of the light curve is subject to long term variations; the seasonal changes in the UV are especially strong in the HeII λ 1640 line. Furthermore, the August 1993 data display orbit-to-orbit fluctuations of the UV flux, similar to those seen in the optical photometry of Beuermann et al. (1995). The most striking feature of the UV light curve of RX J0019 is the secondary minimum at $\Phi \approx 0.6$ seen in the August 1993 data. It was observed in both cameras, with smaller amplitude at short wavelengths (SWP: ~10%, LWP: ~25%). Recent photometry of RX J0019 (Will & Barwig 1996) reveals that the secondary minimum is present also at optical wavelengths, but that it is of highly variable nature.

2.3 The UV Spectrum

From the 2200 Å absorption in the observed LWP spectra a reddening of $E(B-V) = 0.10 \pm 0.03$ is derived, corresponding to $N_\mathrm{H} \approx 5 \times 10^{20}$ atoms cm^{-2} which is compatible to the absorption column determined from the ROSAT data (Beuermann et al. 1995) and to the galactic column density. Fig. 3 shows dereddened average spectra from the August 1993 run. The UV continuum of RX J0019 is characterised by a strong flux decline below 1600 Å and can be approximately described by a 19 000 K blackbody. This turnover is somewhat surprising for a system with presumably disc-dominated emission. The difference spectrum $(\Phi = 0.5) - (\Phi = 0.0)$ can be fitted by a 16 000 K blackbody. The only prominent UV emission line is HeII λ 1640, indicating that the ionizing X-ray radiation in RX J0019 is of lower temperature than in the LMC SSS. At IUE resolution, no velocity variation is observed, limiting the radial velocity to ≤ 500 km/s.

Closer inspection of the IUE spectra of RX J0019 reveals a surprising multitude of absorption features. In Fig. 4 we compare average spectra from two different IUE runs. The similarity of these two spectra indicates that most of the observed features may be real. We identify a number of low ionization species (CI $\lambda\lambda$ 1270-88, CII λ 1335, OI $\lambda\lambda$ 1302-06, SiII $\lambda\lambda$ 1260/65, 1304/09 and SII $\lambda\lambda$ 1251/54/59; note that the feature seen at ~1325 Å is due to a reseau mark). Table 2 lists the equivalent widths of the most prominent absorption features measured from the IUE spectra and compares them to the values expected for interstellar absorption at $d \approx 2$ kpc and $N_\mathrm{H} \approx 4 \times 10^{20}$ (compiled from Shull & van Steenberg (1985), van Steenberg & Shull (1988), Mauche et al. (1989), Burks et al. (1994) and Morton (1975, 1978)). The fact that some lines (e.g. SiII $\lambda\lambda$ 1260/65, MgII λ 2800 and the complex around $\lambda = 1300$ Å) have equivalent widths much larger than what is expected from interstellar absorption and that they show an orbital variation of the equivalent widths is a strong indication that part of the absorption is intrinsic to the system. However, these metal absorption lines are much weaker than in hot (20 000-30 000 K) stellar atmospheres (e.g. Heck et al. 1984). In addition to the low-ionization species, the presence of several high ionization absorptions, e.g. NV $\lambda\lambda$ 1238,43 and SiIV $\lambda\lambda$ 1394,1403,

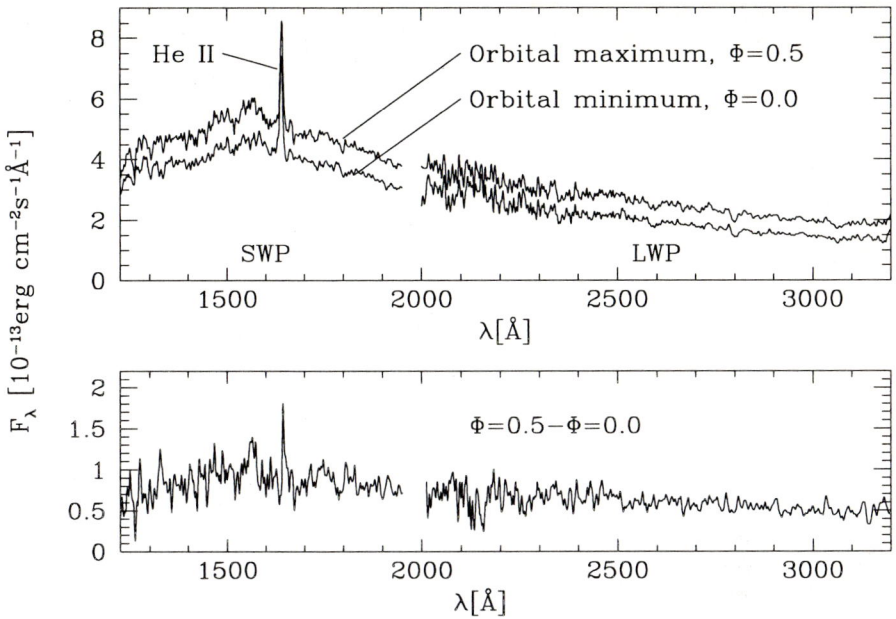

Fig. 3. Average UV spectra of RX J0019 in August 1993. Top: spectra at orbital maximum and minimum. Data have been dereddened with $E(B-V) = 0.1$. Bottom: difference spectrum orbital maximum minus orbital minimum.

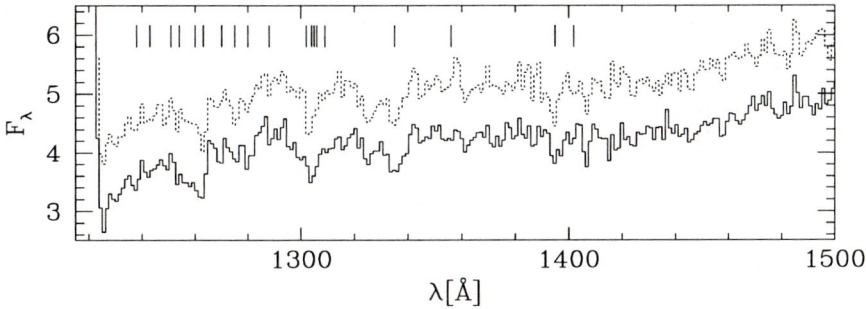

Fig. 4. Absorption features in RX J0019. Solid: average of the 15 spectra from August 1993. Dotted: average of the 5 spectra obtained in July 1992, shifted by one unit. The ticks show the positions of identified features (see text).

indicates intrinsic absorption (although these lines could be due to circumstellar absorption, e.g. Holberg et al. 1995). High resolution UV spectroscopy will allow to separate interstellar and intrinsic features and will also allow to constrain the system geometry from velocity variations of the intrinsic absorption.

Table 2. Equivalent widths [mÅ] of some absorption features as measured from the IUE spectra of RX J0019 and as expected by interstellar absorption.

λ [Å]	July 92	Jan. 93	Aug. 83	ISM
~ 1260	1400 ± 300	1700 ± 300	1850 ± 300 ($\Phi = 0.0$)	SiII, SII, CI.
			2550 ± 200 ($\Phi = 0.5$)	$\Sigma \approx 1150$
~ 1280	800 ± 200	500 ± 200	600 ± 200	CI $\approx$ 300
~ 1300	1300 ± 300	1600 ± 300	2700 ± 300 ($\Phi = 0.0$)	OI, SiII
			3500 ± 400 ($\Phi = 0.5$)	$\Sigma \approx 600$
~ 1335	1100 ± 200	1200 ± 200	1250 ± 200	CII $\approx$ 700
~ 1527	$\leq$ 700	$\leq$ 700	700 ± 200	SiII $\approx$ 500
~ 1560	600 ± 150	$\leq$ 500	600 ± 150	CI $\approx$ 150
~ 1808	$\leq$ 400	$\leq$ 300	$\leq$ 300	SiII $\approx$ 400
~ 2800	2600 ± 300	2200 ± 300	2850 ± 250 ($\Phi = 0.0$)	MgII $\approx$ 1000
			3450 ± 250 ($\Phi = 0.5$)	

3 Discussion

3.1 The UV Light Curve

From the radial velocity of the optical HeII lines and from the depth of the orbital modulation, Beuermann et al. (1995) derived an inclination of $i \approx 20°$ assuming that the optical modulation is due to the varying aspect of the secondary. Now, with the secondary minimum revealed, the light curve of RX J0019 looks reminiscent to that of CAL 87, a high-inclination system. There are, however, some important differences to note: Ingress and egress of the optical primary minimum (Beuermann et al. 1995) are less steep than in CAL 87, and the orbital modulation of RX J0019 ($\Delta m_V \approx 0.5$ mag) is smaller than in CAL 87 ($\Delta m_V \geq 1.5$ mag, Schmidtke et al. (1993)).

Recently, Schandl et al. (1996a,b) successfully fitted the optical light curve of CAL 87. Their model includes a heated secondary and an irradiated accretion disc with an illuminated high rim composed of cold, clumpy matter. This rim turned out to be an important ingredient to reproduce the observed light curve. Schandl et al. (1996a,b) applied their model also to the optical light curve of RX J0019, predicting $i = 60° - 70°$. It is, however, not clear if this model can explain the very deep secondary minimum with the obstruction of the secondary star by the accretion disc (rim).

Another problem concerns the X-ray observations of CAL 87 and RX J0019. The blackbody analyses of the ROSAT data of CAL 87 by Schmidtke et al. (1993) result in a high absorption column ($N_H \sim 10^{22}$ atoms/cm^2) and low (observed) luminosity ($L_X \approx 10^{36}$ erg s^{-1}), compatible with the model of Schandl et al. (1996a,b) where the central X-ray source is hidden from direct view. On the other hand, van Teeseling et al. (1996) show that the ROSAT observations of CAL 87 can be fitted with a hot, compact white dwarf. As for RX J0019, there is no sign for significant intrinsic absorption in the X-ray data.

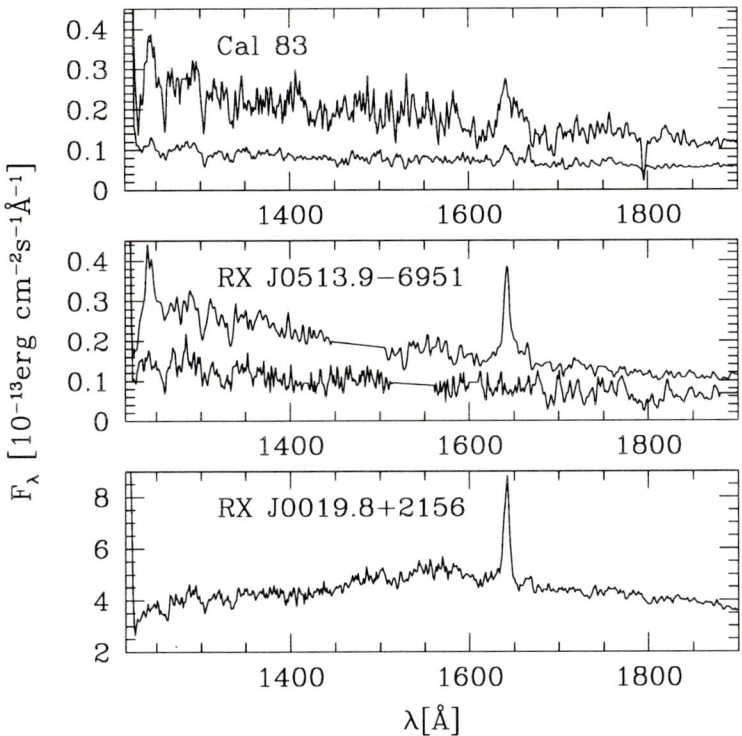

Fig. 5. UV spectra of CAL 83, RX J0513 and RX J0019. The spectra of CAL 83 have been discussed by Crampton et al. (1990) and Bianchi & Pakull (1988), the high state spectrum of RX J0513 by Pakull et al. (1993). In the RX J0513 spectra, cosmic hits have been removed.

3.2 The UV Spectrum

In non-magnetic CV's with a high mass transfer rate (e.g. IX Vel, Long et al. 1994) the accretion disc is the dominant light source at short wavelengths, with a spectrum rising up to far UV wavelengths. In SSS, the hot (a few 100 000 K) and luminous ($\sim 10^{38}$ erg s^{-1}) burning white dwarf should cause an additional heating of the accretion disc, resulting in a very blue continuum.

In Fig. 5, we compare all UV observations of SSS obtained so far. The LMC sources CAL 83 and RX J0513 have almost identical spectra, with a rising continuum as expected from a hot accretion disc. Furthermore, they exhibit UV high and low states with an amplitude of $\sim$ 50%, very similar to the variations observed at optical wavelengths (Reinsch et al. 1996b,c; Southwell et al. 1996). The main difference between the UV appearance of RX J0019 and the LMC sources is the flux turnover below 1600 Å and the lack of high excitation emission lines (e.g NV λ 1240). A reasonable fit for the spectrum at $\lambda \geq 1600$ Å can be achieved using the simple model for an irradiated accretion disc and secondary star described by Reinsch et al. (1996b,c) with the following parameters: $M_{\rm wd} = 1.0 M_\odot$, $M_{\rm sec} = 1.6 M_\odot$, $i = 20°$, $d = 2\,{\rm kpc}$, $L_{\rm wd} = 3.5 \times 10^{36}\,{\rm erg\,s^{-1}}$ and $\dot{M} = 10^{-8} M_\odot/{\rm yr}$. These values should be considered only of indicative character as the model of a flat disc and the assumption of blackbody emission is surely to simple and as the inclination might be higher. However, it is evident that also at higher inclinations, a standard accretion disc with an $\dot{M}$ needed for stable burning ($\sim 3 \times 10^{-7} M_\odot/{\rm yr}$) produces too much UV flux and can not reproduce the observed flux turnover. A speculative model that could account for these feature is a disc truncated in the inner, hot regions; e.g. by the magnetic field of the white dwarf.

Acknowledgment: We thank S. Schandl for interesting discussions about modeling the light curves and spectra of SSS. This work was supported in part by the DARA under project number 50 OR 9210.

References

Beuermann K., Reinsch K., Barwig H., Burwitz V., de Martino D. et al., 1995, A&A 294, L1
Bianchi L., Pakull M.W., 1988, in A Decade of UV Astronomy with IUE, p. 145
Burks G.S., Bartko F., Shull J.M., Stocke J.T., Sachs E.R., Burbidge E.M., Cohen R.D., Junkkarinen V.T., Harms R.J., Massa D., 1994, ApJ 437, 630
Crampton D., Cowley A.P., Hutchings J.B., Schmidtke P.C., Thompson I.B., Liebert J., 1987, ApJ 321, 745
Fujimoto M.Y., 1982, ApJ 257, 767
Greiner J., Wenzel W., 1995, A&A 294, L5
Heck A., Egret D., Jascheck M., Jascheck C., 1984, ESA SP-1052
Holberg J.B., Bruhweiler F.C., Andersen J., 1995, ApJ 443, 753
Long K.S., Wade R.A., Blair W.P., Davidson A.F., Hubeny I., 1994, ApJ 426, 704
Mauche C.W., Raymond J.C., Córdova F., 1988, ApJ 335, 829
Morton D.C., 1975, ApJ 197, 85 & 1978, ApJ 222, 863
Pakull M.W., Motch C., Bianchi L., Thomas H.-C., et al., 1993, A&A 278, L39
Reinsch K., van Teeseling A., Beuermann K., Thomas H.-C., 1996a, in Röntgenstrahlung from the Universe, ed. Zimmermann H.U., et al., p. 183
Reinsch K., van Teeseling A., Beuermann K., Abott T.M.C., 1996b, this volume p. 173
Reinsch K., van Teeseling A., Beuermann K., Abott T.M.C., 1996c, A&A (in press)
Schandl S., Meyer-Hofmeister E., Meyer F., 1996a, this volume p. 53
Schandl S., Meyer-Hofmeister E., Meyer F., 1996b, A&A (subm.)
Schmidtke P.C., McGrath T.K., Cowley A.P., Frattare L.M., 1993, PASP 105, 863
Shull J.M., van Steenberg M.E., 1985, ApJ 294, 599
Southwell K.A., Livio M., Charles P.A., et al., 1996, this volume p. 165
van den Heuvel E.P.J., Bhattacharya D., Nomoto K., Rappaport S.A., 1992, A&A 262, 97
van Steenberg M.E., Shull J.M., 1988, ApJS 67, 225
van Teeseling A., Heise J., Kahabka P., 1996, IAU Symp. 165, 445
Will T., Barwig H., 1996, this volume p. 99

Optical Spectroscopy of RX J0439.8−6809 and 1E 0035.4−7230

A. van Teeseling[1], K. Reinsch[1], K. Beuermann[1], H.-C. Thomas[2], M.W. Pakull[3]

[1] Universitäts-Sternwarte Göttingen, Geismarlandstr. 11, 37083 Göttingen, Germany
[2] Max-Planck-Institut für Astrophysik, Postfach 1603, 85740 Garching, Germany
[3] Observatoire de Strasbourg, URA 1280 du CNRS, F-6700 Strasbourg, France

Abstract. We have identified RX J0439.8−6809 with a very blue $B = 21.5$ object. We find no evidence for X-ray or optical variability. The optical spectrum of RX J0439.8−6809 does not show any absorption or emission features, and suggests that the optical flux is the Rayleigh-Jeans tail of the soft X-ray component. RX J0439.8−6809 may be a short-period accreting binary or a $\sim 1 M_\odot$ post-AGB star.

We also present phase-resolved optical spectroscopy of the 4.1 hr binary supersoft X-ray source 1E 0035.4−7230. We detect strongly variable He II $\lambda 4686$ emission and Balmer absorption lines. From radial velocity measurements we infer that a significant fraction of the hydrogen absorption lines originate on the donor star.

1 RX J0439.8−6809

1.1 Introduction

RX J0439.8−6809 has been discovered with ROSAT by Greiner et al. (1994). From the X-ray data they derived a blackbody temperature of $kT_{bb} = 20 \pm 10\,\text{eV}$ and an absorption column density of $N_H \sim 4 \times 10^{20}\,\text{cm}^{-2}$, making RX J0439.8−6809 one of the softest supersoft X-ray sources. Its direction to the Large Magellanic Cloud, its absorption column, and the absence of objects brighter than $V \sim 19$ in the X-ray error circle indicated an LMC membership. Using the accurate X-ray position from a long-term monitoring campaign, we have identified RX J0439.8−6809 with a very blue and faint object. Here, we summarize the results of follow-up optical spectroscopy and photometry (Van Teeseling et al. 1996).

1.2 Optical Identification

From a long-term monitoring campaign with the ROSAT HRI we have obtained an average X-ray position of RX J0439.8−6809 with a statistical 1σ error of $\sim 2''$. The HRI X-ray position for epoch 2000 is: $\alpha = 4^h 39^m 49\overset{s}{.}6$, $\delta = -68°09'01''$. Figure 1 shows U and V images of the field containing RX J0439.8−6809 together with the X-ray error circle. There is only one object in the X-ray error circle (star 8 of Greiner et al. 1994), which we have identified with the optical counterpart of RX J0439.8−6809. This identification is supported by the extreme blueness of this faint object.

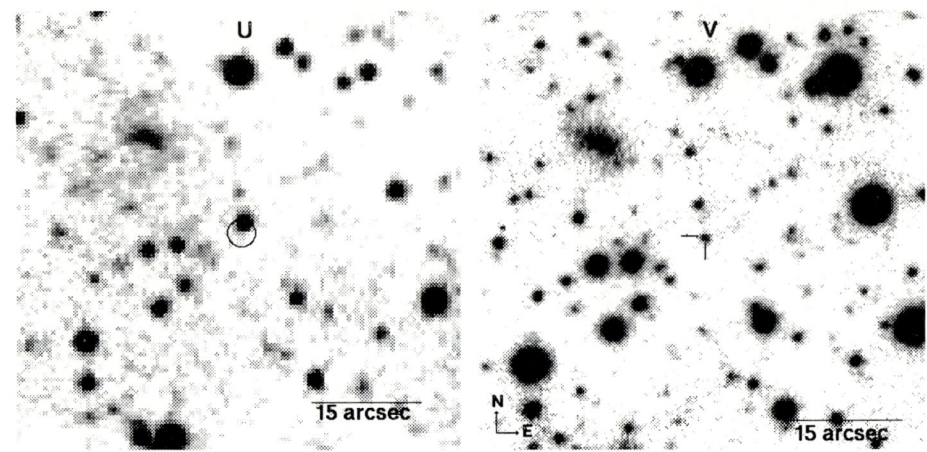

Fig. 1. U and V images of the field containing RX J0439.8–6809. The average HRI X-ray error circle with an error radius of $2''$ is shown in the U image.

Using Landolt standard stars, we have determined the magnitudes of the optical counterpart: $V = 21.74 \pm 0.09$, $B = 21.50 \pm 0.04$, and $U = 20.40 \pm 0.10$. With $E_{B-V} = 0.1$ this gives: $V_0 \approx 21.4$, $(B-V)_0 \approx -0.3$, and $(U-B)_0 \approx -1.2$. If RX J0439.8–6809 is located in the LMC the absolute visual magnitude is $M_V \sim 2.9$.

1.3 Variability and Optical Spectrum

During our X-ray monitoring, the field of RX J0439.8–6809 has been observed almost weekly in B, V, and R with the Dutch 90 cm telescope at La Silla. Neither the X-ray light curve nor the optical light curves of RX J0439.8–6809 show significant variability. We can set a 3σ upper limit to the X-ray variability of $\sim 20\%$. To get a firm upper limit on the optical variability, we have obtained seven more B images with 10 min exposures at the 2.2m telescope at La Silla (Fig. 2), from which we can set a 3σ upper limit to the B variability of ~ 0.2 mag. The optical counterpart of RX J0439.8–6809 is also visible with apparently the same brightness as in 1995 on an ESO J plate taken in December 1975, which excludes strong variability on a time scale of 20 years.

In October 1995 we have obtained an optical spectrum of RX J0439.8–6809 with the 3.6m telescope at La Silla. The exposure was 6×30 min. The spectral resolution is ~ 6 Å per pixel (FWHM resolution of ~ 17 Å) and the wavelength range covered is 3750–6950 Å. The exposures were obtained during photometric conditions with a seeing of $\sim 1.5''$. The average flux-calibrated spectrum is shown in Fig. 3. The optical spectrum of RX J0439.8–6809 reveals a very blue continuum, from which we infer $V \sim 21.9$ and $B - V \sim -0.3$ consistent with the magnitudes inferred from our photometry. The optical spectrum has a blackbody

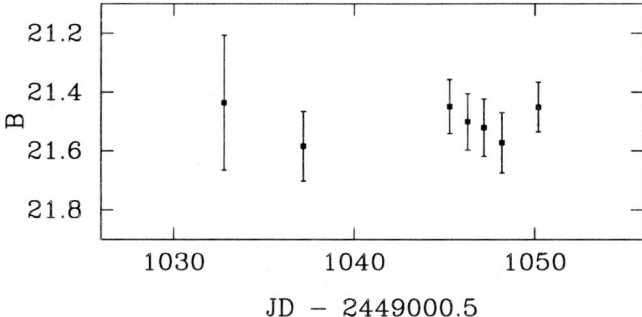

Fig. 2. The B light curve of RX J0439.8−6809 observed with the 2.2m telescope at La Silla.

temperature of $T_{opt} \geq 10^5$ K and suggests that the optical flux is the Rayleigh-Jeans tail of a very hot spectral component. There are no detectable emission or absorption features in the spectrum. In particular, we do not find any evidence for the presence of He II $\lambda 4686$ Å emission.

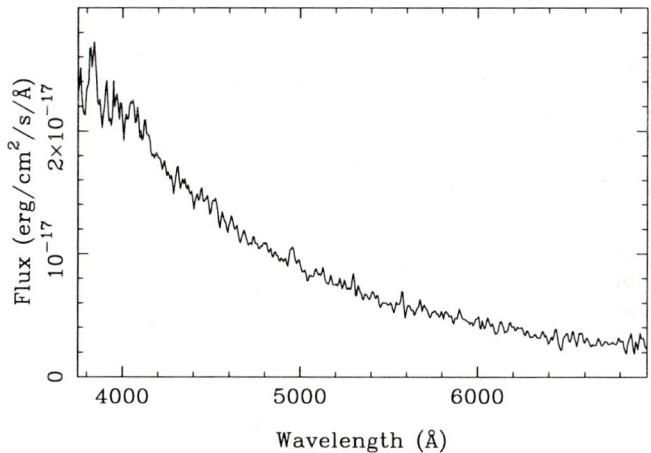

Fig. 3. Mean flux-calibrated spectrum of the optical counterpart of RX J0439.8−6809.

1.4 The Nature of RX J0439.8−6809

The intrinsic colours and absolute visual magnitude of the optical counterpart of RX J0439.8−6809 are not unusual for a low-mass X-ray binary (cf. Van Paradijs & McClintock 1995). However, it is difficult to hide a main sequence star in RX J0439.8−6809 which is sufficiently massive to drive steady or recurrent nuclear burning on a white dwarf (Van den Heuvel et al. 1992; Kahabka 1995).

If RX J0439.8–6809 is located in the LMC at a distance of ∼ 50 kpc, the donor star has to be less luminous than approximately a G0 V star. If RX J0439.8–6809 contains an irradiated accretion disk, the absolute visual magnitude of RX J0439.8–6809 suggests a very short orbital period which would exclude a main-sequence donor star and support a helium-burning or degenerate donor star (Van Paradijs & McClintock 1994). However, a helium-burning star would probably be too luminous and a degenerate donor star may not be able to provide the necessary high accretion rate.

The optical faintness of RX J0439.8–6809, our strong upper limit on the variability, and the absence of He II λ4686 Å emission lead us to an alternative model. If we assume that the optical flux is the Rayleigh-Jeans tail of the soft X-ray component we can constrain the spectral parameters to a narrow range. Independent of whether we use a blackbody spectrum or a $\log g = 7$ LTE model atmosphere spectrum, we find an absorption column of $N_H \approx 4 \times 10^{20}$ cm^{-2}, an effective temperature of $T_{\rm eff} \approx 270\,000$ K, a radius of $R \approx 5 \times 10^9 (d/50{\rm kpc})$ cm, and a corresponding luminosity of $L \approx 1 \times 10^{38} (d/50{\rm kpc})^2$ erg s^{-1}. Figure 4 shows a blackbody spectrum and a $\log g = 7$ model atmosphere spectrum together with the optical spectrum and the defolded ROSAT spectra.

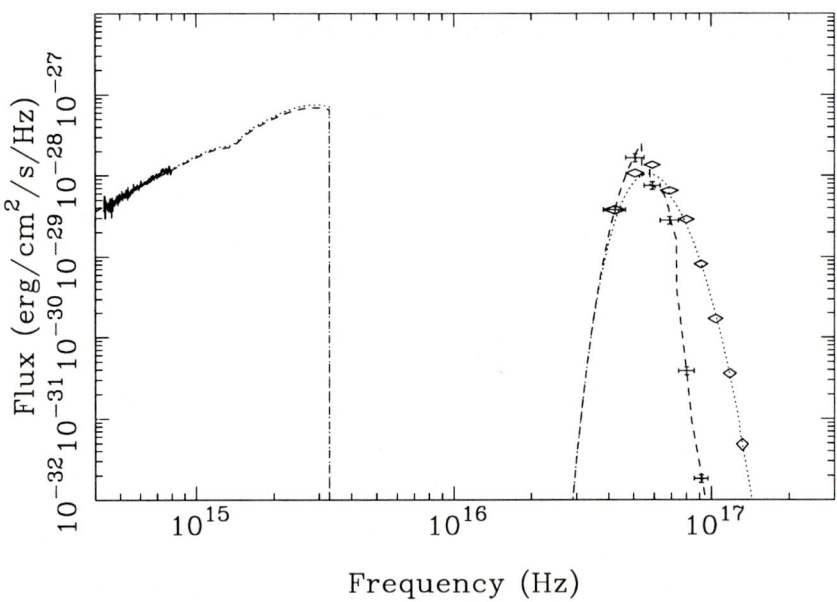

Fig. 4. The dashed line is the absorbed spectrum of a $\log g = 7$ LTE atmosphere with $T_{\rm eff} = 270\,000$ K and a metal abundance of 0.5 solar. The dotted line is an absorbed $T_{\rm bb} = 300\,000$ K blackbody spectrum. For both spectra $N_H = 4 \times 10^{20}$ cm^{-2}. Also shown are the observed optical spectrum and the defolded X-ray spectrum of RX J0439.8–6809, using the blackbody spectrum (diamonds) and the model atmosphere spectrum (crosses).

The inferred spectral parameters are consistent with a $\sim 1 M_\odot$ star near the turn-over from the horizontal shell-burning track to the white-dwarf cooling track (e.g. Vassiliadis & Wood 1994). The evolutionary time scale of such a star would be of the order of a few years and would predict the existence of a planetary nebula around RX J0439.8–6809. However, no nebula could be detected around RX J0439.8–6809 (Remillard et al. 1995), which could be explained if RX J0439.8–6809 is suffering a late helium shell flash (e.g. Iben 1984) and has re-entered the high-luminosity quiescent helium-burning phase. It is unclear, however, whether the lack of variability is consistent with the evolutionary time scale of such a star.

2 1E 0035.4–7230

2.1 Introduction

1E 0035.4–7230 has been identified with a 20th mag variable blue star (Orio et al. 1994). Schmidtke et al. (1996) showed that 1E 0035.4–7230 undergoes nearly sinusoidal $\Delta V \sim 0.3$ variations on a period of $\sim 4.1\,\mathrm{hr}$. This orbital modulation is probably due to the changing aspect of the X-ray heated donor star. Schmidtke et al. (1996) also noticed the presence of weak variable He II $\lambda 4686$ emission and an absorption feature at $\sim 4868\,\mathrm{\AA}$. Here, we present some preliminary results of phase-resolved spectroscopy of 1E 0035.4–7230.

2.2 Phase-Resolved Spectroscopy

In October 1995 we obtained 12 spectra of 1E 0035.4–7230 with EFOSC1 at the 3.6m telescope at La Silla. Each spectrum has an exposure of 30 min, a spectral resolution of $\sim 6\,\mathrm{\AA}$ per pixel (FWHM resolution of $\sim 17\,\mathrm{\AA}$), and covers the 3750–6950 Å region. We have minimized the loss of light due to differential atmospheric refraction by rotating the slit for each exposure perpendicular to the horizon. To obtain an accurate wavelength calibration, each spectrum was followed by a He-lamp observation with the telescope pointed to the same direction. Most exposures were obtained during photometric conditions with a seeing of $\sim 1.5''$. However, the flux-level of the average spectrum, which is shown in Fig. 5, is slightly reduced due to obscuring clouds during two exposures.

The average spectrum shows clear He II $\lambda 4686$ emission and perhaps also weak He II $\lambda 5412$ and O VI $\lambda 5291$ emission. Surprisingly, Hβ is in absorption, but no other Balmer lines are visible in the average spectrum. Both the He II $\lambda 4686$ emission and the Hβ absorption are strongly variable in strength and wavelength. In some individual spectra also the other Balmer lines are clearly in absorption (Fig. 5). The FWHM of the normalized average He II $\lambda 4686$ and Hβ are $\sim 18\,\mathrm{\AA}$ and $\sim 22\,\mathrm{\AA}$, respectively. Both average features are consistent with a narrow line which varies in wavelength. However, in some spectra the FWHM of Hβ seems to be significantly larger (see e.g. Fig. 5).

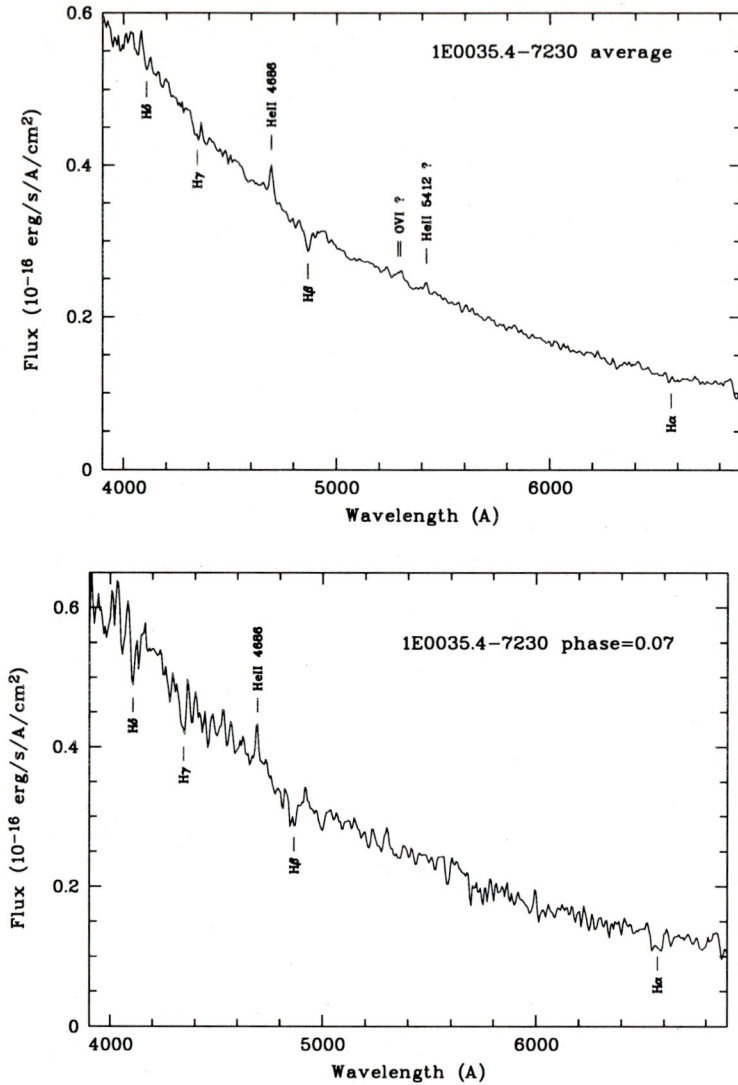

Fig. 5. Average flux-calibrated spectrum of 1E 0035.4–7230 (top panel) and flux-calibrated spectrum for photometric phase 0.07.

We have measured the radial velocity of He II λ4686 and Hβ in each of the 12 normalized spectra by cross-correlating the spectra with the normalized average spectrum and by fitting Gaussian profiles to the lines. Both methods give almost identical results. The He II λ4686 radial velocity is strongly variable, but we could not correlate this variability to the photometric phase. The Hβ radial velocity, however, can be fitted with a sine with the period fixed to the orbital pe-

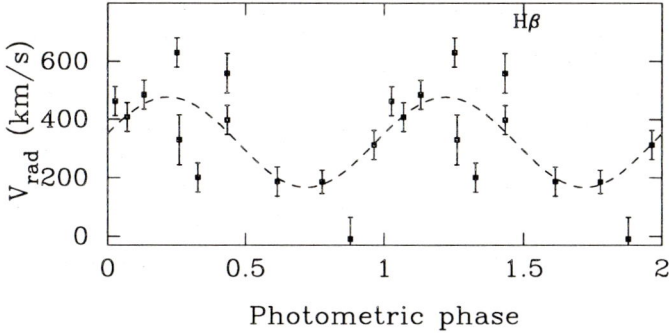

Fig. 6. Radial velocity of Hβ as a function of photometric phase. Each data point is plotted twice. The dashed line is a sine fit to the data.

riod. We have used the photometric ephemeris given by Schmidtke et al. (1996), where phase 0 corresponds to minimum optical light. Leaving the phase as a free parameter, we find a phase offset of the blue-to-red zero crossing compared to minimum optical light of only -0.03, and a coincidence probability of 4%. When one of the zero points of the sine is fixed to photometric phase 0, the probability of getting such a good fit with these data is only 0.5%. When the photometric modulation is due to the changing aspect of the X-ray heated donor star, the Hβ radial velocities imply that a significant part of the Hβ absorption originates on the donor star. However, the large FWHM of Hβ in some spectra suggests that the Hβ absorption is contaminated with a variable absorption component from the accretion disk. This would imply that the real radial velocity amplitude is larger than shown in Fig. 6. We also note that the mean velocity of Hβ is significantly higher than the mean SMC velocity.

2.3 Discussion

In a 4.1 hr binary with a white dwarf and a Roche-Lobe filling quasi-main-sequence star, the secondary star has to be a late K star or an M star. Using a white-dwarf luminosity of $5\,10^{36}$ erg s^{-1}, an $0.8 M_\odot$ white dwarf, and a donor star with the size of an M0 V star, we have calculated the temperature distribution on the surface of the donor star. The spectrum of the heated side of the secondary has been calculated by summing observed stellar spectra. The result depends on the amount of shielding by the accretion disk. Fig. 7 shows the calculated spectra for a flat disk and for an accretion disk with a half angular thickness of 10°. The observed photometric modulation and the average equivalent width of Hβ are consistent with a half angular thickness of $\sim 7°$. However, the model predicts much stronger Hγ and Hδ absorption. The weakness of Hγ and Hδ in the observed spectra may be due to a flatter temperature gradient in the irradiated atmosphere of the donor star (cf. Van Teeseling et al. 1994), or to a much redder spectrum of the heated side of the secondary.

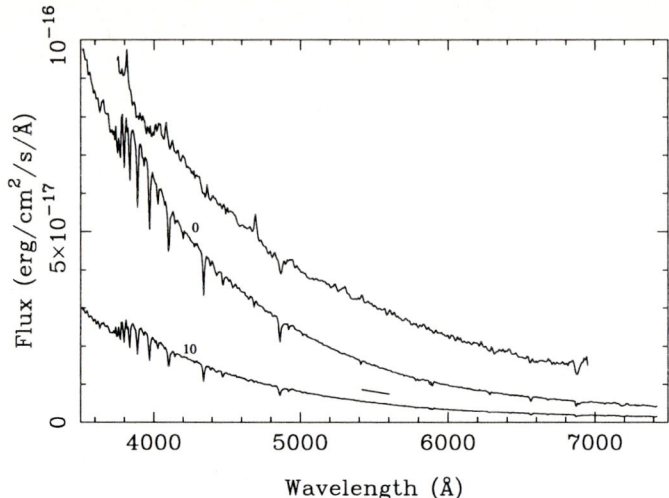

Fig. 7. Average spectrum of 1E 0035.4–7230 scaled to $V = 20.2$. The lower two spectra are the calculated spectra of the heated side of the donor star for no shielding by the accretion disk and for partial shielding by an accretion disk with a half angular thickness of $10°$. The small line is the flux level of the heated side inferred from the photometric modulation of $\Delta V \sim 0.3$, assuming zero contribution at photometric phase zero.

Acknowledgement: This research was supported by the DARA under grant 50-OR 92 10.

References

Greiner J., Hasinger G., Thomas H.-C., 1994, A&A 281, L61
Iben I., 1984, ApJ 277, 333
Remillard R.A., Rappaport S., Macri L.M., 1995, ApJ 439, 646
Kahabka P., 1995, A&A 304, 277
Orio M., Della Valle M., Massone G., Ögelman H., 1994, A&A 289, L11
Schmidtke P.C., Cowley A. P., McGrath T.K., Hutchings J.B., Crampton D., 1996, AJ 111, 788
Van den Heuvel E.P.J., Bhattacharya D., Nomoto K., Rappaport S.A., 1992, A&A 262, 97
Van Paradijs J., McClintock J.E., 1994, A&A 290, 133
Van Paradijs J., McClintock J.E., 1995, in X-ray Binaries, W.H.G. Lewin et al. (eds.), Cambridge University Press
Van Teeseling A., Reinsch K., Beuermann K., 1996, A&A 307, L49
Van Teeseling A., Heise J., Paerels F., 1994, A&A 281, 119
Vassiliadis E., Wood P.R., 1994, ApJS 92, 125

Photometric Observations of Supersoft Sources in the LMC

P.C. Schmidtke, A.P. Cowley

Arizona State Univ., Dept. of Physics & Astronomy, Tempe, AZ, 85287-1504 USA

1 Introduction

The Magellanic Clouds are an ideal laboratory to study supersoft X-ray sources (SSS). The low-energy response of *ROSAT* detectors has offered an unprecedented opportunity to study the soft X-ray flux from objects with blackbody temperatures in the range $kT \sim 10 - 100$ eV. This radiation is only partially attenuated because interstellar absorption in the direction of the Clouds is low. Since the distance is known, properties of individual sources can readily be compared not only to models but also to each other. Furthermore, the LMC and SMC are sufficiently close that ground-based photometry and spectroscopy can be used to study the optical counterparts. As part of a program to identify SSS in the Clouds by their color and/or variability, UBV images of the fields of yet unidentified sources have been obtained with the 0.9-m telescope at Cerro Tololo Inter-American Observatory. Here we report the observations for two sources in the LMC: RX J0439.8–6809 and RX J0550.0–7151.

2 RX J0439.8–6809

CCD images of RX J0439.8–6809 were taken on 1995 November 29 under excellent seeing conditions and are presented in Fig. 1. The marked star corresponds to star #8 of Greiner et al. (1994) and has very blue colors ($V_{mean} \sim 21.6$, $B - V = -0.29 \pm 0.05$, $U - B = -1.25 \pm 0.10$). It is nearly coincident with the ROSAT HRI position given by van Teeseling et al. (1996), who have independently identified the same star as the optical counterpart.

To investigate the optical variability of this source, a set of 33 V images was taken over seven consecutive nights in 1995 November, yielding the light curve (on an instrumental scale) shown in Fig. 2. Variability, even within a single night, is apparent. We have tested the scatter of the observed magnitudes and concluded that the distribution of their deviations from the mean value greatly exceeds that expected from statistical errors alone. (The scatter of data for a nearby field star of similar brightness shows a normal distribution.) Although van Teeseling et al. (1996) found no variability greater than 0.2 mag in B, their measurements have much larger error bars, and individual observations were taken at widely spaced intervals (days or weeks apart).

Because RX J0439.8−6809 appears to vary on short time scale, we calculated a periodogram of the differential magnitudes for periods from 0.8 to 4 days. A portion of this periodogram is displayed in Fig. 3. The two strong peaks with periods near 0.1403 and 0.1637 days are one-day aliases of each other. Tests of the data set indicate that the shorter of these periods is significant with 90% confidence. Since at present we cannot discriminate between the two periods, a folded light curve for each is presented in Figs. 4 and 5. Both curves show a low-amplitude ($\Delta V \sim 0.15$ mag peak-to-peak), nearly sinusoidal variation with one maximum and one minimum through a cycle. Hence, the light curve for RX J0439.8−6809 closely resembles that of the supersoft source 1E 0035.4−7230 (Schmidtke et al. 1996), which is identified with a faint blue star in the SMC ($V_{mean} \sim 20.4$, $B - V = -0.15$, and $U - B = -1.06$). A comparison of the two sources shows the light curve for 1E 0035.4−7230 has a slightly larger amplitude (0.3 mag in V) and longer period (0.1719 days) than present in RX J0439.8−6809.

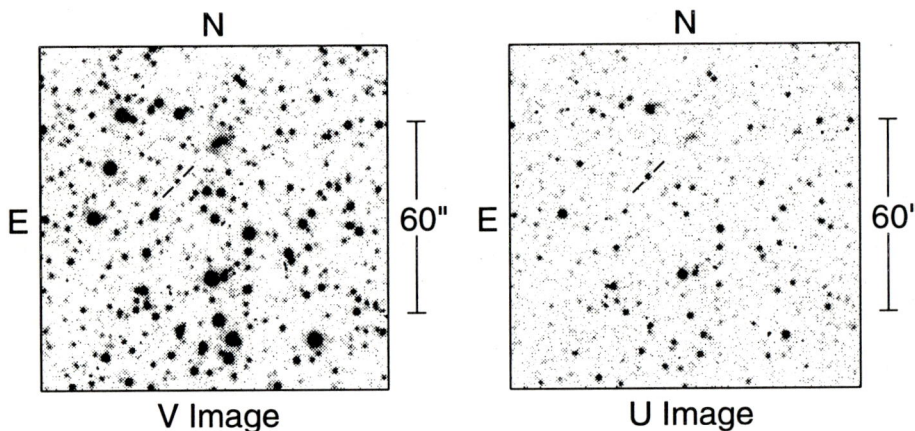

Fig. 1. V and B image of RX J0439.8−6809.

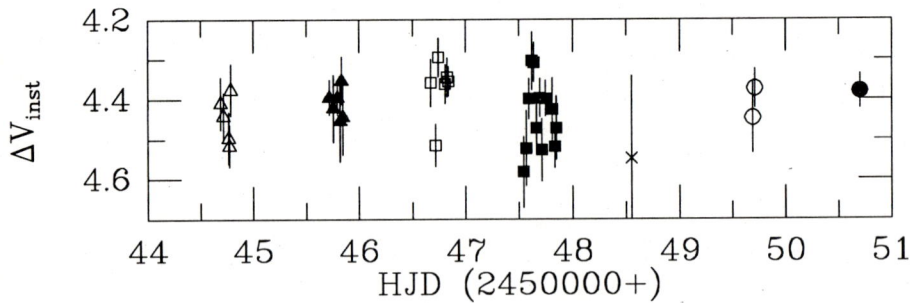

Fig. 2. V light curve for RX J0439.8−6809 in 1995 November.

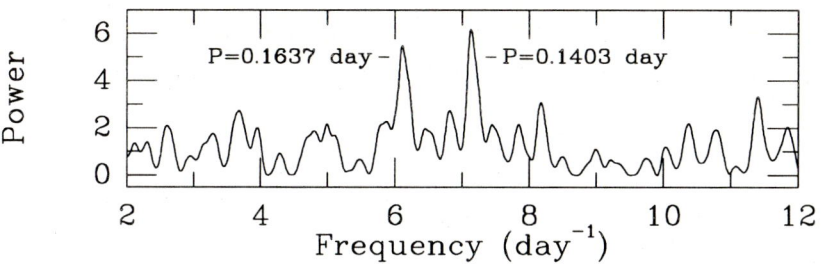

Fig. 3. Periodogram of V magnitudes for RX J0439.8–6809.

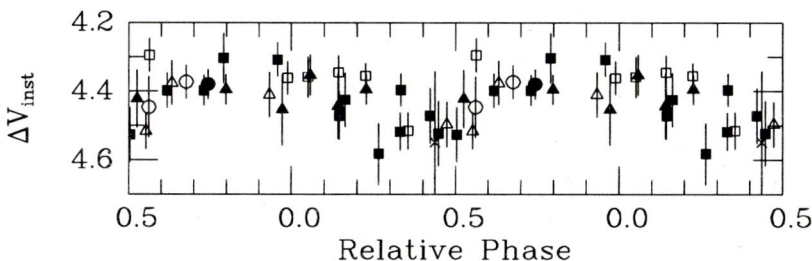

Fig. 4. V light curve for RX J0439.8–6809 folded on 0.1403 day period.

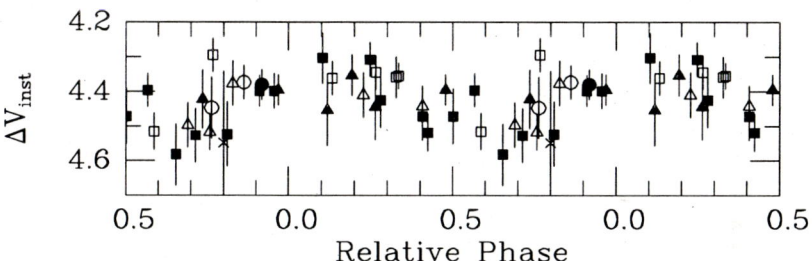

Fig. 5. V light curve for RX J0439.8–6809 folded on 0.1637 day period.

3 RX J0550.0–7151

RX J0550.0–7151 is a supersoft source in the LMC which was detected in the ROSAT PSPC field centered on CAL 87 (Cowley et al. 1993). Since RX J0550.0–7151 lies far off axis in the PSPC images, the point-spread function is extremely large, and the measured coordinates are poorly known. We obtained a well-centered HRI image in 1995, in which a weak source was present which we presumed was the supersoft source. At the HRI position there is a 13th magnitude star, which appeared to be the optical counterpart (Schmidtke & Cowley 1995). We subsequently have learned that in the ROSAT All-Sky Survey two close sources are present (Thomas 1996). These sources were not resolved in the early PSPC pointed observations. The HRI source for which we have found the optical counterpart is not the supersoft source, but rather the other, nearby X-ray source seen in the RASS data. Thus, we explicitly state that the optical counterpart of the supersoft source RX J0550.0−7151 has not yet been found, and this source is not present in either more recent PSPC nor in our HRI images, so it must be highly variable.

The source which we did detect should be referred to as RX J0549.8−7150. It has a HRI count rate of 0.008 sec^{-1} and coincides with a $V = 13.53$, foreground star. A CCD image of the field is shown in Fig. 6, with the HRI position indicated by a +. The measured colors ($B - V = +1.45$, $U - B = +0.86$) indicate a late-type star with a UV excess. Although it has been suggested that the source might be a symbiotic star or composite system, the colors are most similar to other ROSAT-selected dMe stars (see Schmidtke et al. 1994). Recently, Charles et al. (1996) reported that the spectrum of the optical counterpart has strong, narrow Balmer emission superposed on the spectrum of a cool star, in agreement with a dMe identification.

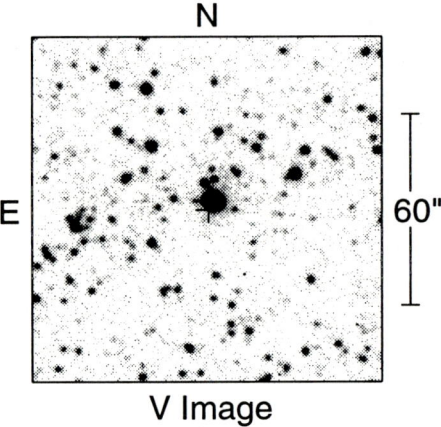

Fig. 6. V image of RX J0549.8–7150.

References

Charles P.A., Southwell K.A., O'Donoghue D., 1996, IAU Circ. 6305
Cowley A.P., Schmidtke P.C., Hutchings J.B., Crampton D., McGrath T.K., 1993, ApJ 418, L63
Greiner J., Hasinger G., Thomas H.-C., 1994, A&A 281, L61
Schmidtke P.C., Cowley A.P., Frattare L.M., McGrath T.K., Hutchings J.B., Crampton D., 1994, PASP 106, 843
Schmidtke P.C., Cowley A.P., 1995, IAU Circ. 6278
Schmidtke P.C., Cowley A.P., McGrath T.K., Hutchings J.B., Crampton D., 1996, AJ 111, 788
Thomas H.-C., 1996 (private comm.)
van Teeseling A., Reinsch K., Beuermann K., 1996, A&A (in press)

Optical Light Curve and Binary Period of the Supersoft X-Ray Transient RX J0513.9–6951

C. Motch, M.W. Pakull

CNRS, Observatoire de Strasbourg, 11 rue de l'Université, F-67000 Strasbourg

1 Introduction

The orbital period is a key parameter for our understanding of the dynamics and of the evolutionary status of contact binary systems. Steady-state hydrogen burning on the surface of a white dwarf requires accretion rates of few 10^{-7} $M_\odot$/yr which for main sequence stars can be reached if the donor stars masses are in the range of 1 - 3 $M_\odot$ and if the orbital periods lie between 8 h and 1.4 d (Rappaport et al. 1994). Much smaller orbital periods are possible, however, for *ex-nova* supersoft X-ray emitters in the constant-bolometric phase. The best-known case is GQ Mus which has $P_{orb} = 85.5$ min (Diaz & Steiner 1990). Orbital periods as long as ≈ 3.8 d may also exist if the mass donor star is slightly evolved (e.g. RX J0925.7–4758; Motch et al. 1994, Motch, 1996).

2 The Recurrent Supersoft Source RX J0513.9–6951

This LMC source (Schaeidt et al. 1993, Pakull et al. 1993) closely resembles the SSS prototype CAL 83 ($P_{orb} = 1.04$ d) in its X-ray - UV - optical luminosities and spectra, albeit with interesting differences. CAL 83 is a quasi-permanent X-ray source, but infrequently exhibits an optically bright state with exceptionally strong emission lines. Such an optical/UV spectrum is also displayed by RX J0513.9–6951, but here during the usual X-ray off state. However, there is some indication that the optical brightness *decreases* during X-ray outbursts (Pakull et al. 1993). Given their close resemblance one might expect similar masses and orbital periods for these two systems. From radial velocity measurements, Crampton et al. (1996) suggested an orbital cycle of 0.76 d, but their observations did not exclude alias periods near 0.32 and 0.43 d.

3 New Photometry

We have carried out CCD photometry of RX J0513.9–6951 during 6 consecutive nights from 1992 December 15 to 20 UT. We used the 90cm Dutch telescope located at ESO, La Silla, equipped with the RCA #5 CCD yielding a pixel size of 0″.51. CCD frames were acquired through a B filter with an integration time of 5 min. All CCD images were corrected for flat-field and bias using standard MIDAS procedures. After rejection of all images with seeing larger than 2″.5 a total of 103 frames remained. We performed differential photometry using a diaphragm of diameter 6″.2. The extraction area was chosen to be large enough to contain the faint companion located 1″.3 away from RX J0513.9–6951 (Crampton et al. 1996) and small enough to exclude light from the couple of stars located at 6″.9 in the north-east direction. We used as main comparison star the bright object located 50″ away from the X-ray source in the south-west direction. Our photometric calibration carried out in April 1992 (Pakull et al. 1993) gives B = 14.74 for this object. From the rms scatter of three other comparison stars with brightness similar to RX J0513.9–6951 we estimate that the error applicable to the differential photometry of RX J0513.9–6951 is $\approx$ 0.025 mag. The average brightness of B = 16.58 is consistent with our previous CCD photometry (Pakull et al. 1993) and with the magnitudes reported by Crampton et al. (1996) indicating that the system was optically bright and probably X-ray inactive at the time.

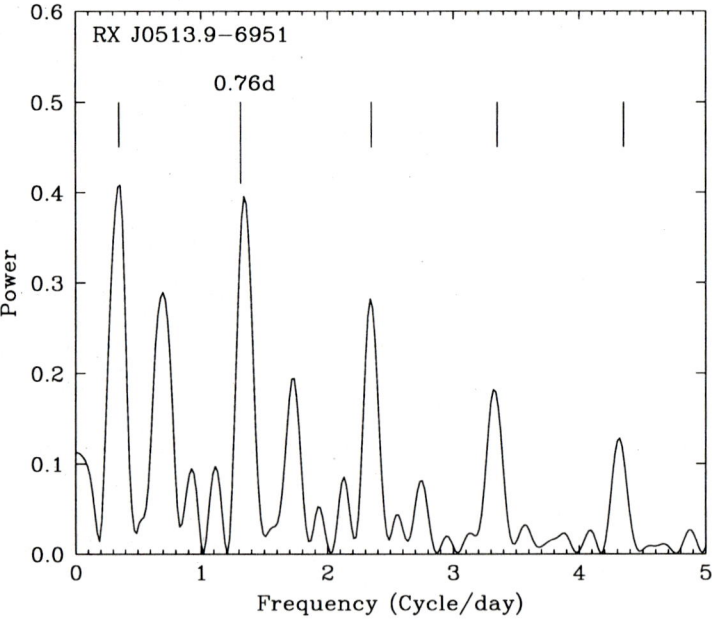

Fig. 1. Power spectrum of the B band data of RX J0513.9–6951. The 0.76 d orbital period proposed by Crampton et al. (1996) from spectroscopic data corresponds to one of the possible 1-day alias peaks.

A least square periodogram (Lomb 1976, Scargle 1982) of our B photometric time series is shown in Fig. 1. There are clear signatures of two sets of 1-day alias periodicities, the most significant one being coincident with that found by Crampton et al. (1996). The same set of alias is present in the minimum string length periodogram (Dworetsky 1983). No time series from any other secondary comparison star (or from seeing data) displays a similar signal. The relatively short length of the observing run and the small amplitude of the orbital modulation does not allow to constrain with large accuracy the best frequency. However, periods in the range of 0.745 ± 0.015 d are acceptable. Assuming a red noise distribution, we estimate at $\leq 10^{-4}$ the probability to find a power that strong at 0.745 d. The coincidence in period of our peak with with that reported by Crampton et al. (1996) (P = 0.76 d) is therefore interesting and in view of these results we conclude that the most probable orbital period is close to 0.76 d or one of its 1-day alias.

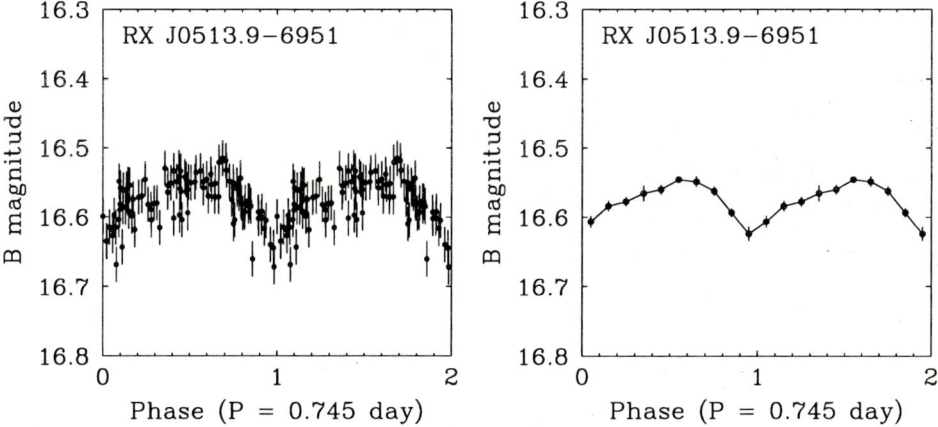

Fig. 2. B band photometric measurements folded with the best period of the periodogram. Time of phase 0.0 (minimum) is JD $2,448,971.737 \pm 0.024$. Left panel: individual data points. Right panel: averaged in 10 bins of phase.

The resulting light curve folded with a best period of 0.745 d is shown in Fig. 2. The full amplitude of the best fit sine wave is 0.059 mag. and the maximum range of variation is ≈ 0.1 mag. Although we note some intrinsic scatter, the light curve appears not to be perfectly sinusoidal, but shows some indication for a relative short phase ($\Delta\phi \approx 0.25$) of minimum light followed by a more gradual rise to maximum light.

4 Conclusion

The optical brightness of RX J0513.9−6951 shows a modulation of 0.1 mag amplitude with a period that is consistent with previously reported radial velocity variations. The light curve displays one relatively narrow minimum per cycle. So far it is not possible to determine the relative phasing between radial velocity and brightness. For the time being we therefore refrain from speculating on possible mechanisms that give rise to the orbital variations. Compared to the light curve of the twin source CAL 83 (Smale et al. 1988) the amplitude is smaller, in agreement with the probable lower system inclination of RX J0513.9−6951.

Acknowledgement: This work is based on observations obtained at the European Southern Observatory, La Silla, Chile with the ESO-Dutch 0.9-m telescope.

References

Crampton D., Hutchings J.B., Cowley A.P., Schmidtke P.C., McGrath T.K., O'-Donoghue D., Harrop-Allin M.K., 1996, ApJ 456, 320
Diaz M., Steiner J., 1990, Rev. Mex. Astron. Af 21, 369
Dworetsky M.M., 1983, MNRAS 203, 917
Lomb N.R., 1976, Ap&SS 39, 447
Motch C., Hasinger G., Pietsch W., 1994, A&A 284, 287
Motch C., 1996, this volume p. 83
Pakull M.W., Motch C., Bianchi L., et al., 1993, A&A 278, L39
Rappaport S., Di Stefano R., Smith J.D., 1994, ApJ 426, 692
Scargle J.D., 1982, ApJ 263, 835
Schaeidt S., Hasinger G., Trümper J., 1993, A&A 284 270, L9
Smale A.P., Corbet R.H.D., Charles P.A., Ilovaisky S.A., Mason K.O., Motch C., Mukai K., Naylor T., Parmar A.N., van der Klis M., van Paradijs J., 1988, MNRAS 233, 51

Non-LTE Model Atmosphere Analysis of the Supersoft X-Ray Source RX J0122.9−7521

K. Werner[1], B. Wolff[2], M.W. Pakull[3], A.P. Cowley[4], P.C. Schmidtke[4], J.B. Hutchings[5], D. Crampton[5]

[1] Universität Potsdam, Lehrstuhl Astrophysik, Germany
[2] Universität Kiel, Institut für Astronomie und Astrophysik, Germany
[3] Observatoire Astronomique, Strasbourg, France
[4] Arizona State University, Department of Physics and Astronomy, Tempe, USA
[5] Dominion Astrophysical Observatory, Victoria, B.C., Canada

Abstract. The supersoft X-ray source RX J0122.9−7521 was discovered in the ROSAT all-sky survey and suspected to be a member of the Small Magellanic Cloud (Kahabka et al. 1994). More recently the source has been optically identified with an extremely hot PG 1159 star (Cowley et al. 1995).
Here we present a NLTE model atmosphere analysis of a pointed ROSAT observation combined with medium resolution optical spectroscopy. RX J0122.9−7521 appears to be the hottest known PG 1159 star. It is compact and relatively massive ($T_{\rm eff}$=180 000 K, $\log g$ =7.5, $M = 0.7\,M_\odot$). In contrast to other stars of this group, the carbon and oxygen abundances in the helium-dominated atmosphere of RX J0122.9−7521 are unusually low. Our results allow a distance determination which clearly rules out SMC membership of RX J0122.9−7521.

1 Introduction

A distinct group of supersoft X-ray sources, the PG 1159 stars, comprises very hot hydrogen-deficient post-AGB stars. These stars represent an evolutionary transition phase between the central stars of planetary nebulae and the white dwarfs (WDs). Their masses are low ($M/M_\odot$=0.5–0.9), and their effective temperatures and luminosities (and hence surface gravities) cover a fairly extended region in the HR diagram (see Fig. 6). Seven out of 27 known PG 1159 stars (including the object under discussion here) were detected with the ROSAT PSPC (see e.g. Werner 1996) and two of these appear in the ROSAT all-sky survey. All ROSAT detected PG 1159 stars have temperatures $T_{\rm eff}\geq$140 000 K, and the X-rays are of thermal origin in the hot photospheres. It could be shown that the non-detection of "cooler" PG 1159 stars is the consequence of a very strong O VI edge near 100Å. The strength of this edge arises from the high oxygen abundance (of the order 10% by mass) in these carbon/helium dominated stars. Further details on these interesting objects may be found in recent reviews by Dreizler et al. (1995) and Werner et al. (1996a).

The first PG 1159 star discovered from its X-ray properties was H 1504+65 (Nousek et al. 1986). The first ROSAT selected PG 1159 star is RX J 2117.1 +3412, which was analyzed with NLTE model atmospheres using PSPC data

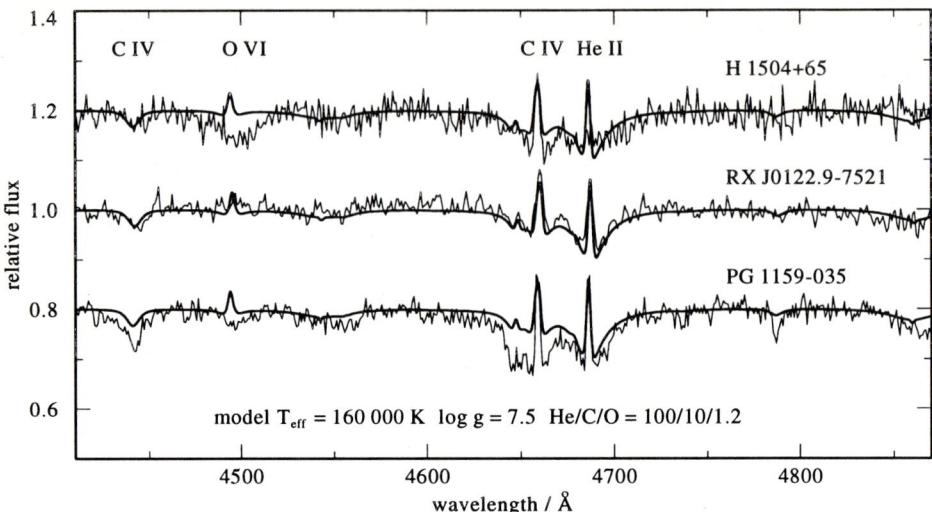

Fig. 1. Optical spectrum of RX J0122.9−7521 (middle) around the characteristic C IV/He II absorption trough region in comparison with spectra from H 1504+65 (top) and the prototype PG 1159−035 (bottom). To facilitate the comparison, overplotted on all three observations is the synthetic spectrum of a model with parameters as given in the panel. The weaker C IV 4660Å and absent O VI 4500Å absorptions in RX J0122.9−7521 are indicative of lower C and O abundances.

as well as optical spectra (Motch et al. 1993). The second ROSAT discovery is RX J0122.9−7521 (Cowley et al. 1995), and we present a similar analysis in this paper. Although RX J0122.9−7521 lies somewhat far from the main body of the Small Magellanic Cloud it was suspected to be a SMC member (Kahabka et al. 1994). Its identification as a PG 1159 star by Cowley et al. 1995 made its SMC membership rather unlikely and we will show that it is in fact a Galactic foreground object. A comparison of visual and X-ray fluxes of both ROSAT discovered stars reveals an unusually high X-ray luminosity for the new PG 1159 star. Although visually fainter by almost 2 magnitudes, the PSPC count rate of RX J0122.9−7521 is more than twice as high as compared to RX J 2117.1+3412 (see Table 1). Our analysis aims at clarifying this behavior.

2 Model Atmospheres and Observations

Synthetic optical and X-ray spectra are computed with a NLTE model atmosphere code developed at Kiel/Bamberg/Potsdam institutes (for numerical details see Dreizler & Werner 1993). The models are plane-parallel and assume hydrostatic and radiative equilibrium. Model atmospheres which are particularly tailored for the analysis of optical and X-ray spectra of PG 1159 stars are described in Werner et al. (1991) and Motch et al. (1993).

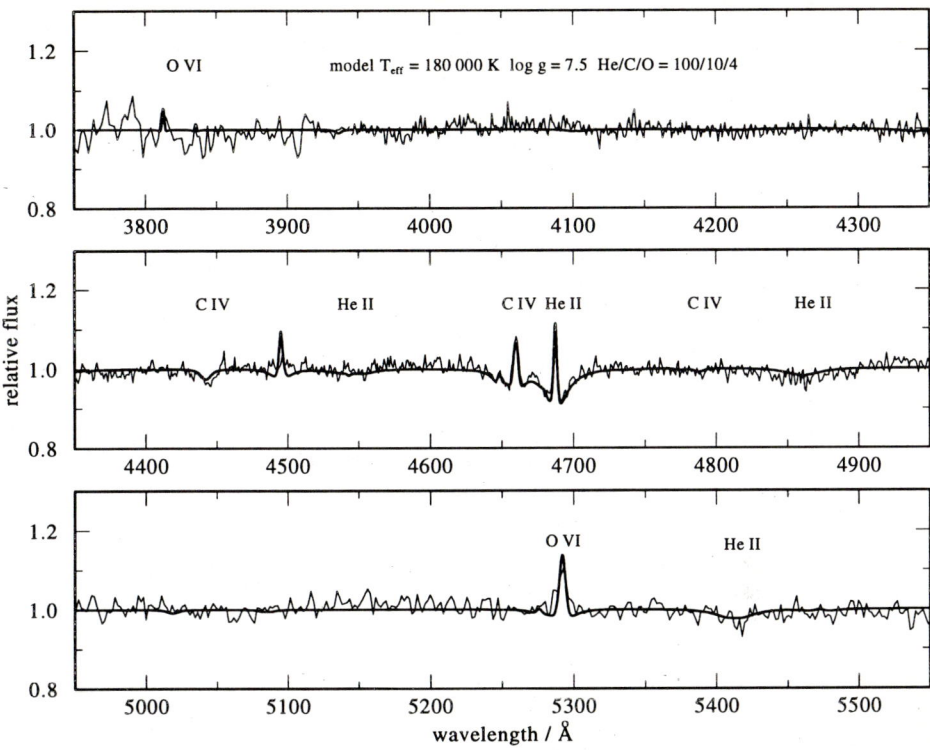

Fig. 2. The complete optical spectrum of RX J0122.9−7521 taken from Cowley et al. (1995). Besides the lines already identified in the close-up in the preceding figure, there is only one additional prominent feature, an emission from O VI 5290Å. Overplotted is the spectrum of the model with the finally adopted parameters.

A hint as to the extraordinarily high X-ray luminosity of RX J0122.9−7521 is immediately evident in Fig. 1 from a comparison of its optical spectrum with those of the prototype star PG 1159−035 and of the above mentioned H 1504+65. It is obvious that the photospheric content of carbon and oxygen in RX J0122.9−7521 is lower than in the comparison stars. Both other stars show a shallow O VI absorption at 4500Å while RX J0122.9−7521 does not, and the depth of the C IV 4660Å absorption complex in RX J0122.9−7521 is much weaker when compared to PG 1159−035. Generally, the appearance of central emission reversals in the C IV 4660Å and He II 4686Å lines indicates that RX J0122.9−7521 is among the hottest members of the PG 1159 group. In fact, a detailed fit to the optical lines (Fig. 2) gives $T_{\rm eff}$=180 000 ± 20 000 K, $\log g$ =7.5±0.5, and an O/He number abundance ratio well below 10% (note that the model predicts an excessively strong O VI 4500Å emission [n=8→10 transitions], which is probably the consequence of a too small model atom). A higher oxygen abundance can be excluded, because otherwise the O VI emission line at 5290Å would be accompanied by shallow absorption wings, as is the case

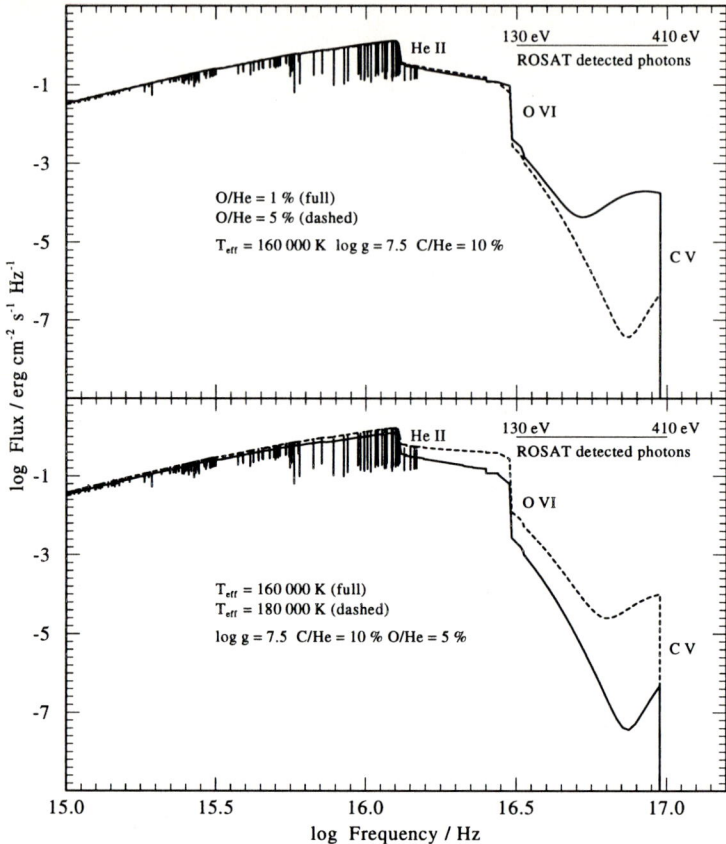

Fig. 3. Top: Flux distribution of two NLTE models with different oxygen abundance. Enhancing the O abundance strengthens the O VI 2p absorption edge at 130 eV and strongly suppresses the secondary flux maximum before the C V ground state edge. The C V edge blocks virtually all photons with energies above 410 eV. Bottom: The influence of the temperature on the soft X-ray flux is extremely strong. Near the C V edge an increase of $T_{\rm eff}$ by 20 000 K produces more than two orders of magnitude more photons.

in the prototype (cf. Werner et al. 1991). The relatively low oxygen abundance is in accordance with the results of our fit to the X-ray spectrum, which will be described below. But in order to understand the basic photospheric features which dominate the X-ray flux let us have a closer look at some models.

In Fig. 3 we compare the fluxes of models which differ in effective temperature and oxygen abundance, demonstrating the effects of these parameters on the emergent spectrum. The figures show the flux distribution from the UV to the soft X-ray region and a horizontal bar indicates the energy range at which photons are detected by the ROSAT PSPC. Two prominent features determine the shape of the soft X-ray flux. The first is a strong O VI edge at 130 eV and

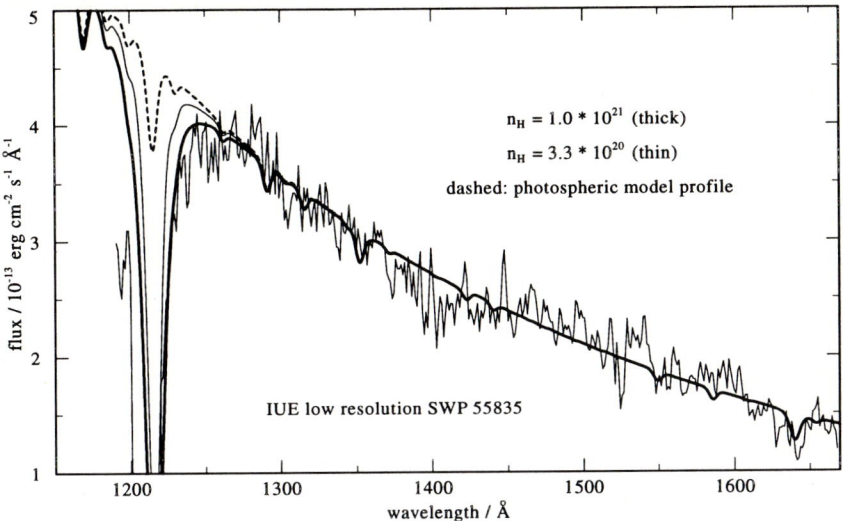

Fig. 4. An attempt to constrain the interstellar hydrogen column density towards RX J0122.9−7521 by a fit to the Ly α red wing fails. See discussion in the text. Due to the high noise, no line features except from interstellar species can be identified.

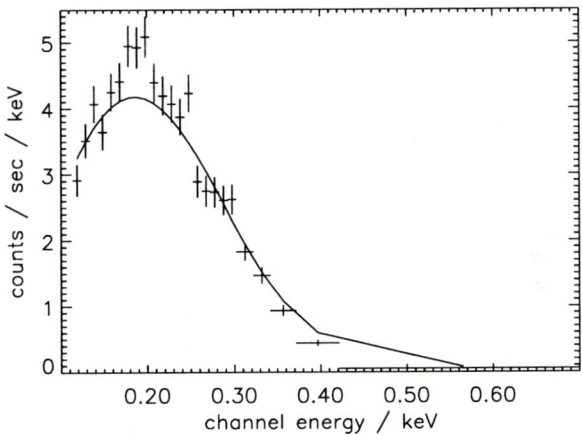

Fig. 5. Best model fit to the observed PSPC pulse height distribution. Model parameters are equal to those for the optical line fit shown in Fig. 2. The resulting interstellar column density amounts to $N_H = 2.6 \cdot 10^{20}$. The pointed observation was performed in Sept./Oct. 1993 and retrieved from the ROSAT archive.

the second is the even stronger C V ground state edge at 410 eV which virtually blocks all the higher energy photons in every model relevant to the present analysis (for a discussion of NLTE models with much higher temperatures, see contributions by Rauch (1996) and Hartmann (1996). Hence, the only feature

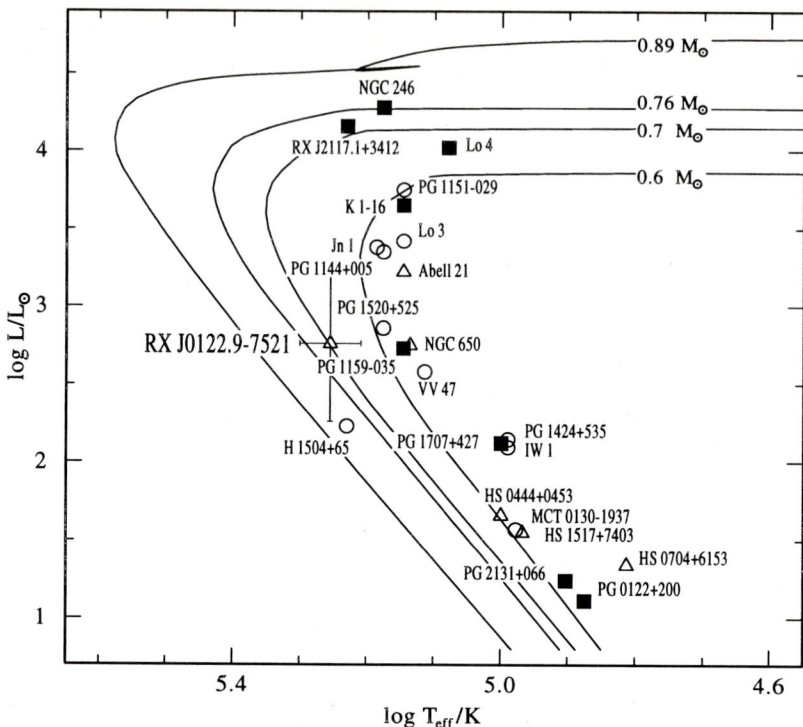

Fig. 6. Position of RX J0122.9−7521 (with error bars) in the HR diagram. It lies on a 0.7 $M_\odot$ post-AGB evolutionary track and represents a descendant of the much more luminous RX J 2117.1+3412. Also shown are positions of previously analyzed PG 1159 stars (different symbols for pulsators □, non-pulsators o, and unknown photometric behavior △) as taken from Werner et al. (1996a). Evolutionary tracks are from Wood & Faulkner (1986).

which remarkably changes the soft X-ray flux is the O VI edge. From Fig. 3 we learn that T_{eff} as well as the O abundance are the two key parameters which determine the strength of the edge. It follows that the analysis of the X-ray data alone gives no useful results, but that the optical spectra have to be used for a sensible interpretation. It should be noted that a change in the surface gravity, within the limits from the optical analysis, leaves the X-ray flux almost unaffected. The same holds for the C/He abundance ratio which is – like $\log g$ – fixed from the optical line profiles.

Interstellar absorption is another unknown parameter which enters the X-ray analysis so it would be most useful to get an independent estimate of the H column density (N_H) towards RX J0122.9−7521. We have obtained a IUE low resolution spectrum hoping to constrain N_H by fitting the Ly α wing. In Fig. 4 we show the IUE spectrum together with the photospheric model flux and the same flux attenuated by two different values for N_H (in cm^{-2}). The lower value ($N_H = 3.3 \cdot 10^{20}$) represents the Galactic column density towards the SMC (Bessell 1991) and hence should be an upper limit for RX J0122.9−7521. It seems that this value is too low to fit to the observed profile and that a value three times higher fits better. This is, of course, unreasonably high and analysis of the X-ray spectra (see below) gives a value of $2.6 \cdot 10^{20}$. We can only speculate about reasons for the failure to get an acceptable fit to Ly α, e.g. the IUE spectrum is rather noisy and it is difficult to define the continuum. As a consequence, the interstellar column towards RX J0122.9−7521 must be treated as an unknown parameter. Fig. 5 shows our best fit model to the observed pulse height distribution. The model (T_{eff}=180 000 K) was selected from a grid which covers T_{eff}=150 000–200 000 K and O/He=0.5–10 (% number ratio). The fit is not perfect but represents a compromise. The model flux is too low by about 20% at the observed peak height (near 0.2 keV), but somewhat too high at the highest observed energies. Increasing T_{eff} improves the low energy fit but at the hard end the flux excess becomes too strong. The opposite holds for decreasing T_{eff}. Although the shape of the flux distribution can be fitted with the adopted oxygen abundance, it could be slightly improved with a higher O/He value (along with a slight increase of T_{eff} in order to compensate for the stronger O VI edge). But then the optical O VI 5290Å emission line develops absorption wings which are not observed.

Table 1. Comparison of results from NLTE analyses of RX J 2117.1+3412 (Motch et al. 1993, Werner et al. 1996b) and of RX J0122.9−7521 (this work)

	RX J 2117.1+3412	RX J0122.9−7521
V-magnitude	13.2	15.4
PSPC ct/s	0.33	0.81
T_{eff}/1000 K	170	180
log g (cgs)	6.0	7.5
He/C/O mass fractions (%)	33/50/17	68/21/11
N_H/cm^{-2}	$3.2 \cdot 10^{20}$	$2.6 \cdot 10^{20}$
log(L/L$_\odot$)	4.16	2.76
log(R/R$_\odot$)	-0.89	-1.6
M/M$_\odot$	0.70	0.70
distance/pc	1400	700

3 Summary

RX J0122.9−7521 is the hottest known PG 1159 star (see Fig. 6), closely followed by RX J 2117.1+3412 (see Motch et al. 1993 and a refined analysis with HST spectra by Werner et al. 1996b). Table 1 summarizes the results of the present analysis and compares the properties of both ROSAT discovered objects. The higher X-ray luminosity of RX J0122.9−7521 is a consequence of both its slightly higher effective temperature and its relatively low oxygen abundance. In addition, the interstellar hydrogen column density is lower, but there remains a discrepancy with the IUE Ly α profile. Comparing the position of both stars in the T_{eff}–$\log g$ diagram reveals very similar masses. However, the deduced total luminosity of RX J0122.9−7521 shows that it is more evolved than RX J 2117.1+3412 (roughly 10^4 years higher post-AGB age). Comparing the measured V magnitude with the model flux clearly rules out SMC membership of RX J0122.9−7521. It yields a distance determination of 700^{+550}_{-310} pc.

We remark that RX J0122.9−7521 has no associated planetary nebula while about 50% of the PG 1159 members are central stars. In addition, it is located outside of the GW Vir instability strip (Werner et al. 1996c), hence we do not expect that RX J0122.9−7521 exhibits photometric variability. In fact, no obvious variability was detected by Cowley et al. (1995) and Bond (1996). Other questions not addressed here concern the possible detection of line profile variations as well as weak N v emission lines in some spectra. These findings deserve further attention.

Acknowledgement: ROSAT data analysis at Kiel and Potsdam is supported by the DARA under grant 50 OR 9409 1. Stefan Jordan kindly supplied his INTEX tool for ROSAT data analysis.

References

Bessell M.S., 1991, A&A 242, L17
Bond H.E., 1996, private communication
Cowley A.P., Schmidtke P.C., Hutchings J.B., Crampton D., 1995, PASP 107, 927
Dreizler S., Werner K., 1993, A&A 278, 199
Dreizler S., Werner K., Heber U., 1995, in White Dwarfs, eds. D. Koester and K. Werner, Lecture Notes in Physics 443, Springer, Berlin, p. 160
Hartmann H.W., Heise J., 1996, this volume p. 6
Kahabka P., Pietsch W., Hasinger G., 1994, A&A 288, 538
Motch C., Werner K., Pakull M.W., 1993, A&A 268, 561
Nousek J.A., Shipman H.L., Holberg J.B., et al. , 1986, ApJ 309, 230
Rauch T., 1996, this volume p. 3
Werner K., 1996, in Röntgenstrahlung from the Universe, eds. H.U. Zimmermann, J.E. Trümper, H. Yorke, MPE Report 263, p. 205
Werner K., Heber U., Hunger K., 1991, A&A 244, 437
Werner K., Dreizler S., Heber U., Rauch T., 1996a, in Hydrogen-Deficient Stars, eds. U. Heber and C.S. Jeffery, The ASP Conference Series, p. 267
Werner K., Dreizler S., Heber U., Rauch T., Fleming T.A., Sion E.M., Vauclair G., 1996b, A&A 307, 860
Werner K., Dreizler S., Heber U., Rauch T., 1996c, in Astrophysics in the Extreme Ultraviolet, IAU Colloquium 152, eds. S. Bowyer and R.F. Malina, Kluwer, p. 229
Wood P.R., Faulkner D.J., 1986, ApJ 307, 659

Implications of Light Metals (Li - Ca) on NLTE Model Atmospheres for Hot Stars

T. Rauch

Universität Kiel, Institut für Astronomie und Astrophysik, Germany
Universität Potsdam, Lehrstuhl Astrophysik, Germany

Abstract. Neither iron line-blanketed LTE model atmospheres nor "standard" NLTE models (including H, He, C, N, O, Ne) can reproduce the observed stellar EUV and X-ray fluxes of hot stars: huge differences appear at energies higher than 54 eV (He II Lyman continuum) which are a result of light metal opacities which have a strong influence on the emergent fluxes of hot stars in this energy range. These metal opacities have not been included in "standard" NLTE model atmosphere calculations so far. We present NLTE model atmospheres including all elements from hydrogen to calcium.
In order to calculate more realistic stellar fluxes, such models should preferentially be used for the EUV and X-ray analysis of very hot stars instead of black bodies which are often used in the analysis of X-ray spectra.

1 Introduction

In the last quarter of this century the numerical models for the interpretation of spectra of hot stars have been continuously improved.

In 1969 Auer & Mihalas presented the complete linearization method for the calculation of "Classical Models". These were limited due to numerical reasons and missing computational capacities — only a handful of NLTE levels and line transitions could be considered.

The rapidly evolving observational techniques provided UV and X-ray spectra (e.g. IUE, Einstein, and EXOSAT satellites) and high-resolution optical spectra (e.g. ESO CASPEC) and have been a challenge for theory: In 1985 Werner & Husfeld presented the newly developed accelerated lambda iteration (ALI) technique for line formation calculations. This method was later applied to the calculation of NLTE model atmospheres (Werner 1986). Presently calculated model atmospheres treat over 200 atomic levels in NLTE and consider more than 10^6 line transitions.

Further progress in observation (e.g. HST, ROSAT, EUVE, ISO, and the ESO VLT) requires adequate models, tailored for an interpretation of the complete wavelength range from the far infrared to the X-ray region.

For "cool" stars ($T_{\text{eff}} <$ 30 kK, spectral type B or later), the fully line blanketed LTE models (Kurucz 1979, 1991) are available, which consider all elements up to the iron group and the blanketing of millions of lines.

The present state-of-the-art NLTE models (Dreizler & Werner 1993, Rauch 1996b) include also all elements up to the iron group (including line blanketing) and permit to perform analyses of very hot stars with realistic models.

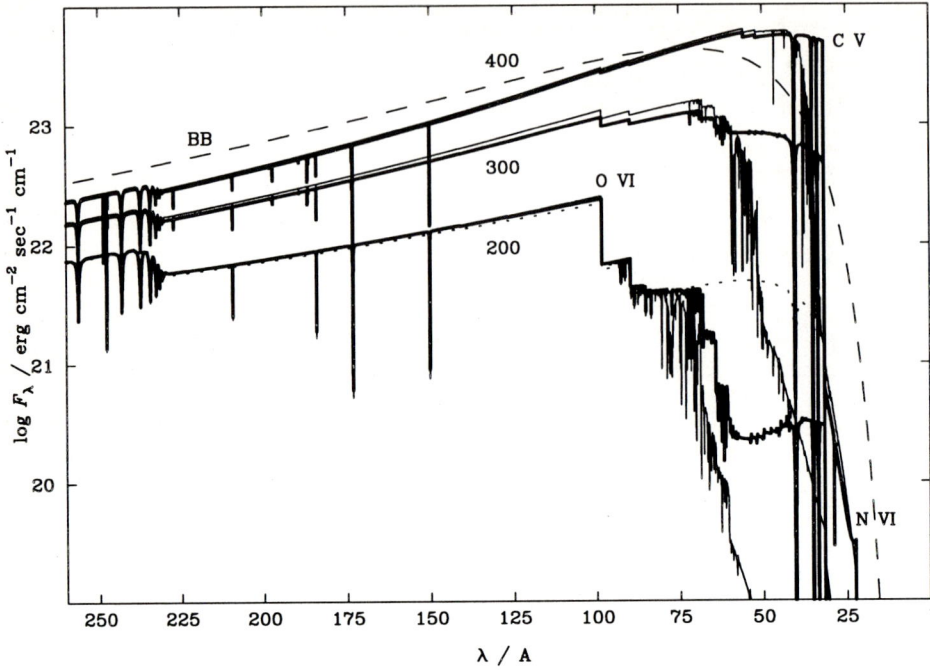

Fig. 1. Emergent fluxes of NLTE model atmospheres with $\log g = 8$, $T_{\text{eff}} = 200$, 300, 400 kK, and solar abundances calculated with H+He+C+N+O (200 kK only, dotted line) and H+He+C+N+O+Ne (———) compared to those of H - Ca (———) models. The dashed line is the flux of a black body (BB) with 400 kK. Note the strong absorption edges of O VI 2p at 98 Å and 2s at 89 Å which almost disappear at $T_{\text{eff}} = 400$ kK due to the changing ionization equilibrium, and the much stronger ground state absorption edge of C V at 31.6 Å which will also disappear towards higher T_{eff} (Fig. 2).

2 NLTE Model Atmospheres

Our model atmospheres are plane parallel, chemically homogeneous, and in radiative and hydrostatic equilibrium. They were calculated using the ALI method.

While the implications of H, He, C, N, and O (Rauch 1993) and of the iron group elements on NLTE model atmospheres have been studied in detail (Dreizler & Werner 1993), the light metals lithium through calcium have been regarded as trace elements and were neglected in model atmosphere calculations so far. For some of them line formation calculations (i.e. fixed atmospheric structure) had been carried out in order to determine photospheric abundances, e.g. Dreizler (1993) for Ne, Mg, Si in sdO stars. However, the "total abundance" of the light metals is comparable to that of the CNO elements and thus, not negligible.

In the construction of our model atoms we have to consider that the number of tractable NLTE levels is restricted due to numerical accuracy (< 200). Thus,

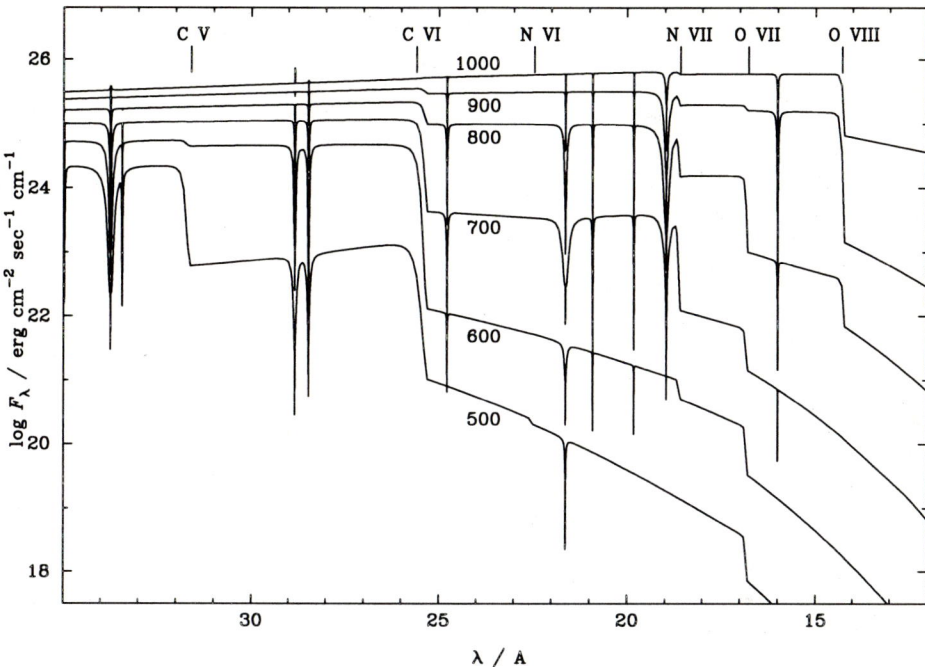

Fig. 2. Emergent fluxes of H+He+C+N+O NLTE model atmospheres (including lines) with $\log g = 9$, $T_{\rm eff}= 500, \ldots, 1000\,\rm kK$, and solar abundances. The wavelengths of the ground state thresholds of C V, C VI, N VI, N VII, O VII, and O VIII are marked. Note that their edges appear always in absorption, and have a maximum at that $T_{\rm eff}$, where the respective ion dominates in the atmosphere due to the ionization equilibrium (e.g. N VII at $\approx 700\,\rm kK$, O VIII at $\approx 900\,\rm kK$).

we used "standard" model atoms for H and He (e.g. Rauch 1993) and small model atoms for C, N, and O. In test calculations, Li, Be, and B were found to be unimportant and neglected. F - Ca were represented by model atoms "as small as possible, as large as necessary" in order to consider all ionization stages that dominate in certain parts of the atmosphere as well as resonance lines. Their photoionization cross-sections were calculated using Opacity Project (cf. Seaton et al. 1994) data.

The changes in the temperature stratification due to the consideration of metals, especially of their resonance lines are demonstrated by Rauch (1996a). The differences in the atmospheric structure have a direct influence on theoretical profiles of e.g. H I and He II lines, which are used in spectral analyses to determine $T_{\rm eff}$, $\log g$, and the photospheric He/H abundance ratio. This is shown in detail by Rauch (1996b).

3 Emergent Flux

In Fig. 1 we show the emergent fluxes of models which are calculated with different T_{eff} and chemical compositions. Obviously, metal opacities reduce the flux in the EUV and X-ray range by orders of magnitude. A comparison to the flux of a black body (BB, at 400 kK, Fig. 1) shows that a BB is unsuitable for analyses of most stars: The BB flux maximum is at ≈ 72 Å while the NLTE flux maximum is at ≈ 50 Å — and at a higher ($\approx 20\%$) flux level. A similar result was found by Heise et al. (1994): Their LTE model atmospheres are also more efficient soft X-ray emitters than BBs but the resulting LTE spectra are much flatter (at a lower flux level than the BBs!).

4 Summary

Emergent fluxes calculated from NLTE model atmospheres which include the "light metals" F - Ca show a drastic decrease of the flux level at energies higher than 126 eV (= O VI 2p continuum) compared to that of "standard" (H+He+C+N+O+Ne) models.

However, at a given T_{eff}, NLTE and LTE model atmospheres have their flux maxima at higher energies than a BB. The flux level of a NLTE model is higher than that of a LTE model (this will be investigated in a forthcoming paper).

For a realistic X-ray analysis of very hot stars the use of NLTE model atmospheres which consider metal opacities is highly recommended.

Acknowledgement: This research was supported by the DARA under grant 50 OR 9409 1.

References

Auer L.H., Mihalas D., 1969, ApJ 158, 641
Dreizler S., 1993, A&A 273, 212
Dreizler S., Werner K., 1993, A&A 278, 199
Heise J., Van Teeseling A., Kahabka P., 1994, A&A 288, L45
Kurucz R.L., 1979, ApJS 40, 1
Kurucz R.L., 1991, in: Stellar Atmospheres: Beyond Classical Models, eds. L. Crivellari, I. Hubeny, D.G. Hummer, NATO ASI Series C, Vol. 341, p. 44
Rauch T., 1993, A&A 276, 171
Rauch T., 1996a, in: Röntgenstrahlung from the Universe, eds. H.U. Zimmermann, J.E. Trümper, H. Yorke, MPE Report 263, p. 63
Rauch T., 1996b (in prep.)
Seaton M.J., Yu Yan, Mihalas D., Pradhan A.K., 1994, MNRAS 266, 805
Werner K., 1986, A&A 161, 177
Werner K., Husfeld D., 1985, A&A 148, 417

Part IV

Intrinsic Variability of SSS

The Long-Term X-Ray Lightcurve of RX J0527.8–6954

J. Greiner[1], R. Schwarz[2], G. Hasinger[2], M. Orio[3]*

[1] Max-Planck-Institut für Extraterrestrische Physik, 85740 Garching, Germany
[2] Astrophysikalisches Institut Potsdam, 14482 Potsdam, Germany
[3] Department of Physics, University of Wisconsin, Madison, WI 53706, USA

Abstract. Supersoft X-ray sources are commonly believed to be stably burning white dwarfs. However, the observations of some supersoft sources show dramatic variability of their X-ray flux on timescales ranging from days to years. Here, we present further observational data of the supersoft X-ray source RX J0527.8–6954 exhibiting a continuous decline over the past 5 yrs. With no clear trend of a concordant temperature decrease this might suggest a evolutionary scenario where the WD leaves the steady burning branch and the combined effect of reduced luminosity and cooling at constant radius produces the observed effect.

1 Introduction

The supersoft X-ray source RX J0527.8–6954 was discovered in the *ROSAT* first light observation (Trümper et al. 1991) of the Large Magellanic Cloud (LMC) in June 1990. It has an extremely soft X-ray spectrum, with spectral parameters (blackbody temperature, absorbing column density) very similar to CAL 83 (Greiner et al. 1991), the SSS prototype (Long et al. 1981). Also, it was readily realized that this source must have brightened up by at least a factor of 10 compared to previous *Einstein* observations when RX J0527.8–6954 was in the field of view but not detected. It was noticed already earlier (Orio and Ögelman 1993, Hasinger 1994) that the countrate of RX J0527.8–6954 had decreased substantially since its discovery.

2 Observational Results

2.1 All-Sky-Survey

As RX J0527.8–6954 is close to the south ecliptic pole, it was scanned during the All-Sky-Survey over a time span of 21 days. The total observation time resulting from 92 individual scans adds up to 1.96 ksec. Due to the scanning mode the source has been observed at all possible off-axis angles with its different widths of the point spread function. For the temporal and spectral analysis we have used an 5′ extraction radius to ensure that no source photons are missed.

---

* On leave from Osservatorio Astronomico di Torino, 10025 Pino Torinese, Italy

No other source down to the 1σ level is within this area. Each photon event has been corrected for its corresponding effective area. The background was determined from a circle 13' off along southern ecliptic latitude with respect to RX J0527.8–6954. The mean countrate was determined to (0.14±0.06) cts/sec. Due to systematic errors of about 20% no definite conclusion can yet be drawn on possible short-term variations of the X-ray flux.

For the spectral fitting the X-ray photons in the amplitude channels 11–240 (though there are almost no photons above channel 50) were binned with a constant signal/noise ratio of 5σ. The fit of a blackbody model results in an effective temperature of kT_{bb} = 40 eV (with the absorbing column fixed at its galactic value), very similar to the results obtained from fitting the PSPC data of the *ROSAT* first light observation (Greiner et al. 1991).

2.2 PSPC Pointings

Several pointings with the *ROSAT* PSPC have been performed on RX J0527.8–6954, starting with the first light observation and continuing with several dedicated pointings. In addition, RX J0527.8–6954 was also in the field of view of a number of other target pointings, mainly within those on the bright supernova remnant N132D which has been used for calibration purposes. We restricted the analysis to pointings with effective exposure times larger than 300 sec and containing RX J0527.8–6954 at less than 50' off-axis angle (Table 1).

Depending on the off-axis angle of RX J0527.8–6954 within the detector, photons have been extracted within 6–10' radius. This extreme size of the extraction circle was chosen because the very soft photons (below channel 20) have a much larger spread in their measured detector coordinates. As usual, the background was determined from a ring well outside the source without having contaminating sources in there. Before subtraction, the background photons were normalized to the same area as the source extraction circle. Since RX J0527.8–6954 is affected by the window support structure in many pointings, we have developed a dedicated procedure to correct for the shadowing/wobbling effect.

As a first step, we took the standard PSPC instrument map together with the effective area table to produce an instantaneous correction map. In the second step these correction maps are added up according to the wobble motion and roll-angle using the attitude table. Both these steps were performed separately in the 11–41 and 42–52 energy channel bands. This energy selection is important because the obscuration of sources (scattering) is energy dependent. We then determine the good time intervals (exposure time) and after multiplication get two exposure maps (in the separate energy bands) containing the effects of vignetting and wobbling. As the next step we determine the relative number of counts of RX J0527.8–6954 in the 11–41 and 42–52 bands and use these as weights for summing the two exposure maps. Finally, the resulting exposure map is used to compute the effective exposure time (given in column 5 of Tab. 1) of RX J0527.8–6954 by averaging over the same location and area as source photons have been extracted.

Table 1. Summary of *ROSAT* observations covering RX J0527.8−6954.

Observation No.	Date	PSPC or HRI	T_{Nom} (sec)	T_{Eff} (sec)	No. of counts	Off-axis angle	Distance to next rib
110173	June 18, 1990	P	2042	1418	296	31'	5'
110176	June 19, 1990	P	2168	1548	292	29'	5'
110074	June 20, 1990	P	754	471	72	28'	9'
110181	June 21, 1990	P	1882	1176	207	46'	15'
110090	June 24, 1990	P	457	310	69	25'	3'
110234	July 6, 1990	H	779	736	19	13'	−
110241	July 7, 1990	H	474	460	12	12'	−
Survey	Oct. 10–31, 1990	P	1965	1357	189	0–55'	−
141800	Dec. 11, 1991	P	1060	541	40	21'	0'
160084	May 5, 1991	P	1757	1040	100	20'	0'
300126	Mar. 5, 1992	P	7802	4972	401	22'	0'
500004	Apr. 5, 1992	P	1100	805	47	20'	1'
400148	Apr. 6, 1992	P	6263	6263	658	1'	20'
300172	May 7–16, 1992	P	6371	3636	403	37	1'
400238	Nov. 26, 1992	H	4063	3788	38	10'	−
400298	Dec. 6, 1992	P	1058	1058	61	1'	20'
300172a	Dec. 16–26, 1992	P	2996	2163	120	36'	10'
400298a	Mar. 11–16, 1993	P	7506	7506	202	1'	20'
500141	Apr. 11, 1993	P	5259	3709	110	19'	1'
141937	Apr. 16, 1993	P	1979	1394	55	21'	0'
141506	June 16, 1993	P	706	446	15	21'	0'
300172b	June 14–27, 1993	P	3882	2630	157	36'	8'
141507	Aug. 24, 1993	P	1334	940	25	22'	0'
201689	Aug. 29/30, 1994	H	8463	8463	16	0'	−
201996	Aug. 10–12, 1995	H	7162	7162	8	0'	−
600782	Oct. 22–24, 1995	H	12872	12588	6	4'	−

2.3 HRI Pointings

There are also a number of HRI pointings which cover RX J0527.8−6954, namely two pointings during the verification phase, one pointing performed in November 1992 in the framework of LMC X-ray source identifications (see Cowley et al. 1993) and three dedicated, on-axis pointings in August 1994, August 1995 and October 1995 (see Tab. 1). We restricted the analysis to pointings with effective exposure times longer than 300 sec and at less than 15' off-axis angle (excluding two verification phase pointings). Source photons have been extracted within 2–3' and were background and vignetting corrected.

We derive a best-fit position ($\pm 5''$) of R.A. (2000.0) = $05^h 27^m 48^s.9$, Decl. (2000.0) = $-69° 54' 09''$ which results from the position averaging of the two individual on-axis HRI pointings in August 1994 and August 1995 (which differ by 4''). The averaging reduces the irreproducible r.m.s. scatter of $\approx 5''$ from the individual pointings due to the fact that the roll angles are different. This new

148 J. Greiner, R. Schwarz, G. Hasinger, M. Orio

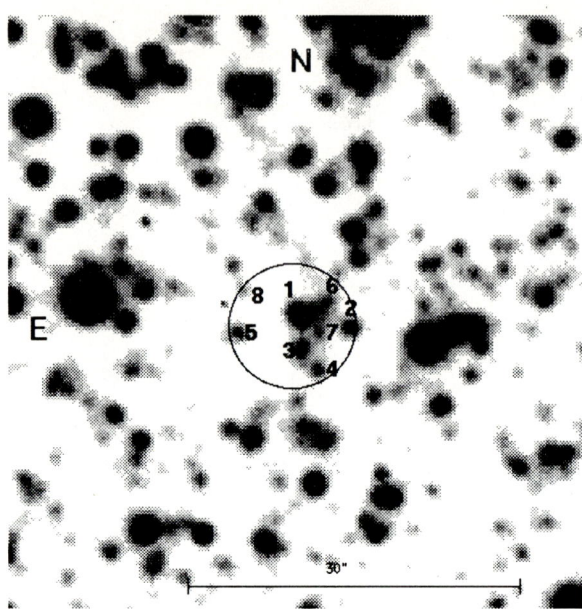

Fig. 1. The X-ray error circle of 5″ radius overplotted on a 10 min B image taken on March 25, 1995 with the ESO 2.2 m telescope at La Silla/Chile (Max-Planck-Institute time). All resolved objects inside the error box are numbered. Object 1 has been spectroscopically identified as B8IV (Cowley et al. 1993). A spectrum of object 2 ($m_B = 18\overset{m}{.}5$) taken on March 28, 1995 (also ESO 2.2 m) suggests spectral type B. Object 6 seems to be variable as compared to the image of Cowley et al. (1993).

position differs by only 2″.5 from that given in Hasinger (1994) which arose from the averaging of several off-axis (4 PSPC and 1 HRI) pointings. Our new position strengthens the conjecture of Hasinger (1994) that the blue object proposed by Cowley et al. (1993) as a strong counterpart candidate is too distant to be a likely counterpart (see Fig. 1). Though this new position is only 6″ from the Cowley et al. (1993) position, it is in the opposite direction with respect to the above mentioned blue object.

In order to compare the HRI intensities of RX J0527.8−6954 with those measured with the PSPC we determined the PSPC/HRI countrate ratio for this supersoft X-ray spectrum in the following two ways: 1) In an empirical approach we selected a few single white dwarfs (WDs) which have been observed with both, the PSPC and HRI, and derive a conversion factor PSPC/HRI = 7.8. This is thought to be an upper limit because such isolated WDs are less absorbed than SSS and thus sample even the lowest PSPC channels. 2) We have used the best fit model of the PSPC spectrum as input for computing the expected HRI countrate using its up-to-date response matrix and effective area. The result is sensitive to the temperature chosen and gives a ratio of PSPC/HRI = 7.7 (7.0) for a blackbody temperature of kT = 35 (40) eV. Thus, we adopt a ratio of 7.5.

2.4 The Lightcurve

Fig. 3 shows the X-ray lightcurve of RX J0527.8–6954 (i.e. the background subtracted counts divided by the effective exposure time) as deduced from 19 *ROSAT* PSPC pointings, six HRI pointings and the All-Sky-Survey data between 1990 and 1994. Two main features can be recognized immediately from this overall 5 year lightcurve: 1) The source has exponentially declined in X-ray intensity since its first *ROSAT* observations in 1990 ($\tau=1.7\pm0.1$ yrs), and 2) there is considerable scatter in the decline which is larger than our estimate of the remaining systematic errors in correcting for the effects of the window support structure.

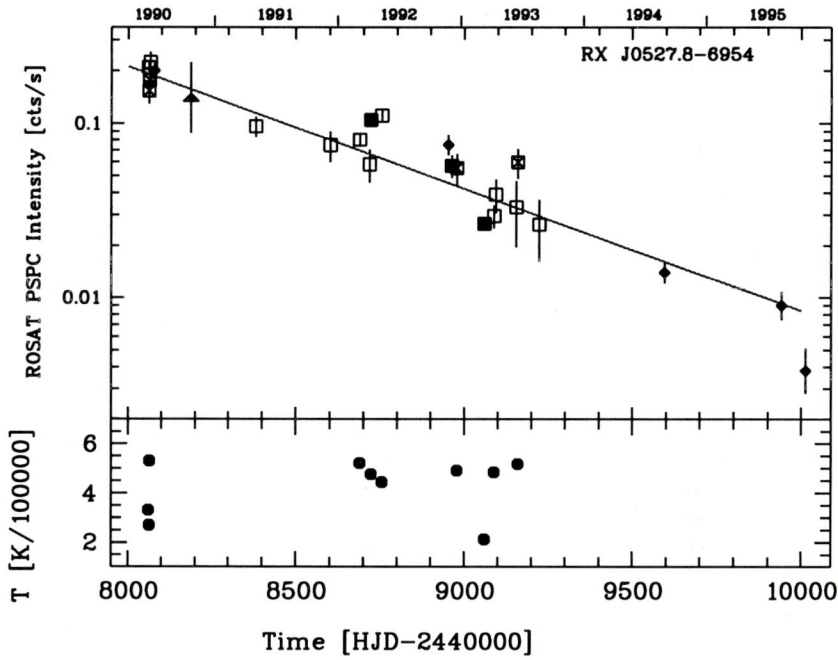

Fig. 2. X-ray lightcurve of RX J0527.8–6954 over the past 5 years as observed with the *ROSAT* satellite. The triangle marks the mean of the scanning observation during the All-Sky-Survey, squares mark PSPC observations (open symbols for off-axis angles larger than 15' and affected by the window support structure, crossed squares for off-axis angles larger than 15' and no obscuration by ribs, and filled symbols for off-axis angles smaller than 15'), and filled hexagons denote HRI observations with the countrates transformed to PSPC rates (see text). Systematic errors (not included in the plotted error bars) are largest for observations marked with open squares and might reach a factor of 2. The solid line is an exponential with $\tau=1.7$ years. The lower panel shows the best-fit blackbody temperature (while N_H was fixed) derived from the PSPC observations. The low temperature point at HJD = 9060 (ROR 400298a) and the spread of best-fit temperatures during the All-Sky-Survey are possibly caused by inadequately corrected gain differences at various off-axis angles.

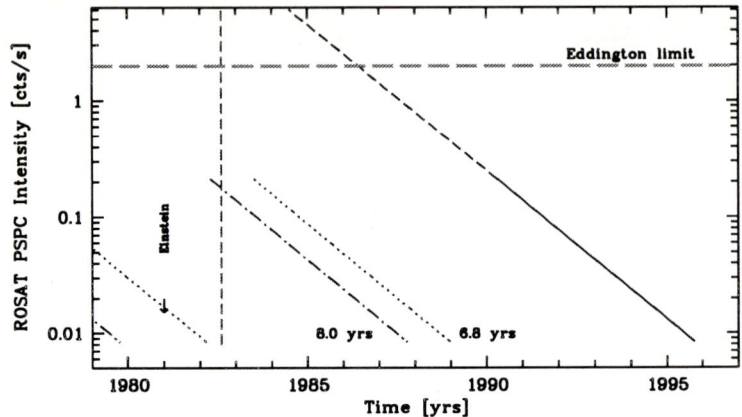

Fig. 3. Schematic presentation of the possible timescales using the $\tau=1.7$ yrs exponential of Fig. 3 and the *Einstein* upper limit. For a repeating source with sawtooth profile the repitition time scale is constrained to 6.8–8 yrs.

Surprisingly, the last HRI observation in October 1995 shows RX J0527.8–6954 to be considerably fainter than the extrapolation of the exponential decline. The X-ray intensity has dropped by a factor of 2.5 within two months. We have checked the housekeeping data of this pointing for any anomalous behaviour in detector properties or background variation - with negative result. Unfortunately, the only steady X-ray source (the SNR N132D) is at 20' off-axis and thus cannot be used as an observational calibration. With the data products of this observation being so recent, further checks of the instrument performance are certainly necessary. If this intensity drop indeed is real, one might speculate on having caught the source during the switch-off.

The total X-ray intensity amplitude between maximum and minimum observed *ROSAT* intensity is a factor of 50 within these 5 years. This is about a factor five larger than the estimated amplitude deduced from the *Einstein* non-detection. At the present X-ray intensity (and during all the last two years) the source would have been invisible again for the *Einstein* observatory.

The decay time up to now is >5.5 yrs. Extrapolating this decay law back and using the *Einstein* upper limit allows to constrain the recurrence time to 6.8–8.0 yrs, if the rise time is very fast (see Fig. 3). The other alternative, a recurrence time of 13.5–15 yrs, would imply that the source exceeded the Eddington limit by a factor of a few for several years.

Though the number of counts detected during the individual PSPC pointings is mostly rather low, we investigated the possibility of X-ray spectral changes during the decline. First, we kept the absorbing column fixed at its galactic value and determined the temperature being the only fit parameter. We find no systematic trend of a temperature decrease (lower panel of Fig. 3). Second, we kept the temperature fixed (at 40 eV in the first run and at the best fit value of the two parameter fit in the second run) and checked for changes in N_H, again finding no correlation.

3 Discussion

The most popular model of SSS involves steady nuclear burning on the surface of an accreting WD (van den Heuvel et al. 1992). If $\dot{M}$ exceeds a certain value that depends on the WD mass and other physical parameters, hydrogen burning on the WD surface is stable.

There are several possible phenomena which might explain completely or partially the X-ray variability of RX J0527.8–6954:

1. Changes in $\dot{M}$ within the small range which is necessary for stable hydrogen burning. The luminosity amplitude in this case can be only a factor 2.3–2.7 (see Fig. 9 of Fujimoto 1982 and Figs. 2 and 8 of Iben 1982). However, since the nuclear luminosity depends mainly on the conditions of the burning envelope, the corresponding changes are expected on the accretion time scale, i.e. much longer than observed in RX J0527.8–6954. Therefore, such $\dot{M}$ changes are ruled out as the only cause of the exponential decline.
2. The atmospheric layers are expanding and thus the effective temperature decreases, while the bolometric luminosity remains constant (evolution along the horizontal track in the log(T)- log(L) plane). The expansion could be caused by increased mass transfer from the secondary. This mechanism has already been suggested by Pakull et al. (1993) for RX J0513.9–6951. However, is is unlikely for RX J0527.8–6954 since (1) it works only near the Eddington rate and (2) no temperature decrease has been observed.
3. The gradual decline of the X-ray intensity also might suggest that we observe the decay phase after a shell flash. Thus, it is important to ask when the decay of RX J0527.8–6954 might have started. The recurrence time of hydrogen flashes is inversely proportional to the mass accretion rate and to the white dwarf mass. Since RX J0527.8–6954 has not been detected with *Einstein* observations, the shell flash must have occurred after 1981. Moreover, since with *ROSAT* we witness a decline from the beginning, the hydrogen-burning plateau phase (horizontal track in the H–R diagram) must have been short. Thus, even if the shell flash happened just after the *Einstein* observation then it would still need a considerable fine tuning of WD mass and accretion rate ($\dot{M} \leq 10^{-8} M_\odot$ yr^{-1}) for the WD to be on the declining portion of the evolutionary track within only 10 years. Moreover, a shell flash with such properties would be accompanied by an optical brightening of the object to $M_V \simeq -5$. At the LMC distance this would correspond to a visual brightness of 14th magnitude which very certainly would not have passed undetected.
4. The WD is cooling at constant radius after a weak flash. Due to the strong sensitivity of the *ROSAT* countrate on the temperature the X-ray amplitude of a factor of 10 does not necessarily translate into a factor of 10 change in bolometric luminosity (even if the latter is assumed to be dominated by the soft X-rays). Thus, the WD can remain within the stable burning range (with its factor 2–3 reduction in bolometric luminosity) while the concordant temperature decrease will shift the Wien tail out of the *ROSAT* window. If the cooling alone would account for the X-ray intensity decay, then the

observed intensity amplitude corresponds to a temperature amplitude of at least two (depending on the absolute temperature). This translates into a cooling time of the order of five years. We can assume that the shortest time for the white dwarf cooling is only slightly longer than the time necessary to burn all the hydrogen envelope mass once there is no mass transfer at all, like it is usually supposed after a nova outburst (Starrfield, priv. comm.). This can be as short as 5 years for a white dwarf of 1 $M_\odot$ (Kato & Hachisu, 1994).

Thus, it seems likely that the variability of RX J0527.8–6954 is due an earlier hydrogen flash on a rather massive white dwarf (m$\geq$ 1.1$M_\odot$) accreting at a high rate (10^{-7} $M_\odot$/yr) as described in Iben (1982) and Fujimoto (1982). In an empirical model of recurrent SSS, Kahabka (1995) related the envelope mass to the decay and recurrence times of these sources. The latter can be used in turn as observables to determine the WD mass and its accretion rate. From a preliminary analysis of the X-ray decline of RX J0527.8–6954 Kahabka (1995) adopted a value of 5 yrs for the time for return to minimum after a flash, and a recurrence time of 10 yrs. These numbers have changed only moderately with the herewithin reported results. With the new timescales we derive M_{WD}=1.2–1.35 $M_\odot$ and $\dot{M}_{accr}$=2–7$\times 10^{-7}$ $M_\odot$/yr. The rather high WD mass is also compatible with the temperature of 5–6×10^5 K and the derived luminosity.

Acknowledgements: JG is supported by the Deutsche Agentur für Raumfahrtangelegenheiten (DARA) GmbH under contract FKZ 50 OR 9201. We are extremely grateful to Y.-H. Chu for providing the data of the October 1995 HRI observation. We thank C. Motch for valuable comments. The *ROSAT* project is supported by the German Bundesministerium für Bildung, Forschung, Wissenschaft und Technologie (BMBW/DARA) and the Max-Planck-Society.

References

Cowley A.P., Schmidtke P.C., Hutchings J.B., et al. 1993, ApJ 418, L63
Fujimoto M.Y., 1982, ApJ 257, 767
Greiner J., Hasinger G., Kahabka P. 1991, A&A 246, L17
Hasinger G., 1994, Reviews in Modern Astronomy 7, 129
Iben I., 1982, ApJ 259, 244
Kahabka P., 1995, A&A 304, 227
Kato M., Hachisu I., 1994, ApJ 437, 802
Long K.S., Helfand D.J., Grabelsky D.A., 1981, ApJ 248, 925
Orio M., Ögelman H., 1993, A&A 273, L56
Paczynski B., Zytkow A., 1978, ApJ 222, 604
Pakull M.W., Motch C., Bianchi L., et al. 1993, A&A 278, L39
Trümper J., Hasinger G., Aschenbach B., et al. 1991, Nat 349, 579
van den Heuvel E.P.J., Bhattacharya D., Nomoto K., Rappaport S.A., 1992, A&A 262, 97

Interpretation of the Long-Term Optical Variations of RX J0019.8+2156

F. Meyer, E. Meyer-Hofmeister

MPI für Astrophysik, 85740 Garching, Germany

Abstract. The optical light curve of this supersoft source by Greiner and Wenzel (1995) provides information on the evolution over 100 years. We interpret the long-term changes as resulting from different stages of burning and cooling of the hot white dwarf. We show, that the inclination under which RX J0019.8+2156 is seen must be around 60°. A part of the variation in the optical light can be understood as a signature of a varying height of the accretion disk rim.

1 Introduction

The detection of luminous supersoft X-ray sources was one of the major achievements of the observation with ROSAT. An important subgroup are the binary sources. The physics of these objects, especially the accretion of mass onto the white dwarf connected with nuclear burning is of high interest for our understanding of binary evolution. The variability in X-rays during the years of X-ray satellites was studied and used to draw conclusions on the white dwarf masses (Kahabka 1995). The X-ray observations usually cover only very short time intervals on the order of hours, separated by months or years from the last data. In contrast to this, the 100 years light curve of RX J0019.8+2156 (henceforth RX J0019) compiled by Greiner and Wenzel (1995) provides us with an extraordinary wealth of information. We analyse the optical light curve in terms of alternating phases of nuclear burning and cooling (Kahabka 1995), superimposed by orbital variation and the variation in height and occasional disappearance of a light reprocessing accretion disk rim, possibly caused by corresponding decrease of the mass overflow rate. This is quantified by computing light curves following the investigation of Schandl et al. (1996a,b) for CAL 87.

2 Phases of White Dwarf Nuclear Burning and Cooling

In Fig. 1 we show the optical light curve of RX J0019 (Greiner and Wenzel 1995), one triangle point more resulting from observation in 1994 was added (Wenzel 1996). The scatter from orbital variations in the light curve is reduced if average values are given as all filled diamonds and triangles.

This light curve shows quasiperiodic abrupt rises by ≈ 0.5 magnitudes followed by subsequent decay over periods of 20–30 yrs. If the optical light comes from irradiation of disk and secondary star by the white dwarf and lies in the

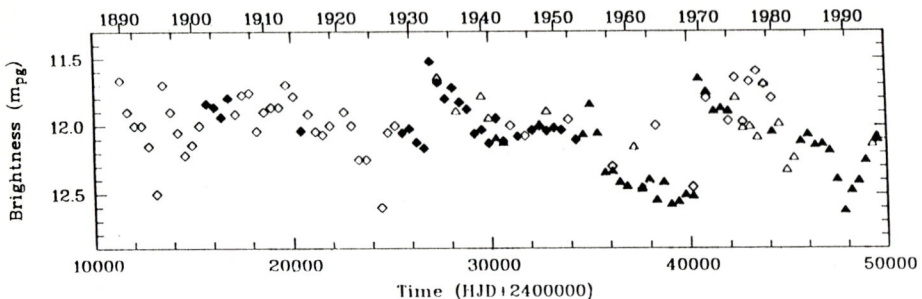

Fig. 1. Optical light curve of RX J0019.8+2156 derived from photographic plates at Harvard (diamonds) and Sonneberg Observatory (triangles), respectively. Filled symbols contain more than nine individual measurements. (Reproduced from Greiner & Wenzel (1995) and complemented by the mean brightness for 1994 (Wenzel 1996).)

Rayleigh-Jeans part of the black body spectrum then a variation by 0.5 mag in the optical corresponds to a variation by a factor 6.3 in the luminosity of the irradiating white dwarf. With an estimated X-ray luminosity at present of $10^{36.8}$ erg/s (Beuermann et al. 1995) this would indicate peak luminosities of order $10^{37.6}$ erg/s, appropriate for steady hydrogen burning on white dwarfs of between 0.6 and 1 $M_\odot$ (Fujimoto 1982). This suggests repeated hydrogen flashes as the cause of the quasiperiodic behaviour (Kahabka 1995). In the high flash state accumulated hydrogen is burnt while in the low state fresh material is accreted. Recurrence times as short as 30 years however would require white dwarf masses larger than 1.1 $M_\odot$ (Fujimoto 1982) at slight variance with the above mass range. We now note that the apparent reignition condition where compressional heating of accreted material alone triggers the new outburst is not realized in our case here. This is indicated by a comparision of the observed X-ray luminosity, $10^{36.8}$ erg/s, with that released by accretion of e.g. 10^{-7} $M_\odot/yr$ on a 1 $M_\odot$ white dwarf, $10^{36.2}$ erg/s. The factor 4 between these values suggests that the white dwarf layers are still cooling when the new flash occurs.

This is not unreasonable. In the burning zone temperatures typically are 10^8 K which during the time of burning heat up the white dwarf material below. This heat is released during the cooling phase. As a diffusive process the cooling luminosity then falls off as $(t/t_0)^{-1/2}$, and this seems to agree loosely with the visual light curve which fluctuates around an intermediate plateau at $m_{pg} \simeq 12$ where the decrease $\Delta m_{pg} \simeq 2.5 \cdot \frac{1}{4} \log(t/t_0)^{1/2} = 0.31 \log t/t_0$ becomes imperceptably on short terms when t/t_0 is large. The cooling law relates the cooling time t_0 of the burnt hydrogen layer to the decrease over the total cooling time t, $t_0 = t \cdot 10^{-\Delta m/0.31} = 1.1$ yr. for e.g. $t = 30$ yr and $\Delta m = 0.44$. The finally slow cooling also can explain the small, if at all, decrease in the X-ray count rate from 2.1 cts/sec in the ROSAT All-Sky-Survey 1990 to 2.0 cts/sec in pointed ROSAT observations 1992 and 1993 (Beuermann et al. 1995), predicted to be of order 5% after 30 yr cooling.

If the above description holds for RX J0019 then the apparent excursions by another $\Delta m \simeq 0.5$ to even lower brightness can not be due to changes in the irradiating white dwarf luminosity. This is also obvious from the ROSAT observations mentioned above that showed no significant variation while the visual light varied by 0.5 mag during the same interval. We must then attribute these variations to a change in the reprocessing surfaces and in particular to the one element that may conceivably vary, the outer disk rim, possibly caused by a spray of optically opaque blobs originating at the impact of the accretion stream on the accretion disk, and varying with the mass transfer.

3 Accretion Disk Height Signatures in the Visual Light

3.1 Description of the Model

In a detailed investigation Schandl et al. (1996a,b) showed that an essential contribution to the visual light originates from reprocessing of X-rays at the disk rim. The accretion flow impinging on the disk produces a stream of blobs which move along the outer disk boundary. These blobs are illuminated facing the hot white dwarf. We assume the spectrum to be black body. Based on a comparison with the fit for CAL 87 it was argued already, that the orbital light curve observed by Beuermann et al. (1995) might indicate an inclination much higher than 20° as derived in that work. A new compilation of more recent data by Will and Barwig (1996) clearly document a pronounced secondary minimum and therefore a higher inclination. In this investigation was also found, that the shape of the phase dependent light curve changes within days. This suggests a high variability connected with a varying mass overflow rate.

Assuming that the geometrical height of the spray, that is the area filled with blobs, depends on the overflow rate, we investigate the influence of this height on the shape of the orbital visual light curve. The light curve is computed based on the model of van den Heuvel et al. (1992) for nuclear burning of accreting white dwarfs using a computer code by Schandl et al. (1996a), which includes the contributions of the white dwarf, the disk with a spray at the rim and the secondary consistently. Adopting this description we find for different luminosities of the white dwarf the light curves shown in Fig. 2. These light curves have the characteristic features of the light curve observed for RX J0019. (Now after knowing the results of Will and Barwig (1996) one could work out a detailed fit to these light curves.) The similarity of the curves in Fig. 2 allows the conclusion, that the shape of the orbital light curve does not depend on the white dwarf luminosity. Since the luminosity from the nuclear burning cannot change within months we assume for our analysis a constant white dwarf luminosity.

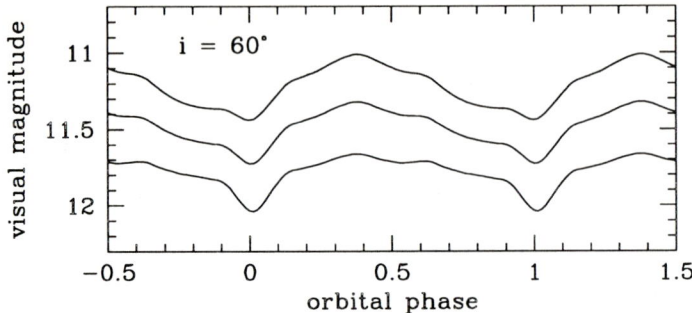

Fig. 2. Theoretical visual light curves computed for a high outer disk rim, $z/r = 0.45$ and three values of the white dwarf luminosity (from top to bottom): 6×10^{37}, 2.2×10^{37}, 6×10^{36} erg/s.

3.2 The Height of the Disk Rim

The minimal disk thickness z is the height of an illuminated disk where the structure is consistently determined with the white dwarf luminosity, no hot spot included. The value z/r, (r distance from white dwarf) increases with size of the disk. The value for the taken geometry is $z/r = 0.11$. As a maximal value we take $z/r = 0.45$ (as found for CAL 87). For that case also the size of the disk was enlarged at the phase where the rim height has its maximum (for details see Schandl et al. 1996a). We study the change of the orbital light curve with z/r increasing from 0.11 to 0.45 and corresponding enlargement of the disk size. In the case of low disk height the contribution of the disk to the total visual light is low, the variation of the secondary seen from either the illuminated or the non-illuminated side is important and produces a deep primary minimum. With increasing disk height the disk contribution becomes more important. The secondary looks less bright from its illuminated side because the high disk rim shadows the secondary star's brightest region. The primary minimum flattens.

We determined from our light curves an average magnitude $\overline{m}$ for the phase interval 0.1 to 0.9 corresponding to the determination of the brightness in the work by Greiner and Wenzel (1995). We show in Figs. 3 and 4 how the average brightness $\overline{m}$ and the depth of the primary minimum change with z/r. For this evaluation two values of white dwarf luminosity were used, first the same as for the analysis of CAL 87, $2.2\,10^{37}$ erg/s and second, $6\,10^{36}$ erg/s. For the observation (Will and Barwig 1996) an inclination of 60° to 70° seems appropriate.

We conclude that a variation of the disk rim height easily produces a scatter of about half a magnitude in the optical light curve. The existence of one recent observation, at very low brightness for phase 1.0 ($12\overset{m}{.}9$ pg at JD 244 9638.47, eclipse minimum E=20954; Wenzel 1996), strengthens the point, that such minima indeed occur and therefore the disk seems to be from time to time without the thickening connected with an impinging accretion flow. This means, that the accretion flow varies from low values to values around $10^{-7} M_\odot$/yr (the latter exspected from the van den Heuvel et al. model).

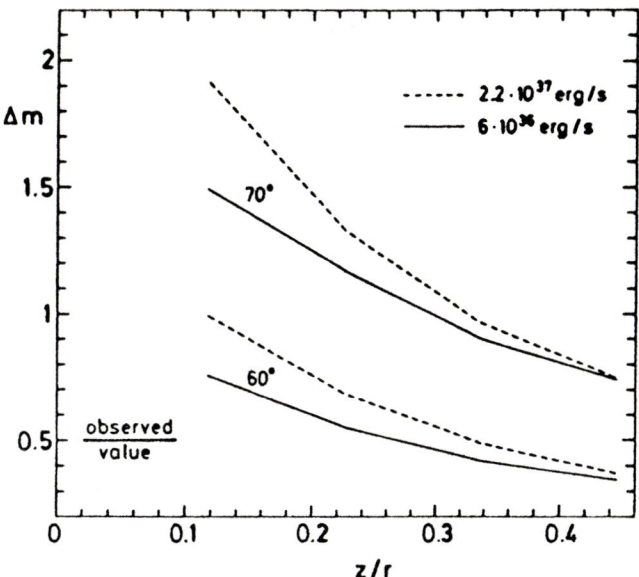

Fig. 3. Depth of primary minimum of the visual light curve of RX J0019 dependent on the chosen maximal height of the outer disk rim (for details of the model see text).

3.3 Long-Term Trends in the Mass Overflow Rate

This scatter seems overlying a long-term trend of decreasing mass overflow rate in the cooling phase. In particular the continuous low state in the years 1960 to 1970 (Fig. 1) indicates a low or non-existent accretion disk rim and lower than usual mass overflow. One could imagine that the decreasing illumination of the secondary might change the mass overflow rate. This can not be excluded since the shadowed L_1-point of the Roche geometry still lies laterally only scale heights away from irradiated regions.

4 Summary

The long-time light curve of RX J0019 shows evidence for quasiperiodic H-flashes where heating of underlying white dwarf layers during the short burning phase stores heat that is gradually released during the off-phase. It aids in reaching reignition under accretion of fresh material and will allow shorter repetition periods than without such heating from below.

The occasional descent of the visual brightness by a further 0.5 magnitudes is interpreted as due to a significant decrease in the height of an accretion disk rim, conceivably due to a transitory or systematic decrease of the mass overflow from the secondary star which on impact sprays matter around the disk to a height z up to $0.45\,r$. Such an accretion disk yields good agreement both for the orbital light curve and its variation.

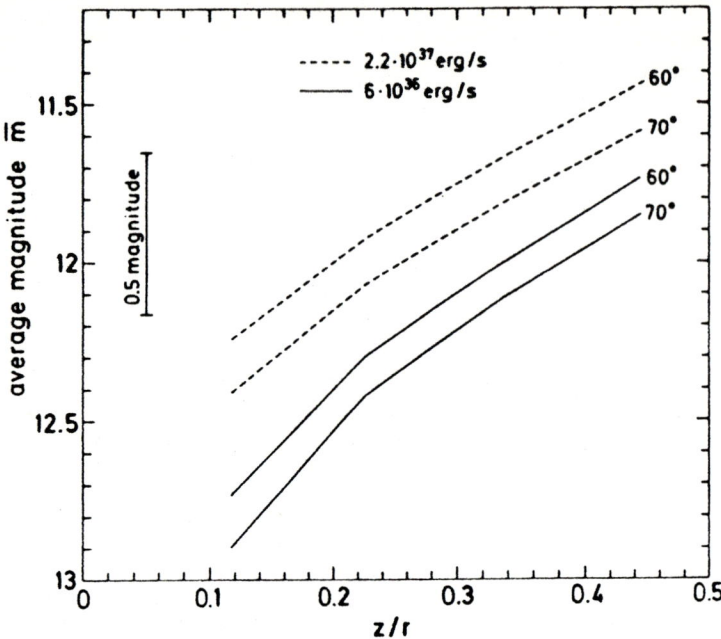

Fig. 4. Average visual magnitude $\overline{m}$ as expected from observations at different phases dependent on the chosen maximal height of the outer disk rim.

Acknowledgement: We thank W. Wenzel for additional information from recent observational data of RX J0019.8+2156 and H. Barwig and T. Will for providing their new orbital light curve of RX J0019 before publication. We are grateful to Susanne Schandl for the computation of light curves with her code, results which were essntial for our conclusions.

References

Beuermann K., Reinsch K., Barwig H., Burwitz V., de Martino D., Mantel K.-H., Pakull M.W., Robinson E.L., Schwope A.D., Thomas H.C., Trümper J. van Teeseling A., Zhang E., 1995, A&A 294, L1
Fujimoto M.Y., 1982, ApJ 257, 767
Greiner J., Wenzel W., 1995, A&A 294, L5
Kahabka P., 1995 A&A 304, 227
Schandl S., Meyer-Hofmeister E., Meyer F., 1996a, A&A (submitted) (MPA report 912, 1995)
Schandl S., Meyer-Hofmeister E., Meyer F., 1996b, this volume p. 53
van den Heuvel E.P.J., Bhattacharya D., Nomoto K., Rappaport S.A., A&A 262, 97
Wenzel W., 1996 (private comm.)
Will T., Barwig H., 1996, this volume p. 99

ROSAT Monitoring of the LMC Supersoft Transient Source RX J0513.9−6951

S.G. Schaeidt

Max-Planck-Institut für extraterrestrische Physik, 85740 Garching, Germany

Abstract. We present the ROSAT X-ray light curve of the LMC supersoft transient source RX J0513.9−6951 from 1990 to 1995, during which 3 X-ray on-states of this source have been observed. From the ROSAT HRI monitoring program in 1994/1995 we can estimate an upper limit for the decline time between X-ray on- and off-state to 7 days. None of the standard scenarios (shrouded neutron star, low-mass X-ray binaries (LMXB), white dwarf with nuclear burning on its surface) can explain the 'fast' X-ray variability of RX J0513.9−6951. It has been proposed, however, that a WD can adapt its radius to varying mass transfer rates on a dynamical time scale.

1 Introduction

The LMC supersoft transient source RX J0513.9−6951 was discovered in its X-ray on-state during the ROSAT All-Sky Survey in 1990 (Schaeidt et al. 1993). The optical counterpart of RX J0513.9−6951 was identified to be a B ∼ 17 mag blue star (Pakull et al. 1993, Cowley et al. 1993). Optical and IUE observations have shown that this object is very similar to CAL 83, the prototype of the supersoft sources (SSS) which is also located in the LMC. Both sources have remarkably similar X-ray spectra.

Although RX J0513.9−6951 seems to be a prototypical SSS, it shows some characteristics, which are not typical for most transient X-ray sources. During its X-ray outburst in the ROSAT survey, there is indication that the optical brightness has declined (Pakull et al. 1993). It was also reported by Cowley et al. (1996) that the optical spectrum shows red- and blueshifted emission components of the H and HeII lines. The radial velocity seems to be of the order of ± 4000 km/s. The existence of the emission components is indicating a strong wind which forces mass loss. The best period measured from the radial velocities of the HeII lines is known to be 0.76 days (Crampton et al. 1996). The same period is now also reported from photometric measurements (Motch et al. 1996, Southwell et al. 1996).

2 ROSAT PSPC Observations

ROSAT PSPC observations were performed during the period from July 1990 to October 1993. The ROSAT All-Sky survey data are presented (see also Schaeidt el al. 1993) and also the pointed observations from the earlier Performance and Verification phase plus a series of pointed observations during the AO-3 period.

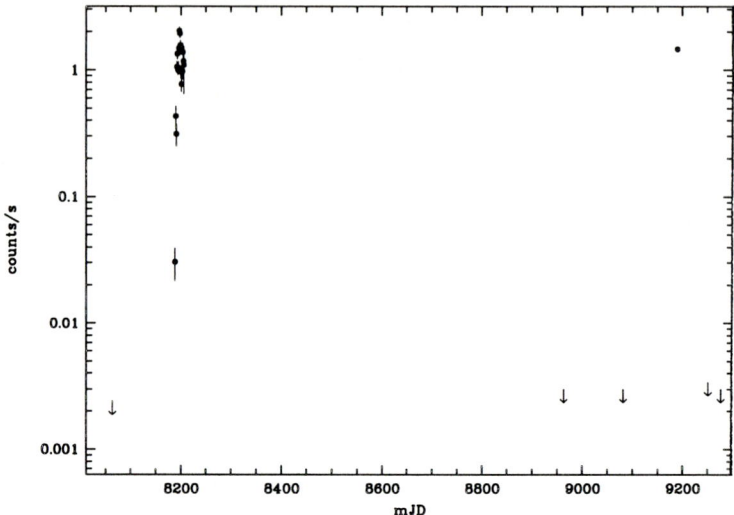

Fig. 1. The ROSAT PSPC lightcurve of RX J0513.9–6951 from July 1990 to November 1993, showing two X-ray on states. The day number is JD-2440000.

2.1 The PSPC Lightcurve

RX J0513.9–6951 was observed during the ROSAT All-Sky survey for 22 days from 1990, October 20th to November 11th with a total exposure of 1849 sec. The corrected light curve for RX J0513.9–6951 is shown in Fig. 1. The count rate increased by a factor of $\sim$ 20 in the first few days and remained high thereafter with an oscillatory behaviour with a timescale of several days. The error bars given in the figure include both the statistical errors and the systematic uncertainty of the correction procedure.

Although the source lies in the direction of the central region of the LMC, it was not detected during the LMC pointings in the ROSAT calibration phase between 1990 June 16 and 25 (Trümper et al. 1991). We have reanalyzed the calibration data to obtain an upper limit for the count rate at the source position found for the survey data (see Fig. 1).

We have also observed RX J0513.9–6951 during the AO-3 period from 1992-1993 several times, while originally only one observation was planned. The only planed observation was scheduled for December 1992, but for certain reasons the requested observation time of 10 ksec was not fulfilled. Therefore this observation was rescheduled in April 1993 with the same result as for the first time. In July 1993 we got finally pointed observations with a reasonable integration time. For the first two observations (Dec 92 and Apr 93) the source was not detected at all, while in the third observation (July 93) the source was again on. In the following two pointed observations the source was again off. For all of these observations we have calculated an upper limit which is consistent with that of the first observation in 1990 (see Fig. 1).

2.2 The PSPC X-Ray Spectrum

For the spectral analysis the background was determined from a ring well outside the source. Both, the source and the background selection were normalized to the same area and finally we subtracted the background counts from the source plus background counts. The 256 amplitude channels have been binned with a signal to noise ratio of 3σ. Because of the very low statistics the first channels (1-7) and the last channels (> 240) have not been included in the fit procedure. We have fitted a blackbody model to the X-ray data. The best fit model yields an effective temperature of $kT_{bb} = 40$ eV and a column density of $N_H = 0.94 \times 10^{21} cm^{-2}$, consistent with LMC membership (Tab. 1). The corresponding bolometric X-ray luminosity is $L_{bol} = 2.3 \times 10^{38}$ erg/s.

This result is consistent with the spectrum for the second detection in July 1993. By analyzing the data in the same way as for the survey, the best – single blackbody – fit model gives an effective temperature of $kT_{bb} = 37$ eV and a column density of $N_H = 1.17 \times 10^{21} cm^{-2}$, which is still consistent with the LMC membership.

Table 1. The spectral parameters of RX J0513.9–6951 for a single blackbody fit model for both the ROSAT survey and the AO-3 X-ray on state. The given count rates are the average over the observation period (X-ray on-state only).

Observation	N_H [$10^{21} \times cm^{-2}$]	T_{bb} [eV]	count rate [PSPC cts/sec]
Survey	0.9	39±7	1.266 ±.374
AO-3	1.1	37±2	1.459 ±.014

3 ROSAT HRI Observations

ROSAT HRI observations are part of a X-ray monitoring program of the supersoft sources in the LMC. We have proposed a monitoring of RX J0513.9–6951 once every week to investigate further the X-ray properties. The monitoring program was also placed in the period of the optical visibility for RX J0513.9–6951, which is from October to March respectively. The HRI monitoring finally started at November 1st in 1994. For the first six observations the source could not be detected. After this serie of observations we have no time coverage with HRI observations for two weeks. Surprisingly, the source was again detected in two further observations which have followed the 'hole'. One week after the last detection, the source was again off. From this observations we can now estimate the turn off time to be smaller than 7 days (see Fig. 2). This turn off time is in the same order as the turn on time, which was observed during the survey. The source stayed off until the end of the monitoring program in March 1995.

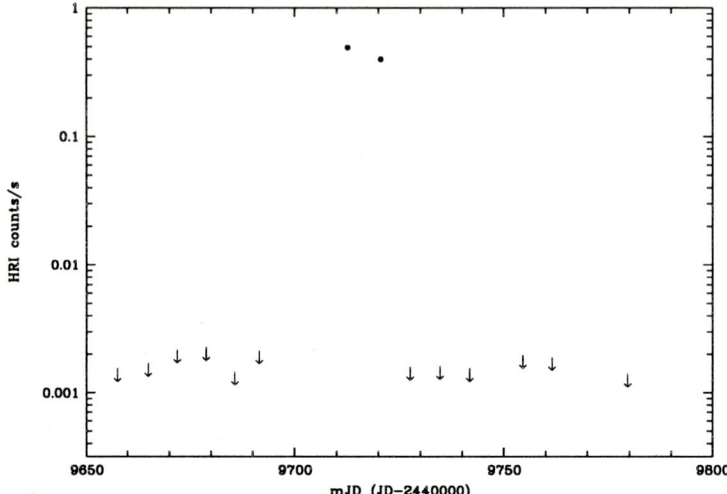

Fig. 2. The ROSAT HRI lightcurve of RX J0513.9–6951 from November, 1994 to March, 1995, showing one X-ray on state over a period of more than 2 weeks.

Comparing the count rate for the HRI X-ray on state of RX J0513.9–6951 to that of the PSPC in the energy range 0.1 – 0.4 keV, which is typically for SSS, we have to apply a factor of $\sim$ 7–8 (Greiner et al. 1996) to correct for the different sensitivity of the HRI. Thus, the HRI count rate corresponds to 3.9 PSPC cts/sec. The highest count rate we have detected during the ROSAT survey was a factor of $\sim$2 lower (Schaeidt et al. 1993).

4 Results and Discussion

For the LMC supersoft X-ray transient source RX J0513.9–6951 we can estimate the duration of the X-ray on-state. From our set of observations during the period from 1990 to 1995 we have detected the switch on time of several days for RX J0513.9–6951 in the ROSAT All Sky survey and now also the decline time from the HRI monitoring program which is in the same order. The total time for which the source will stay in its X-ray on-state can estimate from both observations, the ROSAT survey and the HRI monitoring observations. From the ROSAT survey we have seen, that the source stays in its X-ray on-state for more than two weeks. During the HRI monitoring program we have also seen the source for a period of 2 weeks in its X-ray on-state. Taking into account the gap of observations two weeks before, we can estimate now the period for which RX J0513.9–6951 stays in its X-ray on-state to at least 4 weeks (see Fig. 2).

We have detected for RX J0513.9–6951 in the period from 1990 to 1995 three X-ray on-states. The shortest period between two X-ray on-states was 17 months seen between the AO-3 observations in July, 1993 and the HRI observations in December, 1994.

None of the standard scenarios (shrouded neutron star, low-mass X-ray binaries (LMXB), white dwarf with nuclear burning on its surface) can explain the observed 'fast' X-ray variability of RX J0513.9-6951. Van den Heuvel et al. (1992) have proposed a model to explain the supersoft sources. This model involves a white dwarf in a close binary system, accreting at very high rates (10^{-7} to $4 \cdot 10^{-7}$ $M_\odot$/yr), such that nuclear burning is taking place on the white dwarf surface (a kind of slow nova). The temperatures and blackbody radii measured for SSS fit the prediction for white dwarfs quite naturally. For this model van den Heuvel et al. (1992) predict a time scale of ~100 yrs between cycles of pure accretion and nuclear burning luminosity. But this time scale is to slow to explain the observed 'fast' variability for RX J0513.9-6951.

A more recent interpretation for RX J0513.9-6951 was proposed by Pakull et al. (1993). This is based on the calculations of Kato (1985) which discusses the possibility, that a WD can adapt its radius to varying mass transfer rates on a dynamical time scale. In this scenario we can understand the X-ray outburst in terms of a temporary slightly reduced mass transfer rate. At the same time we would also expect a contraction of the photosphere and therefore a drop of the optical brightness of the white dwarf. A first tendency was reported by Pakull et al. (1993) and now established by Southwell et al. (1996). Pakull et al. (1993) have reported for the first time that the optical counterpart does not have brightened during the X-ray outburst, the quasi-simultaneous optical observations (B band) indicate a declining of the optical brightness by ~0.5 mag.

This view is now established by the recent results obtained using the MACHO data. Southwell et al. (1996) have presented the relative B magnitude lightcurve of RX J0513.9-6951 showing quasi-regular optical drops of ~ 1 mag. Two of the detected X-ray on-states (July 1993 and December 1994) fit well into the interval of the optical drops. The MACHO lightcurve of RX J0513.9-6951 also indicate a quasi-periodic variability on a timescale of ~150 days. Reinsch et al. (1996) have also reported fainter H and HeII emission lines for RX J0513.9-6951 during the December 1993 optical low-state.

The collection of all observations in the optical and X-ray wavelength shows, that the central source in RX J0513.9-6951 is consistent with a white dwarf. But from the present set of observations we can not determine if the X-ray on-state is followed by an optical minimum or vice versa. To solve this question a further program of simultaneous optical and X-ray observations is needed.

5 Summary

We have monitored the LMC supersoft X-ray source RX J0513.9-6951 in the period from 1990 to 1995 with the ROSAT PSPC and HRI. RX J0513.9-6951 was observed during this monitoring program 3 times in its X-ray on-state. Both, from the ROSAT survey data and the HRI data we can estimate an upper limit for the duration of the X-ray on-state to be in the order of 3 to 4 weeks each.

Comparing the optical and X-ray observations of RX J0513.9-6951, its central object is consistent with a white dwarf with nuclear burning on its surface.

Acknowledgement: I would like to thank J. Greiner for further discussions on SSS. The ROSAT project is supported by the Bundesministerium für Bildung, Wissenschaft, Forschung und Technologie (BMBW/DARA) and the Max-Planck-Society.

References

Cowley A.P., Schmidtke P.C., Hutchings J.B. et al. 1993, ApJ 418, L63
Cowley A.P., et al. 1996, in IAU Symp. 165, Compact stars in Binaries, eds. E.P.J. van den Heuvel & J. van Paradijs (Dordrecht: Kluwer), p. 439
Crampton D., Hutchings J.B., Cowley A.P., et al. 1996, ApJ 456, 320
Greiner J., Schwarz R., Hasinger G., Orio M., 1996, A&A (in press)
Kato M., 1985, PASJ 37, 19
Motch C., et al. 1996, this volume p. 83
Pakull M.W., Motch C., Bianchi L., Thomas H.-C., et al., 1993, A&A 278, L39
Reinsch K., van Teeseling A., Beuermann K., Abott T.M.C., 1996, this volume p. 173
Schaeidt S., Hasinger G., Trümper J. 1993, A&A 270, L9
Southwell K.A., Livio M., Charles P.A., et al. 1996, this volume p. 165
Trümper J., et al. 1991, Nature 349, 579
van den Heuvel E.P.J., Bhattacharya D., Nomoto K., Rappaport S.A., 1992, A&A 262, 97

Optical Variability of the LMC Supersoft Source RX J0513.9–6951

K.A. Southwell[1], M. Livio[2], P.A. Charles[1], W. Sutherland[1], C. Alcock[3,4],
R.A. Allsman[5], D. Alves[3], T.S. Axelrod[3,6], D.P. Bennett[3,4], K.H. Cook[3,4],
K.C. Freeman[6], K. Griest[4,7], J. Guern[4,7], M.J. Lehner[4,7], S.L. Marshall[4,8],
B.A. Peterson[6], M.R. Pratt[4,8], P.J. Quinn[6], A.W. Rodgers[6], C.W. Stubbs[4,8],
D.L. Welch[9]

[1] Dept. of Astrophysics, Nuclear Physics Bldg, Keble Road, Oxford OX1 3RH, England
[2] Space Telescope Science Inst., 3700 San Martin Drive, Baltimore, MD 21218, USA
[3] Lawrence Livermore National Laboratory, Livermore, CA 94550
[4] Center for Particle Astrophysics, University of California, Berkeley, CA 94720
[5] Supercomputing Facility, Australian Nat. Univ., Canberra, A.C.T. 0200, Australia
[6] Mt. Stromlo and Siding Spring Observatories, A.C.T. 2611, Australia
[7] Department of Physics, University of California, San Diego, CA 92093
[8] Department of Physics, University of California, Santa Barbara, CA 93106
[9] Dept. of Physics and Astronomy, Mc Master University, Hamilton, Canada, L8S 4M1

Abstract. We present optical spectroscopy and photometry of the LMC supersoft source (SSS) RX J0513.9–6951. Through the exceptional monitoring capabilities of the MACHO project, we show the optical history of this object for a ~ 3 year period. Recurring low states, in which the optical brightness drops by up to a magnitude, are observed at quasi-regular intervals. Analysis of the high state data reveals a small modulation with a semi-amplitude of ~ 0.02 magnitudes at P=$0\overset{d}{.}76278\pm0\overset{d}{.}00005$. By considering all the available data, including optical photometry, spectroscopy and X-ray observations, we suggest a theoretical model to explain the fundamental variations exhibited by this source.

1 Introduction

The LMC source RX J0513.9–6951, discovered in the ROSAT All Sky Survey (Schaeidt et al. 1993), was the first transient, recurrent SSS to be established; three outbursts have so far been detected (Schaeidt, these proceedings). A black body fit yielded $T_{bb} \sim 40$ eV and $L_{bol} \sim 2 \times 10^{38}$ erg s^{-1}.

The source has been optically identified with a V$\sim$ 17 Harvard variable, HV5682 (Cowley et al. 1993; Pakull et al. 1993). This star was observed to display fluctuations of < 0.3 mags recently (Crampton et al. 1996, hereafter C96), but has shown variations of ~ 1 mag historically. Optical spectroscopy (Pakull et al. 1993; C96) revealed a blue continuum with strong H and HeII emission, indicative of an accretion disk, but no evidence of the secondary star. C96 found a "best" period of $\approx 0\overset{d}{.}76$ from the HeII 4686 radial velocities.

2 Spectroscopy

We obtained 10 high (~ 1.3 Å) resolution spectra of RX J0513.9–6951 on the nights 29/11/94 – 1/12/94, using the 3.9 m Anglo-Australian Telescope at Siding Spring, Australia.

2.1 Average Spectrum

We show in Fig. 1 the variance-weighted average of all our blue spectra (which are not flux calibrated). HeII and Hβ are seen strongly in emission, with narrow cores and broad bases. We find marginal evidence for the Nv 4603/4619 Å emission lines reported by C96. The spectral region around 4600 – 4700 Å, shows evidence for the Bowen NIII-CIII $\lambda\lambda$ 4640 – 50 Å complex and probable CIV 4658.3Å, although these lines are contaminated by the underlying HeII emission.

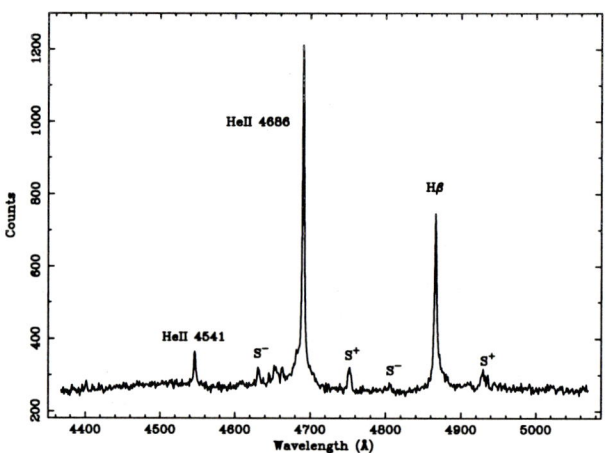

Fig. 1. Average blue spectrum of RX J0513.9–6951. The resolution is ~ 1.3 Å. The principal HeII and H emission features are marked, along with their associated Doppler-shifted components. Note the extended blue wing of HeII 4686 and lack of blue emission in Hβ.

2.2 Evidence for High Velocity Outflows

Emission lines marked S$^+$ and S$^-$ (following the nomenclature of C96) are seen on either side of HeII 4686Å and Hβ. These unusual features were first noted by Pakull (1994 - private communication) and Cowley et al. (1996), and interpreted as high velocity components, suggesting some form of bipolar outflow. The velocities of the S$^+$/S$^-$ components are of order 3800 km s^{-1}, consistent with the measurements of C96.

2.3 Binary Parameters

We folded our radial velocity data obtained from the He II 4686 emission line on the photometric period, P=0$\overset{d}{.}$76278 (see Sec. 3.3). A sinusoidal fit gives a velocity semi-amplitude, K= $14.5 \pm 3\,\mathrm{km\,s^{-1}}$ and a systemic velocity, $\gamma = 297 \pm 1\,\mathrm{km\,s^{-1}}$. The mass function for these parameters is $f(M) = PK^3/2\pi G = 0.0002 \pm 0.0001$ $M_\odot$. From this we conclude that, if the accretor is a white dwarf, the inclination of the system must be very low, $\lesssim 10°$ (see also C96). Furthermore, the implied mass of the companion star is such that it would not fill its Roche lobe if on the main sequence. An evolved donor star is therefore required.

3 Photometry

3.1 The MACHO Project

RX J0513.9–6951 lies in one of the fields surveyed by the MACHO project (see e.g. Alcock *et al.* 1995). Optical photometry is taken using the 1.27 m telescope at Mount Stromlo Observatory, Australia. A dichroic beamsplitter and filters provide simultaneous CCD photometry in two passbands – "red" ($\sim$ 6300-7600 Å) and "blue" ($\sim$ 4500-6300 Å), the latter being approximately equivalent to the Johnson V passband. We are thus afforded a serendipitous opportunity to study the long term behaviour of this source.

The relative "B" magnitude of RX J0513.9–6951 for the period 1992 22 August – 1995 27 November is presented in Fig. 2. Details of the instrumental set-up and data processing may be found in Alcock *et al.* (1995). The absolute calibration of the MACHO fields and transformation to standard passbands is not yet complete, thus the measurements are plotted differentially relative to the observed median. Typical errors are $\sim$ 0.02 mag.

3.2 Optical Variability

The system exhibits pronounced optical variability, with the most dramatic changes occurring on timescales of $\sim$100–200 days. The brightness typically drops by $\sim$0.8–1.0 mag in only $\sim$10 days, after which the light curve maintains a low level for $\sim$30 days. Following the initial drop, the system usually brightens by 0.3–0.4 mag in the first $\sim$8 days, and maintains this plateau level before making the rapid upward transition back to the high state. This pattern of variability is less obvious in the low state at day number $\sim$1540, in which the system appears to undergo a minor outburst, rather than exhibiting the step-like behaviour.

We note that it is probable, both from the recurrence times of the low states and from the decline of the light curve before day $\sim$1320, that a low state was missed in the period $\sim$1320–1370, where there is a gap in the monitoring. Indeed, spectroscopic observations obtained in December 1992 (Reinsch, these proceedings) strongly suggest that RX J0513.9–6951 was in an optical low state during this time. The optical variability is discussed in greater detail in Alcock *et al.* (1996).

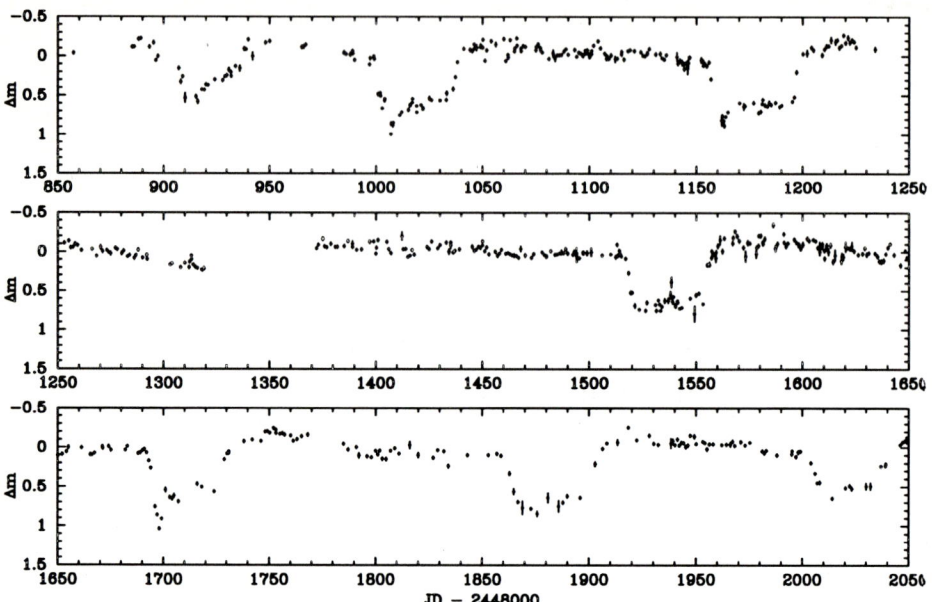

Fig. 2. The MACHO "B" light curve of RX J0513.9−6951, showing quasi-regular magnitude drops of ∼ 1 mag. The day number is JD−2448000.

3.3 The Photometric Period

We searched the high state sections for periodicities by detrending the data and performing a Lomb-Scargle analysis. The resulting periodogram is shown in Fig. 3. We see a dominant peak at P=$0.^{d}76278 \pm 0.^{d}00005$, with lesser power at $3.^{d}22$ and $1.^{d}45$ (the former value is almost certainly a one-day alias). We checked the significance of the peaks by analysing the power spectra of randomly generated datasets, using the sampling intervals of the real data. Our Monte Carlo simulations reveal that the $1.^{d}45$ peak is only significant at the ∼ 1.6σ level. However, the power at P=$0.^{d}76278 \pm 0.^{d}00005$ corresponds to a 5σ detection, leading us to present this as the true orbital period. Independent spectroscopic studies (C96) are consistent with this result, having revealed a candidate period of P≈$0.^{d}76$.

3.4 Orbital Modulation

The detrended high state data were folded on P=$0.^{d}76278$ to examine the form of the orbital modulation. We find that the data are well fitted by a sinusoid of semi-amplitude $0.^{m}0213 \pm 0.^{m}0009$. This is consistent with a low inclination, since the heated face of the donor star would have been expected to produce a larger modulation at higher inclination angles. The phase-averaged, folded light curve is shown in Fig. 4. We derive an ephemeris of T_o = JD 2448857.832(5)+0.76278(5)E, where T_o is the time of maximum optical brightness, and E is an integer.

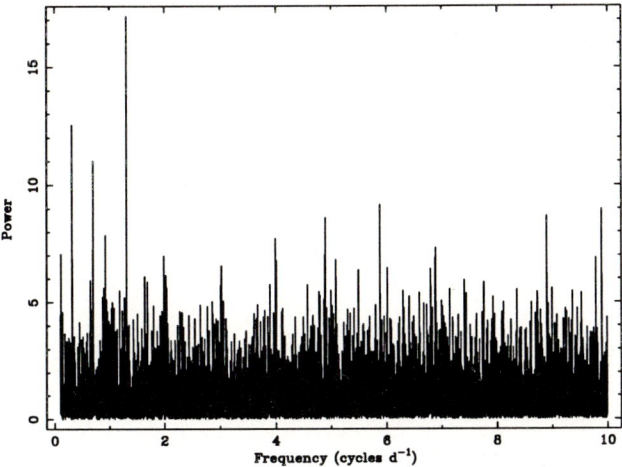

Fig. 3. The Lomb-Scargle periodogram of RX J0513.9–6951 MACHO time series data, excluding the low states. We search a frequency space of $0.1 - 10$ cycles d^{-1} with a resolution of 0.001 cycles d^{-1}. The strongest peak is at $0\overset{d}{.}76278 \pm 0\overset{d}{.}00005$ ($\equiv 1.311$ cycles d^{-1}).

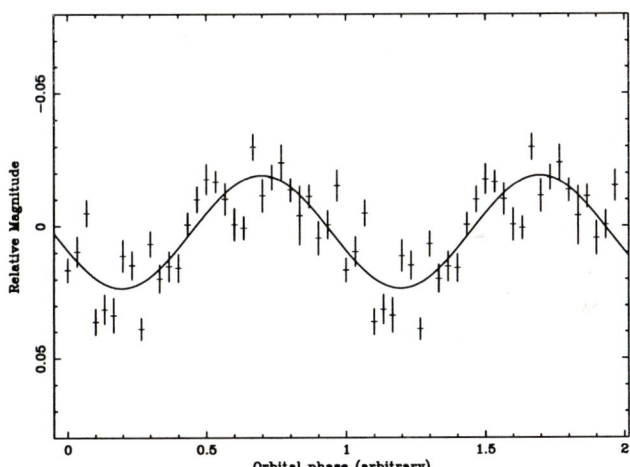

Fig. 4. The light curve of RX J0513.9–6951, folded on a period of $0\overset{d}{.}76278$. The data have been averaged into 30 phase bins, and are fitted with a sinusoid of semi-amplitude 0.0213 mags. Two cycles are plotted for clarity.

4 The Nature of the Compact Object

The observations presented in Sec. 2.2 (see also C96) indicate that RX J0513.9–6951 exhibits bipolar outflows with $V_{\mathrm{bipolar}} \sim 3800\,\mathrm{km\,s^{-1}}$. Since "jet" velocities are always found to be of the order of the escape velocity from the central object, the observed value of V_{bipolar} corresponds to a value of the mass to radius ratio of the compact object of $M/R \sim 40~M_\odot/R_\odot$, which is typical for a white dwarf. Indeed, outflows with velocities of this order have been observed in cataclysmic variables (e.g. Drew et al. 1991).

The observed bolometric luminosity is also consistent with hydrogen shell burning on a white dwarf (see Southwell et al. 1996). We therefore conclude that the compact object in RX J0513.9–6951 is a white dwarf.

5 The Accretion Rate

The accretion rate in this system is probably extremely high, with much of the optical luminosity being generated in the accretion disk. The luminosity of a standard accretion disk is given approximately by (Webbink et al. 1987):

$$M_V^{\text{disk}} \simeq -9.48 - \frac{5}{3} \log \left(\frac{M_{\text{WD}}}{M_\odot} \frac{\dot{M}}{M_\odot \text{yr}^{-1}} \right) - \frac{5}{2} \log (2 \cos i), \qquad (1)$$

where $\dot{M}$ is the accretion rate and i is the inclination angle. If the disk were to generate the entire observed luminosity, then using $M_V = -2$ (e.g. Pakull et al. 1993), for $M_{\text{WD}} = 1 M_\odot$ and $i \simeq 10°$ (see Sec. 2.2), we would have obtained $\dot{M} \simeq 10^{-5} M_\odot~\text{yr}^{-1}$. This accretion rate is of the order of the Eddington value, and it therefore indicates that at least some fraction of the optical light is due to illumination of the disk/secondary (and perhaps nuclear burning on the white dwarf surface).

Another indication that the accretion rate in RX J0513.9–6951 is probably higher than in other similar systems (e.g. CAL 83) comes from the observation of the bipolar outflow (since, for example, jets seem to be associated with episodes of increased accretion rate through the disk in young stellar objects – Reipurth & Heathcote 1993).

6 The Cause of the X-Ray Outbursts

When the X-ray data are examined *together* with the optical data, we find that the X-ray outbursts (for which there were simultaneous optical observations in Jul 1993 and Dec 1994) occurred during optical minima (see Fig. 2), while *no* outbursts were observed during optical high states (see also Pakull et al. 1993). If this represents the rule (rather than being an accident), then it is very difficult to reconcile with a regular thermonuclear flash model which is normally accompanied by radius expansion and an increased optical luminosity. An examination of several possible models for the rise in X-rays led Southwell et al. (1996) to conclude that a contraction of the white dwarf from an expanded state to a steady shell burning phase, as a cause for the X-ray outburst, is consistent with all the available data (see also Pakull et al. 1993). It is important to note that the appearance of a relatively short lived X-ray phase due to contraction of the envelope, during shell burning, has been established observationally for both GQ Mus (Ögelman et al. 1993) and V1974 Cyg (Krautter et al. 1996).

7 The Cause for the Drops in the Accretion Rate

The white dwarf photosphere could contract due to a decrease in the accretion rate (e.g. Kovetz & Prialnik 1994), which may also explain the optical low states. The observed drops in the optical luminosity, by ~ 0.8 mag (Sec. 3.2), if interpreted as a reduction in the accretion rate, correspond to a decrease in $\dot{M}$ by a factor ~ 3. We should note, however, that if the optical luminosity is actually dominated by reprocessed radiation from the accretion disk and/or the secondary star, both being irradiated by the steady burning white dwarf, then the decrease in $\dot{M}$ could be by a larger factor.

The observed drops in optical luminosity in RX J0513.9-6951 resemble a phenomenon exhibited by VY Scl stars and some nova-like variables (e.g. Shafter 1992). The important point here is that RX J0513.9-6951, like the VY Scl stars, experiences only *downward* transitions. Livio & Pringle (1994) suggested a model for VY Scl stars, in which the reduced mass transfer rate is a consequence of a magnetic spot covering the L_1 region; a similar mechanism could be operating in RX J0513.9-6951 (see Southwell et al. 1996 for a detailed discussion).

8 Conclusions

We propose a model consisting of a white dwarf which is probably fairly massive, both because of constraints imposed by the inclination angle (see Sec. 2.3) and by the short X-ray turn-on timescale (see Southwell et al. 1996), which accretes from an *evolved* companion.

The accretion rate is normally very high (perhaps $\sim 10^{-6}$ $M_\odot$ yr^{-1} - e.g. Nomoto 1982). Under these conditions, the white dwarf is slightly inflated and most of the shell luminosity is probably emitted in the UV. Once the accretion rate drops (see Sec. 7), the photosphere contracts slightly (e.g. Kato 1985), raising the effective temperature and thus producing an increase in the X-ray luminosity, which *follows* an optical drop.

It is also possible that the reduction of the accretion rate will actually result in a short thermonuclear flash (e.g. Kovetz & Prialnik 1994). This is because of the fact that steady burning occurs in a rather narrow strip of accretion rates (e.g. Nomoto 1982). The increased X-ray flux irradiates the companion star and, either by inflating material above the secondary's photosphere and causing it to be transferred (e.g. Ritter 1988), or by heating the magnetic spot area (which is generally cooler, e.g. Parker 1979), causes the mass transfer rate to increase again. This produces the step-like behaviour of the optical light curve when the luminosity increases, or the small jump in the optical luminosity in the middle of the low state (around day 1539, Fig. 2). The bipolar outflow also probably becomes stronger in the high mass transfer rate phases (Reinsch, these proceedings).

Assuming that the main ingredients of the model outlined above are correct, we may ask what is the cause for the difference in the X-ray behaviour of RX J0513.9–6951 and similar sources (such as CAL 83). The main difference is probably in the mean mass transfer rate, which is higher for RX J0513.9–6951. Evidence that this is indeed the case is provided both by the brightness of the accretion disk, and by the fact that a bipolar outflow is not observed in CAL 83 (C96).

We note that our model can be tested directly by long-term monitoring of the system in the optical and X-ray regimes. In particular, the model predicts that increases in the X-ray luminosity should *follow* drops (by ~ 1 mag) in the optical luminosity. If this is *not* generally observed, the case for the model in which the rises in X-rays are a consequence of a thermonuclear flash (following the accumulation of a critical mass) would be strengthened.

Acknowledgement: We thank the staff at the AAT and Mt. Stromlo Observatory. KAS acknowledges the hospitality of ST ScI, and support from PPARC. ML acknowledges support from NASA Grants NAGW 2678 and GO-4377.

References

Alcock C., Allsman R.A., Alves D. *et al.* 1995, PRL 74, 2867
Alcock C., Allsman R.A., Alves D. *et al.* 1996, MNRAS (subm.)
Cowley A.P., Schmidtke P.C., Hutchings J.B. *et al.* 1993, ApJ 418, L63
Cowley A.P., Schmidtke P.C., Crampton D. *et al.* 1996, IAU Symp. 165: Compact Stars in Binaries, eds. E.P.J. van den Heuvel, van Paradijs (Dordrecht: Kluwer), p. 439
Crampton D., Hutchings J.B., Cowley A.P. *et al.* 1996, ApJ 456, 320 (C96)
Drew J.E., 1991, MNRAS 250, 144
Kato M. 1985, PASJ 37, 19
Kovetz A., Prialnik D., 1994, ApJ 424, 319
Krautter J., Ögelman H., Starrfield S. *et al.* 1996, ApJ 456, 788
Livio M., Pringle J. E., 1994, ApJ 427, 956
Nomoto K., 1982, ApJ 253, 798
Ögelman H., Orio M., Krautter J. *et al.* 1993, Nature 361, 331
Pakull M.W., Motch C., Bianchi L. *et al.* 1993, A&A 278, L39
Parker E.N., 1979, Cosmical Magnetic Fields (Oxford: Clarendon)
Reipurth B., Heathcote S., 1993, in Astrophysical Jets, eds. D. Burgarella, M. Livio, & C. O'Dea (Cambridge: Cambridge University Press), p. 35
Ritter H., 1988, A&A 202, 93
Shafter A.W., 1992, ApJ 394, 268
Schaeidt S., Hasinger G., Trümper J., 1993, A&A 270, L9
Southwell K.A., Livio M., Charles P.A. *et al.* 1996, ApJ (subm.)
Webbink R.F., Livio M., Truran J.W. *et al.* 1987, ApJ 314, 653

Optical and X-Ray Variability of Supersoft X-Ray Sources

K. Reinsch[1], A. van Teeseling[1], K. Beuermann[1,2], H.-C. Thomas[3]

[1] Universitäts-Sternwarte Göttingen, Geismarlandstr. 11, 37083 Göttingen, Germany
[2] Max–Planck-Institut für Extraterrestrische Physik, 85740 Garching, Germany
[3] Max–Planck-Institut für Astrophysik, Postfach 1603, 85740 Garching, Germany

Abstract. We report new results of ROSAT observations and our optical multi-colour monitoring campaign of supersoft sources in the LMC. During this campaign, we observed an optical low-state of RX J0513.9–6951, lasting ~ 40 days. The decrease of the optical flux, $\Delta V \sim 1$ mag, and the accompanying reddening, $\Delta(B - V) \sim 0.1$–0.2 mag and $\Delta(V - R) \sim 0.1$ mag, can be quantitatively described by variations in the irradiation of the accretion disk by the hot central star. In this simple model, the photospheric radius of a white dwarf with nuclear burning of accreted matter shrinks in response to a temporarily slightly reduced mass-transfer rate. The same model can account for the soft X-ray outburst found in the quasi-simultaneous monitoring with ROSAT.

We also report an improved X-ray position of the transient source RX J0550.0–7151 which excludes the suggested identification of this supersoft source with a bright foreground star. Our optical spectroscopy of all objects within the X-ray error circle is complete to $V \lesssim 19.5$ but has revealed no obvious counterpart.

1 RX J0550.0–7151

1.1 Introduction

RX J0550.0–7151, hereafter called RX0550, is contained in our flux-limited sample of bright, soft, high-galactic latitude sources derived from the ROSAT all-sky survey (RASS) (Beuermann et al., in prep.). The source resides in an obscured part of the LMC and is heavily absorbed. In the RASS, RX0550 has been detected with a countrate of 0.85 PSPC cts/s. The system has also been serendipitously detected with similar countrates in two pointed PSPC observations of CAL 87 (Cowley et al. 1993). During our later pointings with ROSAT (PSPC: Sep. 1993, HRI: May 1994) the system was in an "off state" and undetectable.

1.2 X-Ray Position and Spectrum

The determination of an accurate X-ray position of RX0550 and, hence, its optical identification in the dense LMC field is hampered by the transient nature of the source. All pointed ROSAT observations of the field either contain RX0550 far off-axis or in off state. The only exception was the RASS during which xxx photons from RX0550 have been detected. From a Gaussian fit to the distribution of the photon positions we derive the following position for the supersoft

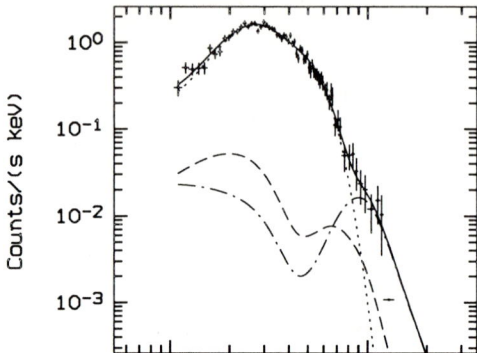

Fig. 1. Blackbody fit to the spectrum of RX0550. The blackbody component has $kT_{bb} = 23 - 30\,\text{eV}$. An unconstrained fit yields $kT_{bb} = 25\,\text{eV}$. The hard component in the spectrum is well described by a two-component spectral model derived for source B from the off-state observation of RX0550.

source: R.A. = $05^h50^m00\overset{s}{.}2$, Decl. = $-71°52'09''$, J2000.0 with a 90 % error radius of $8\overset{''}{.}3$. In our off-state observations of RX0550 we find another source with 0.036±0.003 PSPC cts/s and 0.008±0.002 HRI cts/s, respectively, in the vicinity of this position. The HRI position of this source, however, is $\sim 2.9'$ away from the RASS source and we conclude that this object, henceforth source B, does not coincide with the supersoft source. Our finding that we see two different objects near the position of RX0550 is supported by the different hardness ratios HR1 = -0.31 ± 0.03 of RX0550 and HR1 = $+0.59\pm0.06$ of source B.

From our off-state PSPC observation we have determined the spectral parameters of source B. Its 0.1–2.4 keV spectrum is well described by a two-component optically thin thermal plasma, typical for a stellar corona. In the on-state (off-axis) PSPC observation of RX0550, source B is blended by the ~ 20 times stronger flux from RX0550 and becomes detectable as a separate source only above $\sim 1\,\text{keV}$. The contribution of source B perfectly explains the high-energy component in the on-state spectrum of RX0550. Taking source B as a constant contribution to the X-ray spectrum, the 0.1–2.4 keV flux of RX0550 is well fit by a blackbody with $kT \simeq 25\,\text{eV}$ and $N_H = 2.6\,10^{21}\,\text{cm}^{-2}$ (see Fig. 1).

1.3 Optical Identification

From our analysis of the X-ray data, we can exclude the identification of RX0550 with the $V \sim 13.8$ GSC star at R.A. = $05^h49^m46\overset{s}{.}4$, Decl. = $-71°49'35''$, J2000.0 claimed by Schmidtke & Cowley (1995). The latter star, however, coincides with our X-ray source B. Charles & Southwell (1996) reported that this star shows strong Balmer emission superposed on the spectrum of a cool star. This finding is consistent with the coronal-like X-ray emission and our own optical spectrum showing that source B is a dMe-type star.

The 8″.3 X-ray error radius of RX0550 is sufficiently small to exclude any coincidence with a foreground star (e.g. white dwarf), but still not good enough to allow an identification in the dense field of LMC stars (Fig. 2). We have obtained optical spectra for all stars within the error circle (down to $V \lesssim 19.5$) and for brighter ($V \lesssim 18$) nearby stars. None has displayed clear evidence for He II 4686 or Balmer line emission as expected for a supersoft source of the CAL-83 type. Most stars are clearly reddened ($A_V \sim 1$) and there is no particularly blue candidate in the field down to $V \sim 22$. The temporarily observed near-Eddington luminosity and the strong X-ray variability suggest that RX0550 is an accreting binary. It is, therefore, possible that the system currently is in a low accretion state and we see only the companion star in the optical (which may even be one of our candidate stars).

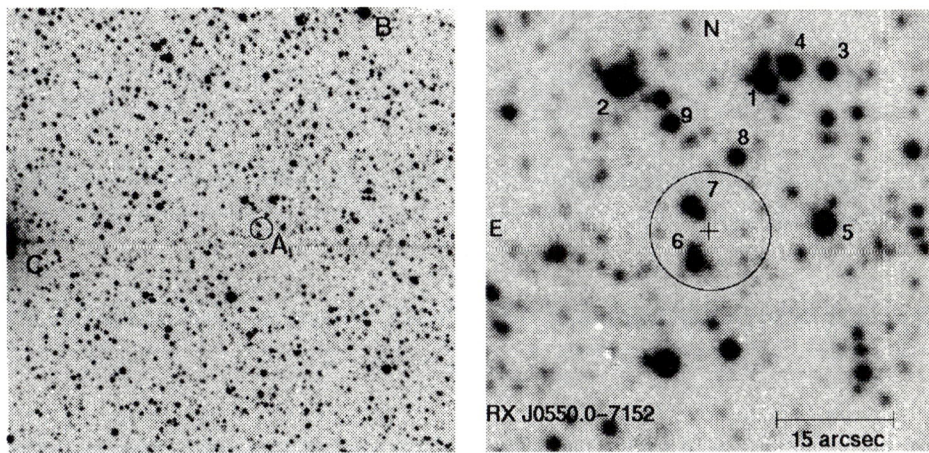

Fig. 2. V-filter CCD image (left) centered on the X-ray position of the supersoft source RX0550. The field of view is about $5' \times 5'$. The RASS position of RX0550 is given by the 8″.3 error circle and labelled with "A". The bright foreground objects "B" and "C" have been detected as faint X-ray sources in the off-state observations of RX0550. The right image is an enlarged version of the central part of the CCD image. Optical spectra have been taken from the possible counterparts labelled "1" to "9"

2 RX J0513.9−6951

2.1 Introduction

RX J0513.9−6951, hereafter called RX0513, has been discovered in the RASS as a transient luminous soft X-ray source (Schaeidt et al. 1993) and has been optically identified with a $B \sim 16.5$ mag blue star in the LMC showing a highly excited emission-line spectrum (Pakull et al. 1993). From historic plates (Leavitt 1908), this object has been known as a variable star, HV 5682. Optical follow-up observations suggest that RX0513 is a high-mass-transfer accreting binary system (Cowley et al. 1993; Pakull et al. 1993) with a probable orbital period of 0.76 days (Crampton et al. 1996). Here, we summarize the results of our systematic optical multi-colour photometric monitoring of RX0513 and consider a simple model which can quantitatively account for the observed colour variation and for the variations of the optical and soft X-ray fluxes (Reinsch et al. 1996).

2.2 Optical Multi-Colour Monitoring

Between October 1994 and March 1995 we have optically monitored RX0513 in three filter bands (B, V, and R) at about weekly intervals with the Dutch 0.9 m telescope at ESO, La Silla. In addition, we have used 92 V-filter CCD direct images of RX0513 which have been obtained with EFOSC2 at the ESO/MPI 2.2 m telescope at La Silla as a by-product of five spectroscopic observing runs between December 1992 and February 1994. Details of the data reduction and analysis are given in Reinsch et al. (1996). The light curves and colour variations of RX0513 are shown in Figs. 3 and 4. The error bars of the V magnitudes and of the $B-V$ and $V-R$ colours correspond to the statistical errors of the differential photometry. The uncertainties in the absolute calibrations are somewhat larger (~ 0.1 mag).

During our monitoring campaign (Fig. 3), RX0513 varied less than 0.3 mag except for one steep decline. In December 1994, RX0513 faded by 0.9 mag within ~ 10 days and slowly returned to its high state within ~ 30 days. The optical low state was accompanied by a colour change, $\Delta(B-V) \sim 0.1$–0.2 mag and $\Delta(V-R) \sim 0.1$ mag. Immediately after the low state, the optical flux remained at a ~ 0.2 mag higher level compared to the pre-minimum brightness for another ~ 50 days. The optical low state was accompanied by an X-ray outburst which has been detected in the quasi-simultaneous monitoring of RX0513 with the ROSAT HRI (Schaeidt et al. 1996). Due to a gap in the X-ray data it is, unfortunately, not clear whether the X-ray increase follows the optical decline or occurs simultaneously.

In December 1993, we have partly covered an event with similiar characteristics (Fig. 4). During five nights of our spectroscopic observing run, RX0513 began to fade at about the same rate as observed one year later. Again, we found that two months later the system was ~ 0.2 mag above the pre-minimum brightness. During the decline, the equivalent widths and intensities of the He II and Balmer emission lines became significantly smaller (Fig. 5). This behaviour is consistent with the reddening of RX0513. At the same time, the P-Cygni absorption features which are very prominent and extend to radial velocities of ~ 3000 km/s in the optical high state, became weaker.

2.3 Modelling of the Optical/X-Ray Variability

The time scales of the optical variability and of the X-ray turn-on of RX0513 are much shorter than the time scales of the long-term variability observed in other

Optical and X-Ray Variability of Supersoft X-Ray Sources   177

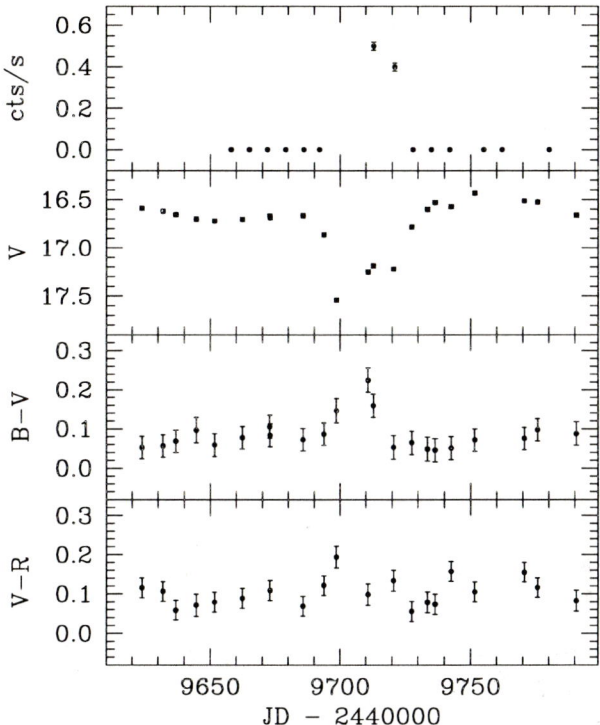

Fig. 3. V, $B-V$, and $V-R$ light curves of RX0513 obtained with the Dutch 0.9 m telescope at La Silla during our optical monitoring campaign (October 1994 to March 1995). The X-ray light curve derived from the quasi-simultaneous monitoring with the ROSAT HRI (Schaeidt et al. 1996) is shown on top.

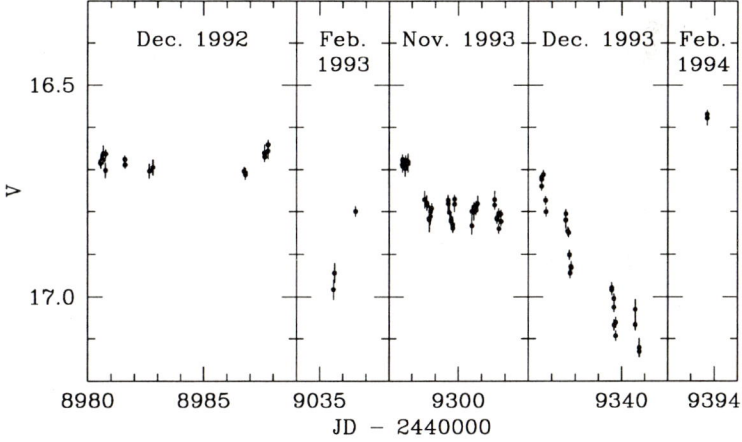

Fig. 4. V-filter CCD photometry of RX0513 obtained with the ESO/MPI 2.2 m telescope between December 1992 and February 1994

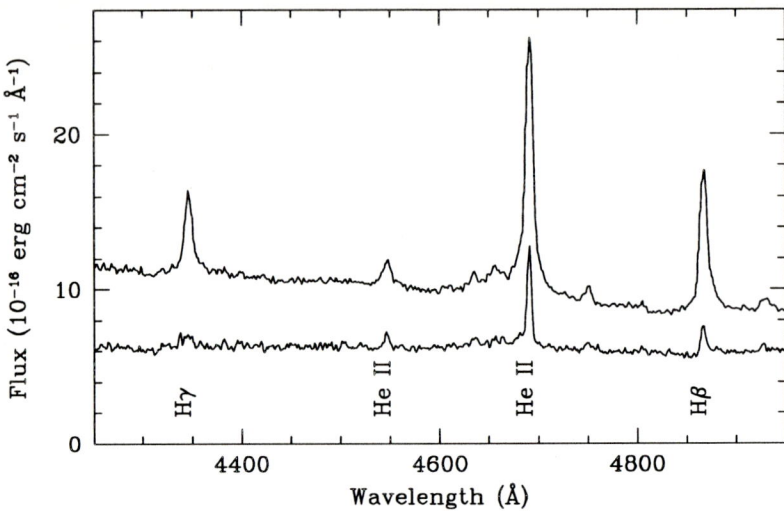

Fig. 5. Optical spectra of RX0513 obtained with the ESO/MPI 2.2 m telescope. a) high state ($V \sim 16.7$, December 1992), b) near a low state ($V \sim 17.1$, December 1993)

SSS, e.g. RX J0019.8+2156 (Beuermann et al. 1995; Greiner and Wenzel 1995) and RX J0527.8-6954 (Greiner et al. 1996). From the MACHO data, Southwell et al. (1996) derive a recurrence time scale for the optical dips $t_{\rm recur} \sim 100\text{--}200$ days. Such a short time scale is difficult to explain by recurrent hydrogen burning as proposed by Kahabka (1995) to explain the variability of other SSS.

Pakull et al. (1993) suggested that the increase in X-ray flux of RX0513 observed during the RASS was due to a temporarily slightly reduced mass transfer rate. In the optical high state, when the accretion rate approaches the Eddington rate, the photosphere of the accreting white dwarf would considerably expand (Kato 1985). With almost constant luminosity the effective temperature would drop and the accreting star would become unobservable in X-rays (cf. Heise et al. 1994). When the accretion rate becomes slightly lower, the photosphere contracts and the white dwarf becomes a supersoft X-ray source again. An important consequence of this model is that the amount of disk illumination depends on the size of the envelope of the accreting white dwarf: when its radius is small, the disk is less illuminated and fainter. Another consequence of the expansion is that the mass-loss from the white dwarf increases (Kato 1985). This explains the prominent P-Cygni absorption in the optical high state.

We have estimated whether an expanding white dwarf could explain the observed optical and X-ray variability and changing $B-V$ (for details see Reinsch et al. 1996). In our simple model, we consider a white dwarf with nuclear surface burning (van den Heuvel et al. 1992) surrounded by a flat standard disk (Shakura & Sunyaev 1973). The hot central star irradiates the disk and the secondary star. As the average angle under which the irradiating flux hits the disk is small, the illumination strongly depends on the radius of the central source.

Table 1. Estimated optical magnitudes of the irradiated disk (V_{disk}), the accreting white dwarf (V_1), the dark side of the secondary (V_{2d}), the heated side of the secondary (V_{2h}), and the total optical magnitude for the optical high and low state. Also given are the luminosity L, radius R, and temperature T of the accreting white dwarf, the ROSAT PSPC X-ray count rate, and the $B-V$ colour of the disk

optical state	PSPC (cts/s)	R (cm)	T (K)	L (erg s^{-1})	$(B-V)_{\text{disk}}$	V_{disk}	V_1	V_{2d}	V_{2h}	V_{tot}
low	2	10^9	$6.04\,10^5$	$9.5\,10^{37}$	−0.15	18.5	25	20.3	18.84	17.8
high	0	$3\,10^{10}$	$1.14\,10^5$	$1.1\,10^{38}$	−0.25	17.4	19.2	20.3	18.76	16.9

For the X-ray on-state/optical low state we have fixed the radius of the accreting star at $R = 10^9$ cm. Using LTE model atmospheres (van Teeseling et al. 1994) and the PSPC X-ray spectrum obtained in July 1993 we find an acceptable fit with $n_{\text{H}} = 7\,10^{20}$ cm^{-2}, $T = 6\,10^5$ K, and a corresponding $L = 9.5\,10^{37}$ erg s^{-1}. For the X-ray off-state/optical high state we use the luminosity core-mass relation $L \approx 6\,10^4\,(M/M_\odot - 0.5)\,L_\odot$ for the horizontal shell-burning track (Paczyński 1970) along which the photosphere of the white dwarf expands with constant luminosity (Iben 1982). Furthermore, we assume a low orbital inclination as suggested by Crampton et al. (1996) and use an LMC distance modulus of 18.5. The optical magnitude of the disk is then fully determined by the radius of the accreting star and the outer disk radius.

In Table 1 we have given the estimated optical magnitudes of the different contributing components, where we have neglected absorption in the optical. It is surprising how well the observed $\Delta V \sim 1$ and $\Delta(B-V) \sim 0.1$ are explained with the simple model of a contracting and expanding photosphere of the accreting and nuclear burning white dwarf, although the observed system appears to be somewhat brighter and redder. This may be due to neglecting the finite disk height, using blackbody spectra, and the uncertainties in the system parameters and the absolute calibration of the observed magnitudes.

The transition to the optical low state was accompanied by a decrease in the optical emission line strengths. This behaviour can easily be explained by the differences in the irradiation geometry suggested above if the lines originate from photoionization. The number of ionizing photons intercepted by either the (flat) disk, the wind, or the secondary star drops with increasing temperature causing the X-ray bright/optical low state to display line emission at a strongly reduced level.

We conclude that the simple model of a contracting and expanding photosphere of a white dwarf accreting near the Eddington rate can quantitatively explain the observed anticyclical optical/ X-ray variability. According to our calculations it is possible that the total luminosity decreases by less than only 20% during an optical low state. Our argumentation that the optical low states are caused by varying irradiation does not depend on the particular cause for the drop in the accretion rate.

Acknowledgement: We thank all observers who observed RX0513 in service mode at the Dutch 90 cm telescope. This research was supported by the DARA under grant 50 OR 92 10.

References

Beuermann K., Reinsch, K., Barwig H., et al., 1995, A&A 294, L1
Charles P.A., Southwell K.A., 1996, IAUC 6305
Cowley A.P., Schmidtke P.C., Hutchings J.B., Crampton D., McGrath T.K., 1993, ApJ 418, L63
Crampton D., Hutchings J.B., Cowley A.P., et al., 1996, ApJ 456, 320
Greiner J., Schwarz R., Hasinger G., Orio M., 1996, A&A (in press)
Greiner J., Wenzel W., 1995, A&A 294, L5
Heise J., van Teeseling A., Kahabka P., 1994, A&A 288, L45
Iben I. Jr. 1982, ApJ 259, 244
Kahabka P., 1995, A&A 304, 227
Kato M., 1985, PASJ 37, 19
Leavitt H.S., 1908, Harvard Ann. 60, 87
Paczyński B., 1970, Acta Astr. 20, 47
Pakull M.W., Motch C., Bianchi L., et al., 1993, A&A 278, L39
Reinsch K., van Teeseling A., Beuermann K., Abbott T.M.C., 1996, A&A (in press)
Schaeidt S., Hasinger G., Trümper J., 1993, A&A 270, L9
Schaeidt S., et al., 1996, this volume p. 159
Schmidtke P.C., Cowley A.P., 1995, IAUC 6278
Shakura N.I., Sunyaev R.A., 1973, A&A 24, 337
Southwell K.A., Livio M., Charles P.A., et al., 1996, this volume p. 165
van den Heuvel E.P.J., Bhattacharya D., Nomoto K., Rappaport S.A., 1992, A&A 262, 97
van Teeseling A., Heise J., Paerels F., 1994, A&A 281, 119

Part V

Supersoft Sources as SN Ia Progenitors

Type Ia Supernovae and Supersoft X-Ray Sources

M. Livio

Space Telescope Science Institute, 3700 San Martin Drive, Baltimore, MD 21218

Abstract. The question of the possible progenitors of supernovae Type Ia (SNe Ia) is examined in the light of the existence of supersoft X-ray sources. It is argued that SNe Ia are thermonuclear explosions of accreting C-O white dwarfs. The existing observational evidence favors somewhat models in which the exploding star ignites carbon upon reaching the Chandrasekhar mass. A careful examination of all the potential progenitor classes reveals that when realization frequencies are combined with a variety of observational charcteristics, no *single* class emerges as containing the obvious progenitors. It is argued that coalescing white dwarfs or supersoft X-ray sources are the most likely progenitor systems. A few critical observations which could help identify the progenitors unambiguously are discussed.

1 Characteristics and the Basic Model

The *defining* characteristics in the spectra of supernovae Type Ia (SNe Ia) are the lack of lines of hydrogen and the presence of a strong red Si II absorption feature ($\lambda 6355$ shifted to ~ 6100Å). The following are some of the observational characteristics of the class:

(1) Nearly 90% of all SNe Ia form a homogeneous class in terms of their spectra, light curves and peak absolute magnitudes. The latter are given by $M_B \simeq M_V \simeq -19.75 + 5\log(H_o/50$ km s^{-1} Mpc^{-1}), with a dispersion of $\sigma(M_B) \sim \sigma(M_V) \sim 0.2$ (*e.g.* Tammann & Sandage 1995).

(2) Near maximum light, the spectra are characterized by high velocity (8000–30000 km s^{-1}) intermediate mass elements (O-Ca). In the late, nebular phase, the spectra are dominated by forbidden lines of iron (*e.g.* Kirshner et al. 1993; Wheeler et al. 1995; Ruiz-Lapuente et al. 1995a).

(3) In terms of explosion strength, SNe Ia can be roughly ordered as: SN 1991bg and SN 1992k represent the weakest events, followed by weak events like SN 1986G, followed by about 90% of all SNe Ia which are called "normals," to the stronger than normal events like SN 1991T.

The luminosity function of SNe Ia declines very steeply on the bright side (*e.g.* Vaughan et al. 1995). Since selection effects cannot prevent the discovery of SNe which are brighter than the "normals," this means that the normal SNe Ia are essentially the brightest.

(4) Fairly young populations appear to be most efficient at producing SNe Ia (tend to be associated with spiral arms in spirals; Della Valle & Livio 1994; Bartunov et al. 1994), but an old population ($\tau \gtrsim 4 \times 10^9$ yr) *can also* produce them (SNe Ia occur in ellipticals; Turatto, Cappellaro & Benetti 1994). The

explosion strength appears to be inversely correlated with the age of the stellar population (Ruiz-Lapuente et al. 1995b).

As a consequence of the above points, and in particular the fact that SNe Ia are found also in elliptical galaxies, which shows that SNe Ia cannot be produced by core collapse of massive stars, the currently accepted model for SNe Ia is that they represent *thermonuclear disruptions of mass accreting white dwarfs* (WDs). Thus, the basic ingredient of the model for SNe Ia is the same type of object that is present in cataclysmic variables (CVs) and in supersoft X-ray sources (SSS). The question that needs to be answered is: which binary system or systems are the immediate progenitors of SNe Ia. Before I address this question, I would like to explain the importance of identifying the progenitors.

2 Why is the Identification of the Progenitors of SNe Ia Important?

Identifying the progenitor systems of SNe Ia is important for the following main reasons:

(i) In the absence of agreed upon detailed models for the explosion itself, a knowledge of the initial conditions and of the distribution of material in the vicinity of the exploding star are essential for the understanding of the explosion.

(ii) An identification of the progenitors will help (through the knowledge of the SNe Ia rate) to put constraints on models for binary systems evolution. In particular, it may help constrain the value of the common envelope efficiency parameter α_{CE} (see *e.g.* Iben & Livio 1993; Livio 1995; for reviews), which is presently very poorly known.

(iii) Galaxy evolution depends on the input from supernovae in terms of energy (radiative and kinetic) and nucleosynthetic products and on the evolution of the SN rate with time.

(iv) The use of SNe Ia to determine H_o and q_o requires a knowledge of the evolution of the rate and the luminosity function of SNe Ia with cosmic time. Both of these quantities depend on the nature of the progenitors.

3 Constraints on the Basic Model

The basic model (thermonuclear disruption of an accreting WD) for SNe Ia can be further constrained. In particular, a combination of observations and theoretical calculations can be used in attempts to determine the required composition of the accreting white dwarf and the location in the WD (*e.g.* center vs. shell) and the fuel where ignition first occurs.

3.1 The Composition of the Accreting WD

The accreting WD could be, in principle of He, C-O or O-Ne-Mg composition. However, in the case of a He WD, the composition of the ejected matter would be primarily He and ^{56}Ni, which is inconsistent with the observations (see § 1). In the case of O-Ne-Mg WDs there is no *direct* observational argument which can exclude them as potential progenitors. Theoretical calculations show, however, that the frequency of events that can be expected from such accretors is not high enough to explain the observed (rather deduced) SNe Ia rate (*e.g.* Livio 1993). Furthermore, detailed calculations show that accreting O-Ne-Mg WDs tend more to collapse quietly to form neutron stars, rather than to explode as SNe Ia (*e.g.* Gutierrez *et al.* 1995).

Consequently, it appears most likely that *the accreting WDs that produce SNe Ia are of a C-O composition.*

3.2 Where in the WD and in which Fuel Does Ignition Take Place?

There are presently two classes of models being considered, one in which *carbon* ignites at the WD *center*, when the WD reaches the *Chandrasekhar mass* and the second, in which the accreted layer of *helium* ($\lesssim 0.2 M_\odot$) on top a C-O WD, ignites *off-center* typically at *sub-Chandrasekhar* masses. In the case of the carbon ignitors, numerical simulations have shown that the energetics, light-curve and composition as a function of ejection velocity are all generally quite consistent with observations, although some difficulties remain for the standard models (*e.g.* to produce rise times to blue maximum longer than 15 days; Branch *et al.* 1985; Harkness 1991; Höflich *et al.* 1995; Höflich *et al.* 1996). The fact that these models always explode at the Chandrasekhar mass has always been regarded as a potential explanation for the homogeneity of SNe Ia. The main difficulty at present with this class of models is related to statistics. Namely, it is not yet clear whether WDs in sufficient numbers to explain the SNe Ia rate can reach the Chandrasekhar mass (hydrogen and helium flashes which lead to mass loss make this difficult). A second, related difficulty, lies in the fact that WDs that are initially more massive than $1.2 M_\odot$ (and therefore more likely, perhaps, to reach the Chandrasekhar mass), tend to quietly collapse to form neutron stars rather than to explode (Nomoto & Kondo 1991), although it is presently not clear how firm this result is.

The helium (sub-Chandrasekhar) ignitors may produce (in principle) three different outcomes, two of which are certainly not normal SNe Ia. In the case of the "direct double detonation" (DDD) (namely, one detonation wave propagating outward through the helium and the second inward, through the C-O), the whole WD is burned into ^{56}Ni and He (Nomoto 1982; Woosley, Weaver & Taam 1980). This is inconsistent with the observation of high-velocity intermediate mass elements.

The "single detonation" (SD) case (the helium detonates but the C-O core does not), in which the core may survive as a WD or disintegrate at low velocities, certainly does not correspond to normal SNe Ia (Nomoto 1982; Woosley, Taam & Weaver 1986).

The only sub-Chandrasekhar model which has some chance to produce an explosion resembling a SN Ia is the "indirect double detonation" (IDD) model.

In this scenario, one detonation propagates outward through the helium, while an inward moving pressure wave results in ignition at the center (through the compression of the core), followed by an outward moving detonation (Livne & Glasner 1990; Woosley & Weaver 1994; Livne & Arnett 1995; Höflich & Khokhlov 1996).

Models of this type have been shown to produce light curves which are in agreement with those observed in SNe Ia. The main problems currently still existing with this model are: (1) The highest velocity ejecta have the wrong composition (^{56}Ni and He; *e.g.* Livne & Arnett 1995), this is due to the fact that in these models the intermediate mass elements have a He/Ni rich layer on top. Also, Si is generally absent (*e.g.* Höflich *et al.* 1996). (2) The light curve rises somewhat faster than observed (due to ^{56}Ni heating) and the early colors are inconsistent with observations (Höflich & Khokhlov 1996), (3) One might expect this model to produce a gradual decline on the bright side of the luminosity function, in contradiction to the observed sharp decline. The last point would have been a consequence of the range of WD masses that the IDD model allows (more massive WDs produce brighter SNe). It should be noted, however, that according to Pinto (private communication), their IDDs can produce excellent fits to both the light curves and the spectra.

As a consequence of the above discussion I feel that *carbon ignitors (at the Chandrasekhar mass) are presently favored by the observations*, but that the uncertainties that still exist in the calculations of the helium ignitor IDD models suggest that this model should still be considered.

4 Candidate Progenitor Systems

We now return to the question of which binary systems could be the immediate progenitors of SNe Ia. Branch *et al.* (1995) examined this question in detail using realization frequencies of SNe Ia obtained from a variety of binaries. The calculations were performed with a population synthesis code of the type used by Yungelson *et al.* (1994, 1996). Of the many uncertainties which are involved in these calculations two are by far the most important ones in terms of their effects on the final results: the efficiency parameter in common envelope evolution, α_{CE}, and the fraction η of the accreted mass, which a WD is able to retain. The efficiency parameter α_{CE} describes the fraction of the gravitational energy change due to spiralling-in inside the common envelope which is deposited into envelope ejection. Since other sources of energy (*e.g.* recombination energy) may be involved, α_{CE} could in principle be even larger than 1 (see Iben & Livio 1993; Livio 1995 for reviews). At present, possible values of α_{CE} are not reliably constrained neither observationally nor theoretically, although the most recent theoretical calculation seems to indicate a value $\alpha_{CE} \simeq 1$ (Rasio & Livio 1996). A similar uncertainty exists in relation to the accumulation ratio η. Different numerical codes result in rather different fractions of the accreted mass which remains atop the WD following shell flashes (*e.g.* Starrfield *et al.* 1992; Prialnik

& Shara 1995). Furthermore, the accumulation ratio in a combination of hydrogen and helium flashes is even less certain (e.g. Iben & Tutukov 1994; Kato, Saio & Hachisu 1989).

Acknowledging the fact that these (and other) uncertainties exist, Branch et al. (1995) adopted the following conservative assumption. A class of objects is *not* considered a major contributor to the SNe Ia rate, if the realization frequencies for young and old populations (respectively) satisfy: $\nu(\sim 10^8 \text{ yr}) < 10^{-4} \text{ yr}^{-1}$, $\nu(\lesssim 10^{10} \text{ yr}) < 10^{-5} \text{ yr}^{-1}$. In Table 1 I give the realization frequencies for all the relevant candidate progenitors for carbon and helium ignitors as obtained by Yungelson et al. (1994, 1996). As can be seen from the table, if one insists on only *one class* of progenitors, then only WD mergers have a sufficiently high predicted frequency among the carbon ignitors. No single class can in fact produce the necessary frequencies among the helium ignitors, although symbiotic systems come close. I would like, however, to make the following comment. In some sense, the permanent supersoft X-ray sources represent what we have always been searching for in SNe Ia progenitors. Namely, WDs that burn steadily, and are therefore capable of growing in mass. We should therefore be very careful before we dismiss this class of potential progenitors on the basis of realization frequencies alone. In fact, the potential of producing the long sought for SNe Ia progenitors gives this class, in my mind, its main importance. The main families which contribute to the supersoft X-ray sources are marked in Table 1 by SSS.

I will therefore examine now in some detail these three classes of objects (merging WDs, symbiotics, supersoft X-ray sources).

WD mergers have been originally suggested as SNe Ia progenitors by Iben & Tutukov (1984) and Webbink (1984). Recent hydrodynamic calculations have shown that when the lighter of the two WDs fills its Roche lobe (the two WDs having been brought together by gravitational radiation), it is totally dissipated within a few orbital periods, to form a thick disk-type configuration (rotationally supported) around the primary (Rasio & Shapiro 1994; Benz et al. 1990). The subsequent evolution of the system is not entirely clear, since it is not known if central or off-center ignition (which may lead to an accretion induced collapse) will occur (e.g. Mochkovitch & Livio 1989, 1990; Mochkovitch et al. 1996). Two points which may give some confidence in the realization frequencies obtained for mergers are: (i) The population synthesis calculations indicate that for populations younger than $\sim 10^{8.5}$ yr, the total (combined) mass of the merging WDs is of order $2.1 M_\odot$, while it converges to the Chandrasekhar mass for older populations (Tutukov & Yungelson 1995). This may be consistent with the inverse correlation found between explosion strength and population age (Ruiz-Lapuente et al. 1995b). (ii) All the recently discovered double WD systems (Marsh 1995; Marsh, Dhillon & Duck 1995) fall into the peak of the period distribution predicted by the population synthesis calculations of Yungelson, Livio, Tutukov & Saffer (1994). On the other hand, it should be realized that some of the paths which lead to the formation of double WD systems involve *two* common envelope phases and thus they are necessarily uncertain (see e.g. Livio 1994).

Table 1. Candidate progenitor systems and their realization frequencies

Type of System	Realization Frequency for Young ($\sim 10^8$ yr) Population	Realization Frequency for Old ($\lesssim 10^{10}$ yr) Population
a. Carbon Ignitors		
Cataclysmic Variable; SSS	2×10^{-5}	—
WD accreting from Roche lobe filling subgiant or giant; SSS	10^{-5}	10^{-6}
Symbiotic System; SSS	10^{-6}	10^{-4}
Helium CV	10^{-4}	—
WD accreting helium from Roche lobe filling giant	10^{-5}	—
Merger of CO + He WDs	10^{-4}	10^{-5}
Merger of CO + CO WDs	10^{-3}	10^{-4}
b. Helium Ignitors		
WD accreting from Roche lobe filling subgiant or giant; SSS ?	—	3×10^{-4}
Symbiotic System; SSS ?	10^{-5}	10^{-3}
Helium CV	10^{-3}	—
Merger of CO + He WDs, or accretion from He WD	?	?

It is important to note that what has been considered until recently as a fatal difficulty for the merger scenario, the tentative detection of hydrogen in the supernova 1990M (Polcaro & Viotti 1991), probably poses no longer a problem, since this detection has been shown to be probably related to absorption in the parent galaxy rather than to the supernova environment (Della Valle, private communication). A difficulty that still exists is statistical in nature; among the 9 double WD systems discovered so far (Saffer, Liebert & Olszewski 1988; Bragaglia et al. 1990; Marsh, Dhillon & Duck 1995; Marsh 1995) not a single one is a SN Ia progenitor candidate, either because the orbital period is too long for the two WDs to merge in a Hubble time, or because the total mass is significantly lower than the Chandrasekhar mass. Other searches for double WD systems have given rather negative results (*e.g.* Robinson & Shafter 1987; Foss, Wade & Green 1991). While Yungelson et al. (1994) have shown that the results of any single one of these searches are still not in conflict with the possibility of producing SNe Ia at a rate compatible with observations ($\sim 3 \times 10^{-3}$ yr^{-1}), the combined negative results of all the searches starts to place uncomfortable constraints on this scenario (see also Renzini 1995).

As noted above, if helium ignitors represent the correct model (which presently appears difficult to accept), then symbiotic systems come the closest to producing the required frequency of SNe Ia. A real observational test for this scenario can come from deep radio observations performed immediately after the explosion. Due to the high wind mass loss rates observed from the giants in symbiotic systems ($\dot{M}_w \sim 10^{-6} M_\odot$ yr^{-1}), the circumstellar density in the vicinity of these systems should be high enough to produce detectable radio emission shortly after the SN explosion. Boffi & Branch (1995) showed that for $\dot{M}_w \sim 10^{-6} M_\odot$ yr^{-1}, the peak of the emission should occur about a week after the explosion and the expected peak flux is about 30 mJy (for a distance of 4 Mpc). Thus, the upper limit to the radio flux from (the rather weak) SN 1986G (Eck et al. 1995), appears to indicate that the progenitor of this supernova at least, is probably not a symbiotic system.

Finally, persistent supersoft X-ray sources, which involve a WD accreting at a rate $\gtrsim 10^{-7} M_\odot$ yr^{-1} from a subgiant companion (van den Heuvel et al. 1992; Rappaport, Di Stefano & Smith 1994; Yungelson et al. 1996) and burning hydrogen stably, include (in principle) the necessary ingredients for a system to reach the Chandrasekhar mass.

However, detailed population synthesis calculations (Yungelson et al. 1996) show that the realization frequencies that are obtained are rather low. This is a consequence of both the detailed history of the mass transfer process (steady burning occurs only in some episodes, while in others, nova outbursts lead to mass ejection) and of helium shell flashes, which lead to the loss of the accumulated helium shell. The rates at which the WDs reach the Chandrasekhar mass for the three families which are the main contributors to the supersoft X-ray sources population are: (i) in CVs, $2.2 \times 10^{-5} yr^{-1}$; in systems with subgiant donors, $0.6 \times 10^{-5} yr^{-1}$ and in systems in which a giant transfers helium via Roche lobe oberflow, $0.7 \times 10^{-5} yr^{-1}$. Thus the total rate, $3.5 \times 10^{-5} yr^{-1}$, is still considerably lower than the observed (rather deduced) one, $\sim 3 \times 10^{-3} yr^{-1}$. It should be noted, that while the rate at which IDDs (from helium ignitors) can (in principle) be obtained from supersoft sources is much closer to the observed SNe Ia rate (see Tab. 1), a detailed inspection of the evolution of these systems shows that most of the mass is accreted at rates that are too high ($\gtrsim 10^{-7} M_\odot yr^{-1}$) to produce successful IDDs (requiring a few $\times 10^{-8} M_\odot yr^{-1}$; Woosley & Weaver 1994).

Ultimately, the real test concerning the viability of supersoft sources (and symbiotic stars) as SNe Ia progenitors may rely on the presence or absence of hydrogen in emission in late time spectra. The point is, that hydrogen ablated from the subgiant donor is expected to end up on the inside of the SN Ia, expanding at low velocities, 200–2000 km s^{-1} (Chagai 1986; Livne et al. 1992). This is due to the fact that the subgiant is first engulfed by the expanding supernova envelope.

5 Conclusions

The following tentative conclusions can be drawn from the discussion presented in this paper.

(1) SNe Ia are *thermonuclear disruptions* of a *mass accreting C-O white dwarf*.
(2) It is not entirely clear yet if carbon is ignited first (in the center), when the white dwarf reaches the Chandrasekhar mass, or if helium is ignited first (off-center), at masses that are lower than the Chandrasekhar mass. Carbon ignition at the Chandrasekhar mass appears presently to be more consistent with observations.
(3) The immediate progenitors of SNe Ia are still unknown. The present "best bets" are: (i) the mergers of binary WDs, which can produce the required statistics, but maybe they lead to accretion induced collapses, rather than to SNe Ia. (ii) The persistent supersoft X-ray sources, which contain the necessary physical ingredients, but appear not to produce the required statistics. However, it is possible that some changes in our understanding of the physics of the mass transfer process (*e.g.* changes in the critical mass ratio which can still produce stable mass transfer; Hachisu, these proceedings) or of mass accumulation, will result in an increased realization frequency.

Acknowledgements: This research has been supported in part by NASA Grants NAGW-2678 and GO-4377 at the Space Telescope Science Institute.

References

Bartunov O.S., Tsvetkov D.Yu., Filimonova I.V., 1994, PASP 106, 1276
Benz W., Bowers R.L., Cameron A.G.W., Press W.H., 1990, ApJ 348, 647
Boffi F., Branch D., 1995, PASP 107, 347
Bragaglia A., Greggio L., Renzini A., D'Odorico S., 1990, ApJ 365, L13
Branch D., Doggett J.B., Nomoto K., Thielemann F.-K., 1985, ApJ 294, 619
Branch D., Livio M., Yungelson L.R., Boffi F.R., Baron E., 1995, PASP 107, 1019
Chagai N.N., 1986, Soviet Astron. 30, 563
Della Valle M., Livio M., 1994, ApJ 423, L31
Eck C., Cowan J.J., Roberts D.A., Boffi F.R., Branch D., 1995, ApJ (submitted)
Foss D., Wade R.A., Green R.F., 1991, ApJ 374, 281
Harkness R.P., 1991, in Supernovae, ed. S.E. Woosley (New York: Springer-Verlag), p. 454
Höflich P., Khokhlov A., 1996, ApJ 457, 500
Höflich P., Khokhlov A., Wheeler J.C., 1995, ApJ 444, 831
Höflich P., Khokhlov A., Wheeler J.C., Nomato K., Thielemann F.K., 1996, in NATO–ASI–Conference on Thermonuclear Supernovae (Dordrecht: Kluwer) (in press)
Iben I. Jr., Tutukov A.V., 1984, ApJS 54, 335
Iben I. Jr., Livio M., 1993, PASP 105, 1373
Iben I. Jr., Tutukov A.V., 1994, ApJ 431, 264
Kato M., Saio H., Hachisu I., 1989, ApJ 340, 509
Kirshner R.P., *et al.* 1993, ApJ 415, 589
Livne E., Arnett D., 1995, ApJ 452, 62

Livne E., Glasner A., 1990, ApJ 361, 244
Livne E., Tuchman Y., Wheeler J.C., 1992, ApJ 399, 665
Livio M., 1993, in Cataclysmic Variables and Related Physics, eds. O. Regev & G. Shaviv (Bristol: Institute of Physics Publishing), p. 57
Livio M., 1994, Mem. S. A. It. 65, 49
Livio M., 1995, in Evolutionary Processes in Binary Stars, eds. R.A.M.J. Wijers et al. (Dordrecht: Kluwer) (in press)
Marsh T.R., 1995, MNRAS 275, L1
Marsh T.R., Dhillon V.S., Duck S.R., 1995, MNRAS 275, 828
Mochkovitch R., et al. 1996 (in prep.)
Mochkovitch R., Livio M., 1989, A&A 209, 111
Mochkovitch R., Livio M., 1990, A&A 236, 378
Nomoto K., 1982, ApJ 257, 780
Nomoto K., Kondo Y., 1991, ApJ 367, L19
Polcaro V.F., Viotti R., 1991, A&A 242, L9
Prialnik D., Shara M.M., 1995, AJ 109, 1735
Rappaport S., Di Stefano, R., Smith J.D., 1994, ApJ 426, 692
Rasio F.A., Shapiro S., 1995, ApJ 438, 887
Rasio F.A., Livio M., 1996, ApJ (submitted)
Renzini A., 1995, in Supernovae, ed. R. McCroy (Cambridge: Cambridge University Press) (in press)
Robinson E.L., Shafter A.W., 1987, ApJ 322, 296
Ruiz-Lapuente P., Kirshner R.P., Phillips M.M., Challis P.M., Schmidt B.P., Filippenko A.V., Wheeler J.C., 1995a, ApJ 439, 60
Ruiz-Lapuente P., Burkert A., Canal R., 1995b, ApJ 447, L69
Saffer R.A., Liebert J., Olszewski E., 1988, ApJ 334, 947
Starrfield S., Shore S.N., Sparks W.M., Sonneborn G., Truran J.W., Politano M., 1992, ApJ 391, L71
Tammannn G.A., Sandage A., 1995, ApJ (in press)
Tutukov A.V., Yungelson L.R., 1995, in Cataclysmic Variables and Related Objects, ed. A. Bianchini, M. Della Valle, M. Orio, ASSL 205, p. 495
Turatto M., Cappellaro E., Benetti S., 1994, AJ 108, 202
van den Heuvel E.P.J., Bhattacharya D., Nomoto K., Rappaport S.A., 1992, A&A 262, 97
Vaughan T.E., Branch D., Miller D.L., Perlmutter S., 1995, ApJ 439, 558
Webbink R.F., 1984, ApJ 277, 355
Wheeler J.C., Harkness R.P., Khokhlov A.V., Höflich P., 1995, Phys. Rep. (in press)
Woosley S.E., Taam R.E., Weaver T.A., 1986, ApJ 301, 601
Woosley S.E., Weaver T.A., 1994, ApJ 423, 371
Woosley S.E., Weaver T.A., Taam R.E., 1980, in Type Ia Supernovae, ed. J.C. Wheeler (Austin: University of Texas), p. 96
Yungelson L.R., Livio M., Tutukov A.V., Saffer R.A., 1994, ApJ 420, 336
Yungelson L.R., Livio M., Truran J.W., Tutukov A.V., Fedorova A.V., 1996, ApJ (in press)

Luminous Supersoft X-Ray Sources as Progenitors of Type Ia Supernovae

R. Di Stefano

Harvard-Smithsonian Center for Astrophysics, Cambridge, MA 02138

Abstract. In some luminous supersoft X-ray sources, hydrogen accretes onto the surface of a white dwarf at rates more-or-less compatible with steady nuclear burning. The white dwarfs in these systems therefore have a good chance to grow in mass. Here we review what is known about the rate of Type Ia supernovae that may be associated with SSSs. Observable consequences of the conjecture that SSSs can be progenitors of Type Ia supernovae are also discussed.

1 Introduction

1.1 The Quest for Type Ia Supernovae and Their Progenitors

Type Ia supernovae can provide important clues about the age and evolution of the Universe. Several searches expected to significantly increase the discovery rate of Type Ia supernovae are underway (see, e.g., Leibundgut et al. 1995, and Perlmutter et al. 1995). The goal of the searches is to use these bright events to measure cosmological parameters, particularly the Hubble constant, H_0, and the deceleration parameter, q_0. The success of these programs depends upon having a good understanding of the characteristics of Type Ia supernova explosions. Of particular interest is the extent to which the maximum flux, light curve profile, and spectral characteristics are uniform among Type Ia supernovae, and the ability to quantify variations. To this end, an understanding of the progenitor systems and of variations among progenitors would be important. Yet the fundamental nature of the progenitors remains mysterious. We don't even know whether the progenitors are all of one type, or whether there may be several different types of progenitor. Livio (1996) has provided us with a comprehensive review of progenitor models. Other recent reviews include those by Wheeler (1996) and Branch et al. 1995.

1.2 Luminous Supersoft X-Ray Sources as Type Ia Progenitors

Rappaport, Di Stefano, & Smith (RDS; 1994) proposed that close-binary supersoft sources (CBSSs) might be Type Ia progenitors. They found that reliable calculations of the rate of Type Ia supernovae that might be associated with CBSSs required a much better understanding of the evolution of the systems than was available at the time. To derive a first estimate they assumed (1) a constant accretion rate, (2) conservative mass transfer, and (3) that the total mass of the white dwarf needed to grow to $1.4 M_\odot$ for a supernova to occur. If it was further assumed that the accretion rate needed to be within the range of rates compatible with steady burning throughout the evolution, the computed rate was less than a tenth of that required. On the other hand, relaxing this condition could yield rates in the requisite range. Thus, although conclusive results were not obtained, the possibility was open that CBSSs could contribute substantially to the rate of Type Ia supernovae. Yungelson et al. (YLTTF; 1996) took a somewhat different approach, and derived supernova rates compatible with the lower limits computed by RDS, as did Canal, Ruiz-Lapuente, & Burkert 1996. Although YLTTF did follow the complete evolution of some systems, neither their calculations nor those of RDS addressed the fundamental problems that prevented a first principle evolution to be carried out for many CBSSs.

Furthermore, neither investigation treated Roche-lobe-filling systems in which the donor was very evolved at the start of mass transfer. Such systems can have mass transfer rates in or near the steady nuclear burning region. Whereas for CBSSs, rates of this magnitude are driven by the thermal time-scale readjustment of the donor, when the donor is more evolved its nuclear evolution can push the mass transfer rate into the requisite region. We will refer to Roche-lobe-filling systems in which (1) $\dot{m}$ can be within the range for steady burning of hydrogen, and (2) the donor is initially more evolved than typical in CBSSs ($m_c(0) > \sim 0.2 M_\odot$), as wide-binary supersoft sources (WBSSs). The appearance of such systems will depend on the mass transfer rate, the mass ejection rate, and on the optical depth profile. They may or may not have the observational characteristics of SSSs or of symbiotics. Whatever the observational signature, WBSSs are characterized by the state of the donor, and the fact that there is an epoch, while the donor fills its Roche lobe, during which the mass transfer rate will allow for the more-or-less steady burning of hydrogen. The systems originally proposed by Whelan and Iben (1973) as Type Ia supernova progenitors, as well as those considered by Hachisu, Kato and Nomoto (HKN; 1996) are subsets of WBSSs.

Di Stefano et al. (DNLWR; 1996) reviewed some of the uncertainties faced in computing the rate of supernovae predicted by the close-binary supersoft model, and began to study the role of mass ejection. Similar work is ongoing for the wide-binary supersoft model (Di Stefano 1996a). This paper will focus on study of the close-binary supersoft sources; the paper by Di Stefano and Nelson (DN, 1996a) serves as a companion paper which includes much of the background touched on more lightly here. Before proceeding with the details of completed and ongoing work, however, it is worth taking a moment to review the context in which the work takes place.

1.3 Promise and Problems

Although white dwarfs that achieve the Chandrasekhar mass, M_C, have long been thought to be progenitors of Type Ia supernovae, viable progenitor models have not been easy to devise. This, in spite of the fact that several varieties of accreting white dwarfs, especially cataclysmic variables (CVs) and symbiotics, have been the subject of intensive research during recent decades. In CVs, for example, the donor is typically a low mass star and, because the accretion rate is low, most or all of the mass it donates can be lost in hydrodynamic events associated with episodes of nuclear burning. The problem with symbiotics is different. Even though the donor may have enough mass to contribute in order to push the white dwarf over the Chandrasekhar limit, and even though the mass accretion rate can be compatible with steady nuclear burning, the mass transfer phase is generally too short-lived for most white dwarfs to reach M_C (Kenyon et al. 1993). Recently, Yungelson et al. (1995) showed that wind-driven symbiotics, in which the donor does not fill its Roche lobe, are likely to make only a negligible contribution to the rate of Type Ia supernovae if the white dwarf needs to achieve the Chandrasekhar mass in order for an explosion to occur.

Given these difficulties, it has been suggested that accreting white dwarfs may become Type Ia supernovae even if they do not reach M_C (see, e.g., Woosley and Weaver 1994). The critical circumstance may instead be the ability to accrete in such a way as to form a helium mantle of $\sim 0.1 - 0.2 M_\odot$ around a C-O white dwarf. There has been good deal of study and discussion about these sub-Chandrasekhar progenitor models in recent years, but a consensus on the likelihood that they constitute a large fraction of the observed Type Ia supernovae has not yet emerged. However, even if it would become clear that reaching the Chandrasekhar mass is not an absolute requirement, this might not much change the rate of supernovae associated with some of the accreting white dwarf models. For example, Yungelson et al. (1995) found that, even if the accretion of as little as $0.1 M_\odot$ could lead to a supernova, symbiotics can account for at most 1/3 of the rate inferred from observations.

It was against this backdrop that luminous supersoft X-ray sources burst onto the scene. CBSSs seem, on the face of it to be perfect candidates for Type Ia supernova progenitors. A significant fraction of the donors are massive enough that they could donate sufficient mass to help their white dwarf companion achieve M_C. And the mass transfer rates can be within the range required for steady nuclear burning. Thus, the white dwarfs can genuinely increase in mass. Although the candidacy of CBSSs thus sounds promising, there are problems as well. In fact, the very features that allow the mass transfer rate to be high enough to be compatible with steady nuclear burning, the fact that the donor may be more massive and also slightly evolved, also makes the candidacy of CBSSs as Type Ia supernova progenitors somewhat problematic. This is because these same features tend to be associated with unstable mass transfer, so that many of the candidate systems risk a common envelope that would likely end the phase of steady accretion onto the white dwarf.

In this paper we will not be able to resolve the uncertainties. Instead we will attempt to clearly delineate them and the steps (both in rate computations and other tests of SSS models) that can be taken to narrow them.

2 Defining the Relevant Rates

It is important to clearly delineate the physical processes whose rates we would like to compute. The first hypothesis we would like to test is that the evolution of SSSs can lead to a rate of Chandrasekhar-mass explosions consistent with the rate of observed Type Ia supernovae. In this scenario, a C-O white dwarf accretes hydrogen from a companion in either a close-binary supersoft source (CBSS) or a wide-binary supersoft source (WBSS). The hydrogen burns to helium, but is likely to burn through to heavier elements before a helium mantle can develop. Thus, if the white dwarf started with an initial mass less than $\sim 1.2 M_\odot$, we are likely to witness a "classic" Chandrasekhar-mass Type Ia supernova explosion of a C-O white dwarf.

A second hypothesis we would like to test is that SSSs could lead to sub-Chandrasekhar-mass explosions. Presumably, this would require that a significant helium mantle would be able to develop, and may therefore be unlikely. Nevertheless we keep track of the numbers of systems in which the white dwarf accretes as much as $\sim 0.2 M_\odot$.

A third hypothesis, is that the explosions are actually triggered in CBSSs and WBSSs in which the binary evolution breaks down, and a common envelope ensues, leading to the merger of the white dwarf with the core of the donor. If the donor has a helium core at the time the common envelope commences, then the merger might lead to something like a sub-Chandrasekhar explosion. If, however, the donor has a C-O core (as could be the case for WBSSs), then the merger process could possibly produce a composite object with mass greater than or equal to M_C.

In practice, we find that events of all three types are associated with the evolution of CBSSs and WBSSs. It is the computation of the relative rate that is complicated by difficulties in computing the fraction of CBSSs and WBSSs that can survive as viable mass transfer binaries without experiencing a common envelope. It is interesting to note, however, that whatever the eventual breakdown of the relative rates, all of these types of events are predicted by the SSS models and should be observed.

3 Recent Work

3.1 Quantifying the Problems

As discussed by DN, the condition that the donor continuously fill its Roche lobe, together with the conservation of angular momentum, leads to an equation for $\dot{m}$, the mass loss rate of the donor of the following generic form.

$$\dot{m}\mathcal{D} = \mathcal{N} \qquad (1)$$

$\mathcal{D}$ has a functional dependence on β, which is itself a function $\dot{m}$. Thus, equation (1) can be viewed as a non-linear equation for $\dot{m}$. There are problems with stability when $\mathcal{D}$ passes through zero and/or is negative. In general $\mathcal{D}$ can be written as $\mathcal{A} + \beta\mathcal{B}$. If $\mathcal{D}$ is negative for all $\beta > 0$, we will say that the system is in class I; systems in class I cannot be evolved using the standard formalism. A system will be said to be in Class II if there is a value of $\beta = \beta_{crit}$, such that $\mathcal{D}$ is positive only for $\beta < \beta_{crit}$; the evolution of systems in class II can be started, but will fail as the rate of mass transfer increases, if β becomes equal to or exceeds β_{crit}. A system will be said to be in Class III if $\mathcal{D}$ is positive for all values of $\beta < 1$; systems in class III can be evolved from start to finish.

Using as input the systems that emerge as CBSS candidates from the population synthesis study of RDS, DNLWR found the following statistics. (1) Across a range of assumptions about the properties of primordial binaries and the value of α, the common envelope ejection factor, the rate at which CBSS candidate systems are formed in a galaxy such as our own is $\sim 0.5 - 1.0$ per century. This is just $\sim 2 - 3$ times as large as the rate of Type Ia supernovae inferred from observations. The rate at which WBSS candidates are formed is more sensitive to input assumptions about α, but can be comparable to the CBSS formation rate. (2) Across the same range of assumptions, we found that between $45 - 72\%$ of all CBSS systems were in class I and therefore could not be evolved. Between $10 - 20\%$ of all systems were in class II; their evolution crashed sometime after beginning, generally as the system approached the steady nuclear burning region. Between $17 - 36\%$ of all systems were in class III and could therefore be fully evolved. The story these statistics tell is somewhat more damning than may be obvious at first, since the systems in class III typically either have a mass ratio, $q = m/M$ (where m is the mass of the donor and M is the mass of the white dwarf), that is small (i.e., not much greater than unity), or else contain donors that are not very evolved. The associated mass transfer rates therefore tend to be small; the system does not spend much time in the steady nuclear burning region, and the white dwarf does not grow significantly. Thus, even though (and in some sense because) systems in class III can be followed, they tend not to be good candidates even for sub-Chandrasekhar Type Ia supernovae. Table 1 illustrates the range of results we obtained.

3.2 The Role of Mass Ejection

Table 1 illustrates two important features. First, for the population synthesis study of RDS, the majority of systems cannot be evolved using the standard formalism; they would seem to be candidates for a phase of common envelope evolution and possible mergers. Second, if the retention factor, β can be small–i.e., if the white dwarf can eject incident material it cannot burn, then a large enough fraction of systems may survive as viable binaries, to allow CBSSs to account for a significant fraction of either Chandrasekhar-mass or sub-Chandrasekhar-mass explosions. This is the point illustrated by the last row of the table, in which an "optimistic" treatment was used: all systems for which $\mathcal{D}$ was less than zero, were artificially saved, until the system parameters changed enough to increase $\mathcal{D}$ above zero. This treatment is not realistic and was designed to give us an

Table 1. Classification of CBSS candidates by retention-factor

		$\Delta M \geq 0.2$				
		Class I	Class II	Class III		
1	CON	0.72	0.10	0.18	0.012	0.10
2	CON	0.67	0.12	0.21	0.009	0.02
3	CON	0.73	0.10	0.17	0.012	0.10
4	CON	0.53	0.15	0.32	0.015	0.16
5	CON	0.51	0.18	0.31	0.016	0.14
6	CON	0.45	0.20	0.36	0.018	0.18
6	OPT	0.45	0.20	0.36	0.55	0.81

Summary of the results of evolutionary calculations. "Set" refers to the data sets of CBSS candidates that emerge from each of 6 population synthesis studies we have carried out along the lines described in RDS. Note that there is relatively little variation among the results derived for different data sets. "Case" refers to the class of evolutionary "treatment" used to evolve the CBSS candidates. There are two classes of treatment, conservative (CON) and optimum (OPT). The numbers in each column represent the average fraction of systems that fall into the category indicated by the column headings. A treatment is characterized by the values of the parameters used in the evolution of the CBSS candidates. These include a_1 and a_2 (see DN), and the value of $\tilde{\xi}_{ad}$. In rows $1-6$, the average of the results for 9 separate conservative treatments is shown. In our standard conservative treatment, $\tilde{\xi}_{ad} = 4$, $a_1 = 2$, and $a_2 = 1$. Although the results for individual treatments are not shown, we note that the results among the conservative treatments are not generally dramatically different for different treatments. The exception is for $\tilde{\xi}_{ad} = 10$. This case tends to maximize the value of $\mathcal{D}$, so that all systems can be evolved; we find however, that the mass transfer rates tend to be so low that no system reaches $1.4 M_\odot$. In row 7, the results for the optimum treatment, which has been applied here only to data set 6, are shown. The evolutionary parameters are the same as those for the standard conservative treatment; when $\mathcal{D} < 0$, however, β is chosen so as to set $\mathcal{D}$ equal to $\mathcal{D}_{min}$. Note that all systems in Class I and some in Class II are candidates for mergers.

upper limit. The fact that the upper limit so-derived is in the range of observed rates, illustrates the significant role played by mass ejection in the computation of the Type Ia supernovae rates.

It has since been discovered (HKN) that there are steady state solutions in which the white dwarf can eject the matter that it cannot burn. This is a potentially important result. Together with several other steps, it should help us to better quantify the Type Ia supernovae rate associated with SSSs. The additional needed developments include the following. (1) A population synthesis study which differs from that of RDS in including the effects of winds prior to the first common envelope phase. This has already been done by YLTTF, and by DN. The results are to move some systems from class I into class II. (2) The inclusion of radiation-driven winds. This allows us to evolve some systems in class II that would otherwise fail. We find, though, that this by itself does not lead to a significant increase in the number of systems in which the white dwarf accretes $\sim 0.2 M_\odot$ or more. (3) Implementing the full non-linear solution to Equation (1). This will allow us to better determine which systems in class II can actually be saved (Di Stefano 1996a). (4) Explicitly including a common envelope phase for those systems in which the binary evolution fails. This will allow us to better quantify the number of merger events expected. (5) Completing a full population synthesis study, including evolution, for wide-binary supersoft sources.

Work along these lines is underway and should help us to narrow the uncertainties in computations of the rate of Type Ia supernovae associated with luminous supersoft X-ray sources.

4 Predictions and Tests of the Model

A promising coincidence of computed and observed rates would not be conclusive evidence that the model is the unique correct Type Ia progenitor model. What would be needed in addition are testable physical predictions that go beyond rate computations. In this section we focus on two types of test. The first is the identification of individual progenitors, and the second is *post facto* study of supernovae and their remnants for "secondary characteristics" (Branch et al. 1995) that may be related to properties of the progenitor.

4.1 Searching for Progenitors

One way to definitively identify progenitors is to observe a system before it experiences an explosion. The problem with this approach is the events are rare. To date, no Type Ia supernova progenitors have been identified. If the rate of events is ~ 0.3 per century per galaxy, we would need to have detailed prior information about ~ 30 galaxies to have a good chance of identifying a progenitor sometime in the next decade. To test the supersoft source progenitor models we therefore ask if the distinctive signatures of SSSs would help us to identify the site of a progenitor in a distant galaxy.

X-Ray Observations As part of the study of the detectability of SSSs, Di Stefano & Rappaport (1994) seeded the Magellanic Clouds and M31 with SSSs drawn from a distribution generated by using the CBSS model. Because steady nuclear burning white dwarfs of higher mass tend to be hotter and more luminous, we found that the sources most likely to be detected in M31 were those with high-mass white dwarfs. This is illustrated in Figure 1.

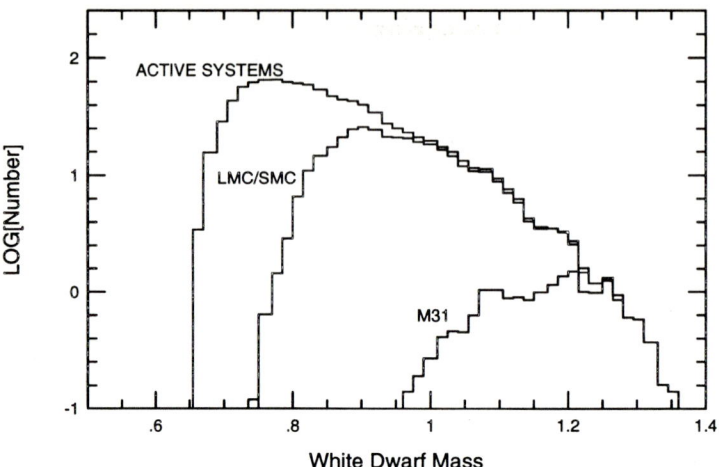

Fig. 1. The uppermost curve represents all active CBSSs as calculated by RDS. The middle (low) curve shows only systems that would likely have been detected by ROSAT in the LMC/SMC (M31).

Note that ROSAT's study of M31 should have detected evidence of all steady nuclear burners with $M > 1.2 M_\odot$ (see Greiner et al. 1996, Supper et al 1995). An important caveat is that the system should not be self-obscured, by a heavy wind, for example. This selection effect, favoring X-ray detection of systems with high-mass white dwarfs, becomes more pronounced as the distance to the host galaxy and/or absorption increases. Deep images of the most distant galaxies in which sources can be detected and resolved by X-ray satellites would therefore seem to provide potentially promising ways to identify possible progenitors. This is especially true if the Chandrasekhar-mass models are correct.

Observations of Supersoft Nebulae The radiation emitted by SSSs is highly ionizing. If the sources are housed in an ISM with a local number density, n, of more than $\sim 1 - 2$ cm^{-3}, they may be expected to exhibit an ionization nebula with high enough surface brightness to be detected, and with distinctive properties (Rappaport et al. 1994; Chiang 1996). The central source is capable of maintaining the ionization of $\mathcal{O}(100) M_\odot$, with, for example, $\sim 2 - 8\%$ of the bolometric luminosity emerging in the $\lambda 5007$ line of [O III]. We will refer to these distinctive nebulae as supersoft nebulae. CAL 83 is associated with a nebula that fits the general expectations computed for a supersoft nebula (Pakull and Motch 1989; Remillard, Rappaport and Macri 1995 [RRM]). At the detection limit of RRM, no other SSS in the Magellanic Clouds exhibits such a nebula. It is unknown what fraction (1) of the sources discovered to date, and (2) of all active SSSs, may be associated with supersoft nebulae.

Di Stefano, Paerels, and Rappaport (1995; DPR) noted that at least some supersoft nebulae should have luminosities in [O III] $\lambda 5007$ comparable to the cut-off of the planetary nebula luminosity function (PNLF) (see Fig. 2).

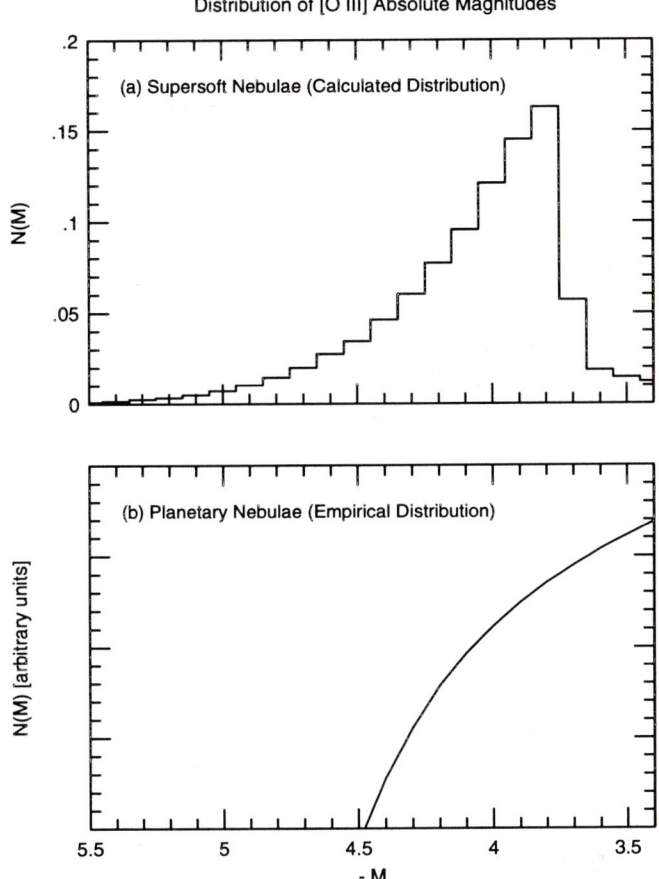

Fig. 2. The top panel shows the supersoft nebula luminosity function (SNLF) in [O III]. The normalization is not known; if, e.g., 10% of all SSSs have supersoft nebulae, then a total of $\mathcal{O}(100)$ supersoft nebulae may be expected to exist in a galaxy such as our own. The bottom panel shows the empirically-derived planetary nebula luminosity function (PNLF).

The PNLF is used to determine extragalactic distances (see, e.g., Jacoby et al. 1992, Jacoby and Ciardullo 1993). Comparison between the SNLF and the PNLF therefore indicates that, if there are significant numbers of supersoft nebulae in distant galaxies, we should be able to detect individual SSSs in galaxies at least as far from us as the Virgo cluster. Planetary nebula surveys have been and are continuing to be carried out for dozens of galaxies. Thus, there is some chance that a coincidence between the location of a nebula and the site of a later Type Ia supernova explosion could be observed during the next decade, if SSSs can be progenitors of Type Ia supernovae (Di Stefano 1996b). Features which can help to distinguish between supersoft nebulae and planetary nebulae have been considered by DPR. It is interesting to note that supersoft nebulae are more

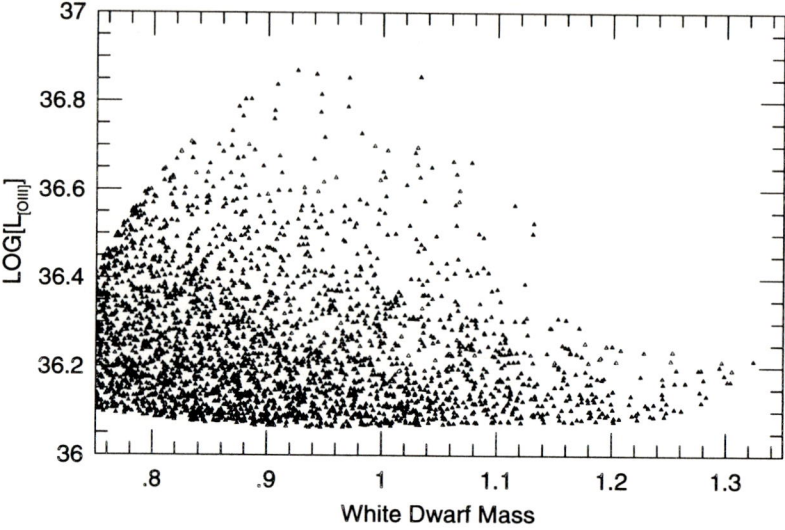

Fig. 3. The distribution of [O III] luminosities as a function of white dwarf mass. (Di Stefano 1996b)

efficient emitters in the [O III] line when the temperature of the central source is moderate, and the mass of the white dwarf is smaller than $1.2 M_\odot$. This is illustrated in Fig. 3.

Thus, while X-ray detection of SSSs in external galaxies is most likely to test and constrain Chandrasekhar-mass models, detection of supersoft nebulae in distant galaxies is most likely to test and constrain sub-Chandrasekhar-mass models.

4.2 Predicting Supernova Characteristics

Observational work that may allow us to eventually identify individual progenitors is exciting, but the returns are necessarily uncertain. On the other hand, we know that ongoing search programs for supernovae will certainly lead to the study of dozens of Type Ia supernovae during the next decade. Thus, if a set of tests to be applied to each observed explosion could be devised to assess the likelihood that the progenitor was a SSS, we might have a better chance of verifying or falsifying the hypothesis that SSSs are the progenitors of a significant fraction of Type Ia supernova explosions.

Branch et al. 1995 discussed a range of so-called "secondary characteristics" of supernovae that could be used to constrain progenitor models. For example, the amount and distribution of circumstellar matter can be checked via radio observations. (See, e.g., Boffi & Branch 1995, Eck et al. 1995.) Evolutionary calculations allow us to compute the total amount of mass ejected by each system and to follow the time history of mass ejection. In ongoing work, both for wide- and close-binary systems, we are therefore tracking mass ejection. Our calculations also allow us to compute the ionization state of any local ISM (as well as that of ejected material) prior to the explosion; post-explosion limits on these quantities are also possible to obtain (see, e.g., Kirshner, Winkler & Chevalier 1987, and Smith et al. 1991).

5 Conclusions

The possibility that supersoft sources are progenitors of Type Ia supernovae is intriguing. There are many hurdles to be gotten over, however, before we can properly assess the situation.

One small hurdle has already been passed. That is, we have established that the pool of close-binary supersoft sources and wide-binary supersoft sources is large enough that, should a substantial fraction of the systems lead to supernovae, the rate of explosions could be comparable to the rate inferred from observations. This result emerges in a straightforward way from population synthesis analyses. It is interesting that the rate at which candidate progenitors are formed is, in most of our simulations, just a few times larger than the required rate. Thus, if less than 0.1 of the candidates could become supernovae, the rate of explosions due to SSSs would constitute only a small fraction of the requisite rate.

The main hurdle, then, is to determine what fraction of the candidates are actually supernova progenitors. This is a difficult problem. Solving it requires making advances in the study of the binary evolution of systems in which a more massive and possibly quite evolved star donates mass to a white dwarf companion. It seems possible that recent and ongoing work may help us to determine the fraction of candidate systems that can survive as viable binaries in which the white dwarf accretes significant mass. Even the binaries that do experience common envelopes are interesting, and determining the rates of all possible outcomes is therefore important.

Whatever the outcome of the rate calculations, the ability to evolve individual systems allows us to compute some features of the post-explosion system related to the total amount of mass ejected or to the state of ionization. Such calculations may help us to constrain the SSS models for Type Ia supernova progenitors. Further, X-ray and nebular observations of galaxies may eventually provide complementary constraints on the progenitor models.

In summary, the status of SSSs as progenitors of Type Ia supernovae is still uncertain. But there are clear lines of investigation that should help us to narrow the uncertainties.

Acknowledgement: I would like to thank Scott Kenyon, Lorne Nelson, Kenichi Nomoto, Saul Rappaport, and J. Craig Wheeler for interesting and useful discussions. This work has been supported in part by NSF under GER-9450087.

References

Boffi F.R., Branch D., 1995, PASP 107, 347
Branch D., Livio M., Yungelson L.R., Boffi F.R., Baron E., 1995, PASP 107, 1019
Canal R., Ruiz-Lapuente P., Burkert A., 1996, ApJ 456, L101
Chiang E., 1996, MIT senior thesis
Di Stefano R., Paerels F., Rappaport S., 1995, ApJ 450, 705
Di Stefano R., Rappaport S., 1994, ApJ 437, 733
Di Stefano R., Nelson L.A., Lee W., Wood T., Rappaport S., 1996, in Thermonuclear Supernovae, Proc. of the NATO ASI workshop, eds. Canal R., Ruiz-Lapuente P. (in press)
Di Stefano R., 1996a, 1996b, (in prep.)
Di Stefano R., Nelson L.A., 1996, this volume p. 3
Eck C.R., et al., 1995, ApJ 451, 53
Greiner J., Supper R., Magnier E.A., 1996, this volume p. 75
Greiner J., Hasinger G., Kahabka P., 1991, A&A 246, L17
Greiner J., Hasinger G., Thomas H.C., 1994, A&A 281, L61
Hachisu I., Kato M., Nomoto K., 1996, this volume p. 205
Jacoby G., Ciardullo R., 1993, in Planetary Nebulae, Proc. IAU Symp. 155, eds. R. Weinberger & A. Acker (Dordrecht: Kluwer)
Jacoby G., et al. 1992, PASP 104, 599
Kahabka P., Pietsch W., Hasinger G., 1994, A&A 288, 538
Kenyon S.J., Livio M., Mikolajewska J., Tout C.A., 1993, ApJ 407, 81
Kirshner R.P., Winkler P.F., Chevalier R.A., 1987, ApJ 315, L135
Leibundgut B. et al. 1995, The Messenger 81, 19
Livio M., 1996, this volume p. 183
Pakull M.W., Motch C., 1989, in Extranuclear Activity in Galaxies, eds. E.J.A. Meurs & R.A.E. Fosbury (Garching: European Southern Observatory)
Perlmutter S., et al. 1995, ApJ 440, L41
Rappaport S., Di Stefano R., Smith J., 1994, ApJ 426, 692
Rappaport S., Chiang E., Kallman T., Malina R., 1994, ApJ 431, 237
Remillard R., Rappaport S., Macri L., 1994, ApJ 439, 646
Schaeidt S., Hasinger G., Trümper J., 1993, A&A 270, L9
Smith C.R., Kirshner R.P., Blair W.P, Winkler P.F., 1991, ApJ 375, 652
Supper R., et al., 1996, (A&A subm.)
Trümper J., Hasinger G., Aschenbach B., Bräuninger H., Briel U.G., et al. 1991, Nature 349, 579
van den Heuvel E.P.J., Bhattacharya D., Nomoto K., Rappaport S.A., 1992, A&A 262, 97
Wheeler J.C., 1996, in Evolutionary Processes in Binary Stars, Proc. of the NATO ASI Workshop (in press)
Whelan J., Iben I. Jr., 1973, ApJ 186, 1007
Woosley S.E., Weaver T.A., 1994, ApJ 423, 371
Yungelson L., Livio M., Tutukov A., Kenyon S.J. 1995, ApJ 447, 656
Yungelson L., Livio M., Truran J., Tutukov A.V., Fedorova A., 1996, ApJ (in press)

A New Model for Progenitors of Type Ia Supernovae and Its Relation to Supersoft X-Ray Sources

I. Hachisu[1], M. Kato[2], K. Nomoto[3]

[1] Dept. of Earth Science and Astronomy, College of Arts and Sciences, University of Tokyo, Tokyo 153, Japan
[2] Dept. of Astronomy, Keio University, Yokohama 223, Japan
[3] Dept. of Astronomy, University of Tokyo, Tokyo 113, Japan

Abstract. We propose a new model for progenitors of Type Ia supernovae and discuss its relation to super soft x-ray sources. The model consists of an accreting white dwarf and a lobe-filling, mass-losing low mass red giant. When the mass accretion rate exceeds a critical value, there is no static envelope solution on the white dwarf. For this case, we find a new strong wind solution, which replaces the static envelope solution. Even if the mass-losing star has a deep convective envelope, the strong wind stabilizes the mass transfer until the mass ratio of the mass-losing star to the mass-accreting white dwarf reaches 1.15, i.e., $q < 1.15$. A part of the transferred matter can be accumulated on the white dwarf at a rate which is limited to $\dot{M}_{\rm cr} = 9.0 \times 10^{-7}(M_{\rm WD}/M_\odot - 0.50)M_\odot$ yr^{-1}, and the rest is blown off by wind. The photospheric temperature is kept around $T \sim 1 \times 10^5 - 2 \times 10^5$ K during the wind phase. After the wind stops, the temperature quickly increases up to $\sim 1 \times 10^6$ K. The white dwarf steadily burns hydrogen and accretes helium so that it can grow up to $1.38 M_\odot$ and explode as a Type Ia supernova. The expected birth rate of this type of supernovae is consistent with the observed rate of Type Ia supernovae.
The hot star may not be observed during the strong wind phase due to self-absorption by wind itself. Strong winds stop when the mass transfer rate decreases below $\dot{M}_{\rm cr}$. Then, it can be observed as a supersoft x-ray source.

1 Introduction

Observations and models strongly indicate that Type Ia supernovae (SNe Ia) are thermonuclear explosions of accreting white dwarfs. Theoretically both the Chandrasekhar mass white dwarf models (Ch) and the sub-Chandrasekhar mass models (sub-Ch) have been considered (e.g., Woosley 1990; Nomoto et al. 1994; Nomoto et al. 1996; Branch et al. 1995 for recent reviews). Though these white dwarf models more or less can account for various observational aspects of SNe Ia, the exact binary evolution that leads to SNe Ia has not been identified. Different evolutionary scenarios have been proposed, neither of them positively proven yet. They include: 1) a double degenerate scenario (DD), i.e., merging of double C+O white dwarfs with a combined mass surpassing the Chandrasekhar mass limit (e.g., Iben & Tutukov 1984; Webbink 1984) and 2) a single degenerate scenario (SD), i.e., accretion of hydrogen via mass transfer from a binary companion at a relatively high rate (e.g., Nomoto 1982). Currently, the issues of *Ch* vs. *sub-Ch* and *DD* vs. *SD* are still debated (e.g., Branch et al. 1995 for a review).

Among the possible combinations, the DD/Ch scenario has not been well supported. Observationally, the search for the DD has discovered only several systems whose combined mass is less than the Chandrasekhar mass (Branch et al. 1995; Renzini 1996 for reviews). Theoretically, the DD has been suggested to lead to accretion-induced-collapse rather than SNe Ia (Nomoto & Iben 1985; Saio & Nomoto 1985).

For the SD scenario, possible observed systems may be symbiotic stars (e.g., Munari & Renzini 1992). Kenyon et al. 1993 have suggested that symbiotics are more likely to lead to the sub-Chandrasekhar mass explosion because the available mass in transfer may not be enough for white dwarfs to reach the Chandrasekhar mass. Renzini 1996 has thus concluded that the SD/sub-Ch combination scores better than the SD/Ch and DD/Ch combinations.

However, photometric and spectroscopic features of SNe Ia are better reproduced by the Ch model than the sub-Ch model (e.g., Nomoto et al. 1994; Nomoto et al. 1996; Branch et al. 1995; Höflich & Khokhlov 1996). Here we shed new light on the SD/Ch scenario and propose a new progenitor model for SNe Ia. Our scenario may also account for why the number of the observed DD is significantly less than the predictions by Iben & Tutukov 1984; Webbink 1984.

In the scenario of Iben & Tutukov 1984 and Webbink 1984, they excluded a close binary system consisting of a mass-accreting white dwarf and a lobe-filling, mass-losing red(sub)-giant, mainly because such a system suffers from unstable mass transfer when the mass ratio of the mass-accreting white dwarf to the mass-losing red giant exceeds 0.79, i.e., $q > 0.79$. However, the advent of new opacities may change all of these pictures because a strong peak in the opacity has been reported at the temperature of $\log T$ (K) ~ 5.2 (Iglesias et al. 1987; Iglesias et al. 1990; Iglesias & Rogers 1991; Iglesias & Rogers 1993; Rogers & Iglesias 1992). This peak in the new opacity is about three times larger than that of the Los Alamos opacity (Cox & Stewart 1970a; Cox & Stewart 1970b; Cox et al. 1973]). Such a large enhancement of the new opacity certainly drives a strong wind on the mass-accreting white dwarf, because the acceleration of envelope matter is directly affected by the opacity value especially when the luminosity is very close to the Eddington luminosity as seen in nova envelopes (Kato & Iben 1992; Kato & Hachisu 1994).

If it is the case, strong winds from the mass-accreting white dwarf change the stability condition up to $q < 1.15$ and are able to open a channel to a Type Ia supernova explosion.

2 Progenitor Model

We assume that the progenitor of Type Ia supernovae is a close binary system consisting initially of a C+O white dwarf with $M_{\rm WD,0} = 0.8 - 1.2 M_\odot$ and a low-mass red(sub)-giant star with $M_{\rm RG,0} = 0.8 - 1.5 M_\odot$ having a helium core of mass $M_{\rm He} = 0.2 - 0.4 M_\odot$. This assumption is consistent with the fact that SNe Ia appear everywhere in spiral galaxies and even in elliptical galaxies (e.g., Tammann 1982; Cappellaro & Turatto 1988). We have followed binary evolution of these systems by using an empirical formulae and obtained the parameter range which can produce an SN Ia.

When the low-mass companion evolves to a red(sub)-giant and fills its inner critical Roche lobe, mass transfer begins from the companion to the white dwarf. If a steady mass transfer is realized, its rate is given by

$$\frac{\dot{M}_2}{M_2} = \left(\frac{\dot{R}_2}{R_2}\right)_{\rm EV} / H(q), \qquad (1)$$

where $(\dot{R}_2/R_2)_{\rm EV}$ represents specifically the evolutionary change in the secondary radius and

$$H(q) = \left(\frac{2}{3} - \frac{0.4q^{2/3} + q^{1/3}/3(1+q^{1/3})}{0.6q^{2/3} + \ln(1+q^{1/3})}\right)(1+q) - 2(1-q), \qquad (2)$$

where q is the mass ratio, $q \equiv M_2/M_1$ (M_1 is the mass of the primary, i.e., the C+O white dwarf component, and M_2 the mass of the secondary, i.e., the low-mass star component). Here we use the empirical formula proposed by Eggleton 1983 as an effective radius of the inner critical Roche lobe. To estimate $(\dot{R}_2/R_2)_{\rm EV}$ we use the empirical formulae proposed by Webbink et al. 1983:

$$\left(\frac{\dot{R}_2}{R_2}\right)_{\rm EV} = (c_1 + 2c_2 y + 3c_3 y^2) \frac{\dot{M}_{\rm He}}{M_{\rm He}}, \qquad (3)$$

where $y \equiv \ln(M_{\rm He}/0.25 M_\odot)$ with the mass of the helium core $M_{\rm He}$. The increasing rate of the helium core mass is calculated by $\dot{M}_{\rm He} = L_2/X\varepsilon_H$, where L_2 is the luminosity of the secondary, X the hydrogen mass fraction, and $\varepsilon_H = 5.987\times 10^{18}$ ergs g^{-1} the nuclear energy release per unit mass. We have used the same parameters as Webbink et al. 1983 for Population I stars.

For a sufficiently large mass of the secondary M_2 (i.e., $q > 0.79$), however, equation (1) gives a positive value of $\dot{M}_2$. This means that the mass transfer proceeds not on an evolutionary time scale but rather on a thermal or dynamical time scale. The gas falls very rapidly onto the white dwarf and forms an extended envelope around the white dwarf (e.g., Nomoto et al. 1979; Iben 1988). This envelope expands to eventually fill the inner and then outer critical Roche lobe. It results in the formation of a common envelope, in which the two cores are spiraling in each other. It forms a very compact binary system consisting of a C+O white dwarf and a helium white dwarf and never produces an SN Ia.

However, the recent version of the opacity changes the situation. In such a large mass accretion rate, hydrogen shell burning ignites soon after the envelope mass exceeds a critical value. Then, the nuclear burning dominates the compressional heating as a source of the luminosity. The luminosity is very close to the Eddington luminosity and, therefore, the envelope structure depends strongly on

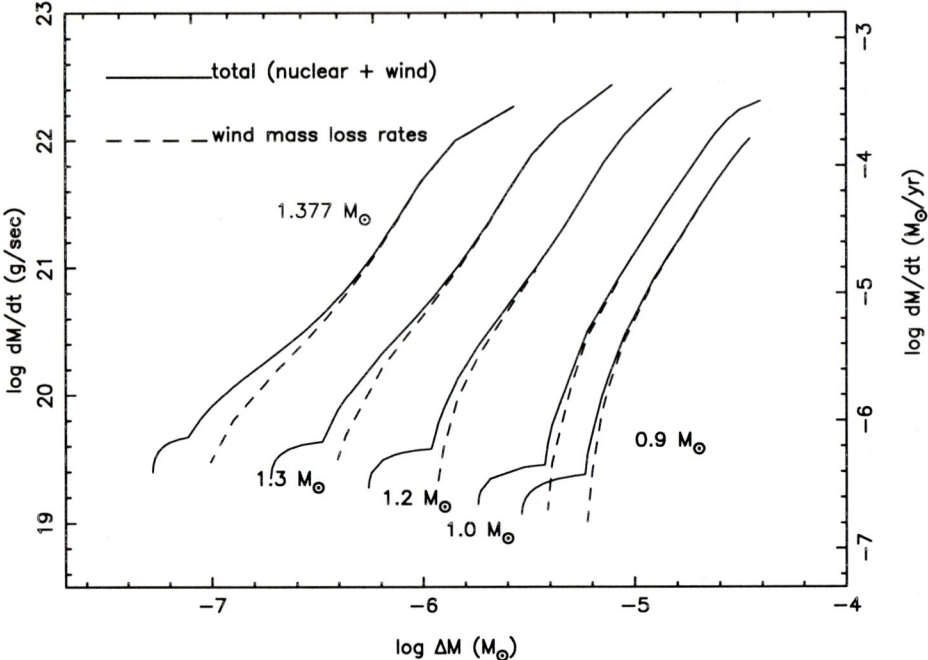

Fig. 1. Wind mass loss rates (dashed line) and total mass consumption rates, i.e., nuclear burning rates plus wind mass loss rates, are plotted against the envelope mass for white dwarfs with masses of 0.9, 1.0, 1.2, 1.3 and 1.377 $M_\odot$. The mass is attached to each line.

the opacity. The new opacity, OPAL opacity, has a very strong peak (three times larger than the old Los Alamos opacity) at $\log T$ (K) ~ 5.2 so that the luminosity easily exceeds the local Eddington luminosity if the photospheric temperature of the envelope is lower than $\sim 200,000$ K. It drives an optically thick wind. Optically thick wind is a continuum-radiation driven wind in which the acceleration occurs deep inside the photosphere. We have shown such solutions in Fig. 1. Wind mass loss rates (dashed lines) are plotted against the envelope mass. A part of the transferred matter is blown off by the wind and the rest is accreted onto the white dwarf after it burns into helium. The total mass consumption rate (nuclear burning plus wind mass loss) is also plotted in the same figure. Here, we assume steady-state and spherical symmetry to calculate structures of mass-losing envelopes and implicitly assume that the white dwarf accretes matter from the equator and blows wind from the other area.

We also plot the relation between the photospheric temperature and the total mass consumption rate in Fig. 2. If the mass transfer rate is given, the wind mass loss rate is determined from Fig. 1 and the photospheric temperature is obtained from Fig. 2. It should be noted that the photospheric temperature is higher than 100,000 K when the mass transfer rate is as small as or smaller than $\sim$

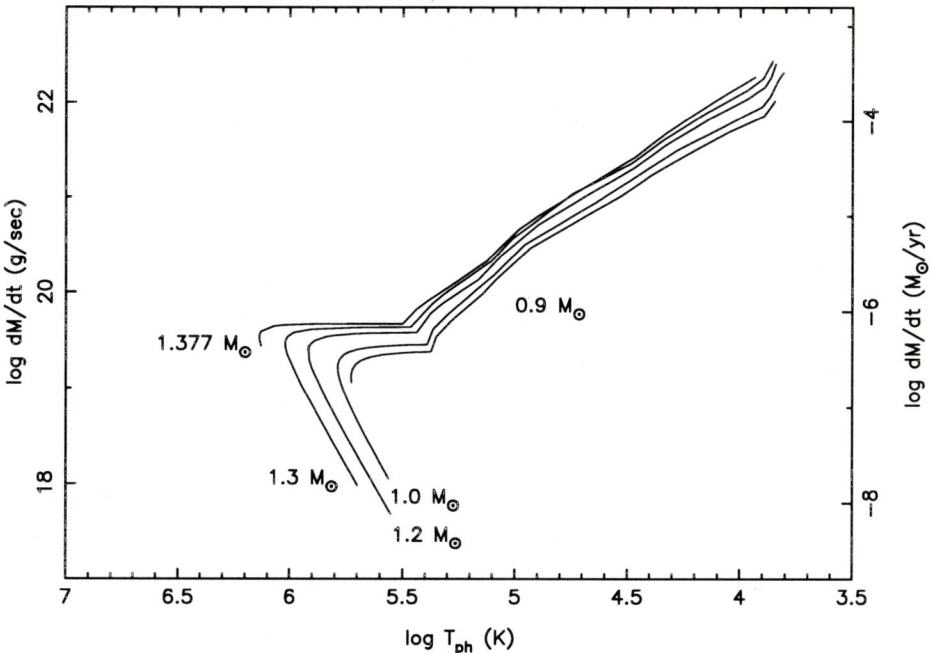

Fig. 2. Total mass consumption rate vs. photospheric temperature for white dwarfs with masses of 0.9, 1.0, 1.2, 1.3 and $1.377 M_\odot$.

$1 \times 10^{-5} M_\odot$ yr^{-1}. Even if the mass transfer rate is as large as $\sim 1 \times 10^{-4} M_\odot$ yr^{-1}, this optically thick wind is able to blow off the transferred matter. For such a high rate, the photospheric radius expands and the photospheric temperature decreases to $\sim 10,000$ K.

The wind velocity is about several hundred to one thousand km s^{-1} when the mass transfer rate is as small as or smaller than $\sim 1 \times 10^{-5} M_\odot$ yr^{-1}. For such a case, specific angular momentum of the wind is estimated as

$$\frac{\dot{J}}{\dot{M}} = \ell a^2 \Omega_{\mathrm{orb}}, \qquad (4)$$

$$\ell = \left(\frac{q}{1+q}\right)^2, \qquad (5)$$

where J and M are the total angular momentum and the total mass of the system, respectively, and a the separation, Ω_{orb} the orbital angular velocity. We assume a circular orbit. The wind carries angular momentum thereby reducing the orbital separation. In this sense, wind helps to destabilize the mass transfer. On the other hand, wind decreases the total mass of the binary so that it increases the orbital separation. This effect stabilizes the mass transfer. Which effect wins determines the stability of mass transfer. These effect of mass transfer can be expressed as

$$\frac{\dot{M_2}}{M_2} = \left(\left(\frac{\dot{R_2}}{R_2}\right)_{EV} - H_1(q)\left(\frac{\dot{M_1}}{M_1}\right)\right)/H_2(q), \tag{6}$$

$$H_1(q) = -\frac{2}{3} + \frac{0.4q^{2/3} + q^{1/3}/3(1+q^{1/3})}{0.6q^{2/3} + \ln(1+q^{1/3})} + \frac{1}{1+q} - 2 + 2\ell\frac{1+q}{q}, \tag{7}$$

$$H_2(q) = \frac{2}{3} - \frac{0.4q^{2/3} + q^{1/3}/3(1+q^{1/3})}{0.6q^{2/3} + \ln(1+q^{1/3})} + \frac{q}{1+q} - 2 + 2\ell(1+q). \tag{8}$$

Function $H_2(q)$ changes its sign at $q = 1.15$. Wind mass loss can stabilize the mass transfer from $q = 0.79$ to $q = 1.15$. It should be noticed that very small specific angular momentum of wind such as $\ell = (q/(1+q))^2 = 0.25$ for $q = 1$ is essential to stabilize the mass transfer.

The white dwarf can accrete the processed matter at a rate of

$$\dot{M}_{cr} = 9.0 \times 10^{-7}\left(\frac{M_{WD}}{M_\odot} - 0.50\right) M_\odot \text{ yr}^{-1}, \tag{9}$$

during the strong wind phase as shown in Figs. 1 and 2. When the mass transfer rate decreases below this critical value, optically thick wind stops. If the mass transfer rate decreases further below $\sim 1.0 \times 10^{-7} M_\odot$ yr^{-1}, hydrogen shell burning becomes unstable to trigger a weak shell flash. In this paper, we simply assume that the white dwarf can accrete all the transferred matter until the mass transfer rate decreases below $0.5 \times 10^{-7} M_\odot$ yr^{-1}.

The system can be specified by three parameters, i.e., the initial white dwarf mass, $M_{WD,0}$, the initial red giant mass, $M_{RG,0}$, and the mass of the helium core of the red giant at the beginning of mass transfer, $M_{RG,0}(He)$, (or, the separation of the binary, a_0). We have followed the evolution of these close binary systems for various sets of parameters. If the mass transfer rate becomes smaller than $0.5 \times 10^{-7} M_\odot$ yr^{-1} before the white dwarf reaches $1.38 M_\odot$, we regard that hydrogen shell flashes on the white dwarf blow off the accumulated matter and the white dwarf never grows any more. We simply assume that all the transferred matter can be processed and then accumulated on the white dwarf when the mass transfer rate is larger than $0.5 \times 10^{-7} M_\odot$ yr^{-1}. The example of such an evolution is plotted in Fig. 3. In this case, the white dwarf can grow up to $1.38 M_\odot$ to induce an SN Ia. The mass of hydrogen envelope on the white dwarf is $\sim 10^{-6} M_\odot$ (Fig. 1) at the explosion, which is too small to be observed. On the other hand, there still remains $\sim 0.1 M_\odot$ hydrogen in the red giant envelope and it may be observed as low velocity components of the H-alpha line. The final outcome of the evolution is summarized in Fig. 4 against these three parameters. The region to induce SN Ia is smaller for smaller $M_{RG,0}(He)$.

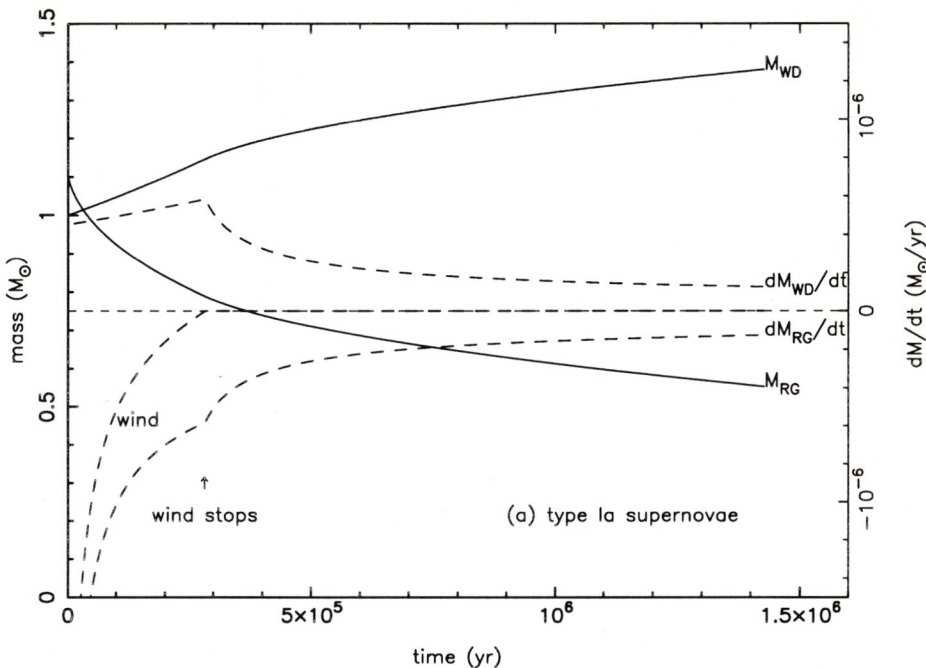

Fig. 3. Time evolution of the system. (a) The white dwarf increases its mass through $1.38 M_\odot$ and explodes as a SN Ia. Solid lines denote the masses of the white dwarf and the red giant companion, respectively. The dashed lines denote the net mass accumulation rate onto the white dwarf, the wind mass loss rate, and the mass transfer rate from the red giant component, respectively from top to bottom.

3 Discussion

The steady hydrogen shell burning converts hydrogen into helium atop the C+O core and increase the mass of the helium layer gradually. When its mass reaches a certain value, helium ignites. For the accretion rate given by equation (9), helium shell burning is unstable to grow to a flash. Once a helium shell flash occurs on relatively massive white dwarfs ($M_{WD} \gtrsim 1.2 M_\odot$), a part of the envelope mass is blown off due to wind mass loss (Kato et al. 1989); however, the ratio of the lost mass to the initial envelope mass is small (less than 10%). The total mass lost from the system would be smaller than $0.02\ M_\odot$.

The effective temperature of the accreting white dwarf is about $\sim 100,000 - 200,0000$ K during the strong wind phase. The progenitor may not be observed due to self-absorption by wind itself. Once the mass transfer rate decreases below the critical value given by equation (9), wind stops and the photospheric temperature quickly increases up to $\sim 1 \times 10^6$ K because its steep dependency on the mass transfer rate as shown in Fig. 2.

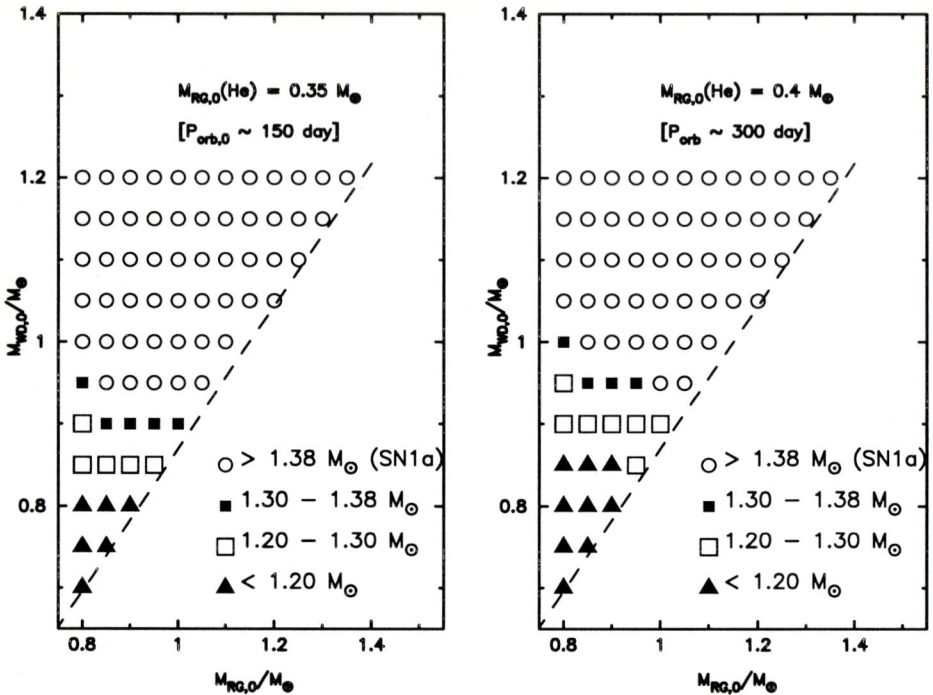

Fig. 4. Outcome of the evolution as a function of the initial mass of the white dwarf, the initial mass of the red giant, and the mass of the helium core at the beginning of mass transfer.

If we assume that $5-8\ M_\odot$ stars leave C+O white dwarfs larger than $0.9\ M_\odot$ (e.g., Weidemann & Koester 198; Iben & Tutukov 1985), the rate of SNe Ia coming through this route is close to the observed rate, i.e., $0.002\ \text{yr}^{-1}$ in our Galaxy. Here, we have estimated the rate by using equation (1) of Iben & Tutukov 1984 and substituting $\Delta \log A = 1$, $q = 0.1 - 0.3$, $M_A = 8$, and $M_B = 5$.

It has been pointed out that the number of the observed DD is significantly less than the predictions by Iben & Tutukov 1984; Webbink 1984. Our scenario suggests that the binary system can avoid the formation of a common envelope even for the unstable mass transfer (i.e., $q > 1.15$ in Case A and Case B mass transfer) if the mass transfer rate does not exceed much larger than $\sim 10^{-4} M_\odot\ \text{yr}^{-1}$. This may resolve the discrepancy between the DD prediction and the DD observation mentioned above.

Acknowledgement: This research has been supported in part by the Grant-in-Aid for Scientific Research (05242102, 05640314, 06233101) and COE research (07CE2002) of the Japanese Ministry of Education, Science and Culture.

References

Branch D., Livio M., Yungelson L.R., Boffi F.R., Baron E., 1995, PASP 107, 717
Cappellaro E., Turatto M. 1988, A&A 190, 10
Cox A.N., King D.S., Tabor J.E., 1973, ApJ 172, 423
Cox A.N., Stewart J., 1970a, ApJS 19, 243
Cox A.N., Stewart J., 1970b, ApJS 19, 261
Eggleton P.P., 1983, ApJ 268, 368
Höflich P., Khokhlov A., 1996, ApJ 457, 500
Iben I. Jr., 1988, ApJ 324, 355
Iben I. Jr., Tutukov A.V., 1984, ApJS 54, 335
Iben I. Jr., Tutukov A.V., 1985, ApJS 58, 661
Iglesias C.A., Rogers F., 1991, ApJ 371, L73
Iglesias C.A., Rogers F., 1993, ApJ 412, 752
Iglesias C.A., Rogers F.J., Wilson B.G., 1987, ApJ 322, L45
Iglesias C.A., Rogers F.J., Wilson B.G., 1990, ApJ 360, 221
Kato M., Hachisu I., 1994, ApJ 437, 802
Kato M., Iben I., 1992, ApJ 394, L47
Kato M., Saio H., Hachisu I., 1989, ApJ 340, 509
Kenyon S.J., Livio M., Mikolajewska J., Tout C.A., 1993, ApJ 407, L81
Munari U., Renzini A., 1992, ApJ 397, L87
Nomoto K., 1982, ApJ 253, 798
Nomoto K., Iben I. Jr., 1985, ApJ 297, 531
Nomoto K., Nariai K., Sugimoto D., 1979, PASJ 31, 287
Nomoto K., Yamaoka H., Shigeyama T., Kumagai S., Tsujimoto T., 1994, in Supernovae (Les Houches, Session LIV), ed. S. Bludman et al. (Elsevier Sci. Publ.), p. 199
Nomoto K., Yamaoka H., Shigeyama T., Iwamoto K., 1996, in IAU Colloquium 145, Supernovae and Supernova Remnants, ed. R. McCray & Z. Wang (Cambridge University Press), p. 49
Renzini A., 1996, in IAU Colloquium 145, Supernovae and Supernova Remnants, ed. R. McCray & Z. Wang (Cambridge University Press), p. 77
Rogers F.J., Iglesias C.A., 1992, ApJS 79, 507
Saio H., Nomoto K., 1985, A&A 150, L21
Sienkiewicz R., 1975, A&A 45, 411
Tammann G.A., 1982, Supernovae: A Survey of Current Research, eds. M.J. Rees and R.J. Stoneham (Dordrecht: Reidel), p. 371
Webbink R.F., 1984, ApJ 277, 355
Webbink R.F., Rappaport S., Savonije G.J., 1983, ApJ 270, 678
Weidemann V., Koester D., 1983, A&A 121, 77
Woosley S.E., 1990, in Supernovae, ed. A.G. Petcheck (Springer Verlag), p. 182

Transient and Recurrent Supersoft Sources as Progenitors of Type Ia Supernovae and of Accretion Induced Collapse

P. Kahabka

Astronomical Institute "Anton Pannekoek" and Center for High Energy Astrophysics, University of Amsterdam, Kruislaan 403, 1098 SJ Amsterdam, The Netherlands

Abstract. Modeling the X-ray light curves of the transient supersoft X-ray sources RX J0045.4+4154 and RX J0513.9-6951 by recurrent shell flashes massive white dwarfs ($M_\odot \sim 1.33 - 1.38\ M_\odot$) accreting at high rates ($\gtrsim 0.1 - 3.\ 10^{-7}\ M_\odot\ yr^{-1}$) are found. These are candidate progenitors of accretion induced collapse (AIC). RX J0019.8+2156 and RX J0527.8-6954 can be considered to be candidate progenitors of SNe Ia.

1 Introduction

The final fate of accreting white dwarfs (WDs) in binary systems is still a question of debate and of detailed theoretical investigations. The most recent calculations spanning a wide parameter range have been presented by Prialnik and Kovetz, 1995. These calculations clearly show that depending on the masses, mass accretion rates and the thermal history (temperature) of the WDs different final fates result: erosion, growth, destruction by an explosion or collapse. The last two events may lead to a type Ia supernova (SN Ia) and an accretion induced collapse (AIC) respectively. The specific conditions which have to apply are at least in part fulfilled in the new and distinct class of supersoft sources established after the extensive *ROSAT* observations (cf. for a review of the observations Hasinger, 1994; Kahabka & Trümper, 1996 and for detailed population synthesis calculations van den Heuvel et al., 1992; Rappaport, Di Stefano & Smith, 1994; Di Stefano & Rappaport, 1994; Yungelson et al., 1996). The aim of this article is to present the present state of knowledge about the properties of supersoft binaries being interesting as possible progenitors for SN Ia and for the AIC. Although the methods used to infer system parameters may be of concern and the parameters themselves may be highly uncertain, such an approach seems to be justified in order to initiate a first direct comparison with predictions from population synthesis calculations. The large uncertainties inherent in the numbers of existent and predicted supersoft binary systems and the still existent uncertainty in the rate of SNe Ia (and the AIC) allow only a qualitative comparison of numbers and rates. It is not the aim of this article to solve any kind of discrepancies. As the observed numbers of identified (or highly probable) binary supersoft sources per host galaxy (Milky Way, LMC, SMC, and M31) are extremely small ($\sim$ 7,5,3, and 1) such an attempt would lack any reliability.

2 Classification Scheme for Supersoft Sources

After optical identifications have been achieved for a large fraction of the supersoft sources discovered with the *ROSAT* and the *EINSTEIN* satellite it became clear they do not form a homogeneous class but comprise several source classes. In this section a classification scheme is set up and in section 3 it is investigated which sources are candidate progenitors of SNe Ia and of AIC. One classification is according to the nature of the companion star, a main-sequence star, a slightly evolved star in or slightly above the main-sequence (subgiant), or a lower mass red giant star. Another classification is in terms of the mass transfer ($\dot{M}$) properties in these binary systems. Steady-state nuclear burning (with the hydrogen being converted into helium at about the same rate as it is accreted) has been predicted to occur only within a narrow range of mass transfer rates ($\sim 1 - 4 \times 10^{-7}\ M_\odot\ yr^{-1}$) for WD masses in the range $0.7 - 1.2\ M_\odot$. Below that rate nuclear burning happens in weak flashes and at still lower rates in strong flashes or even in an explosive event (a nova outburst). Yungelson et al., 1996 proposed to term these classes CV-type, subgiant and symbiotic (donor properties). They proposed a sub-classification into permanent and recurrent sources which would relate to their mass transfer properties. Systems accreting slightly above the stability band of steady nuclear burning and undergoing cyclic (time variable) accretion (cf. Pakull et al. 1993) may constitute an additional sub-class. Class identifiers M,S,G and sub-class identifiers p,c,r are defined in Table 1.

This classification scheme has been applied to the observed supersoft sources which are binaries. From a sample of 26 classical novae only two objects have been found as supersoft sources, Nova Mus 1983 and Nova Cyg 1992 (Ögelman et al. 1993; Krautter et al. 1996). They are found below the CV period gap of $\sim$2-3 hours. Two additional supersoft sources 1E 0035.4-7230 and RX J0439.8-6809 with orbital periods in the range of $\sim$3-4 hours (Schmidtke et al., 1996a, 1996b) are supposed to have main-sequence star donors and possibly lower mass white dwarfs. These possible novae are not considered to be likely candidates for SNe Ia. Three symbiotic systems RR Tel (Jordan et al. 1994), AG Dra (Greiner et al. 1996) and SMC3 (Jordan et al. 1996) are supersoft sources. They are possibly all symbiotic novae. The nature of the donor star is not always very conclusive although it has been proposed to be a giant of spectral type G to M.

Table 1. The classification scheme of supersoft sources.

Component	Class definition	Abbrev.
Donor	Main-sequence star (in part red dwarf dK to dM)	M
	Evolved main- or post main-sequence star (subgiant)	S
	Giant (of low mass $\sim 1 - 3\ M_\odot$)	G
White dwarf	Steady-state burning sources (persistent)	p
	Cyclic accretion (passing above the stability band)	c
	Unstable burning (mild flashes) sources (recurrent)	r

Table 2. Optically identified supersoft binary systems (with possibly evolved main-sequence stars or post main-sequence stars) which are candidate progenitors of SNe Ia and of AIC.

Name	P_{orb}[d]	Ref.
CAL 87	0.442	1-3
RX J0019.8+2156	0.660	4-6
RX J0513.9–6951	0.76	7-12
CAL 83	1.04	1,13-14
RX J0925.7-4758	3.8	15,16

Ref.: (1) Long et al. 1981; (2) Pakull et al. 1988; (3) Schmidtke et al. 1993; (4) Beuermann et al. 1995; (5) Greiner & Wenzel 1995; (6) Gänsicke et al. 1996; (7) Schaeidt et al. 1993; (8) Pakull et al. 1993; (9) Cowley et al. 1993; (10) Cowley et al. 1996; (11) Crampton et al. 1996; (12) Reinsch et al. 1996; (13) Smale et al. 1988; (14) Greiner et al. 1991; (15) Motch et al 1994; (16) Motch 1996. CAL 83

3 Transient and Recurrent Sources

A large fraction of the supersoft sources varies in X-rays with time. The origin of variability on shorter time scales (hours) may be connected to the binary orbit (cf. Kahabka 1996a); the longer time scales have been proposed to be related to unstable nuclear burning at accretion rates slightly below the steady-state burning rate (Kahabka 1995) or to cyclic accretion at rates exceeding temporary the stability limit (Pakull et al. 1993). In Fig. 1 we show the *ROSAT PSPC* X-ray light curves of two transient supersoft sources. For the transient RX J0045.4+4154 in M31 two X-ray outbursts have been detected (White et al. 1995). The LMC transient RX J0513.9-6951 has been discovered in the *ROSAT* all-sky survey during an X-ray outburst and two additional outbursts have been detected during a monitoring campaign (Schaeidt et al. 1993; Schaeidt 1996; Reinsch et al. 1996).

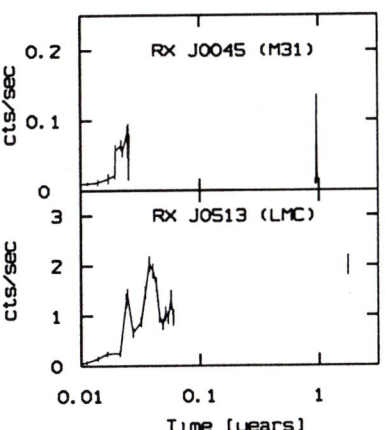

Fig. 1. *ROSAT PSPC* X-ray light curves of the cyclic accreting supersoft source RX J0513.9-6951 in the LMC and the recurrent supersoft source in M31.

3.1 Recurrent Shell Flashes

Unstable nuclear burning has e.g. been modeled in the work of Prialnik and Kovetz, 1995 using hydrodynamic calculations. They derive a gid of models covering the parameter space of supersoft sources (in terms of white dwarf mass, temperature, and mass accretion rate). The envelope masses they derive (critical masses M_{acc}) at the onset of the outburst can be represented in an analytical form by (cf. Kahabka, 1995)

$$\log M_{acc} = 3.532 + 1.654 \, (M_{WD})^{-1.581} \ln(1.4 - M_{WD}), \qquad (1)$$

for an accretion rate of $\dot{M} = 10^{-7} M_\odot \mathrm{yr}^{-1}$ and with the mass of the white dwarf M_{WD}. The observables *outburst recurrence time* t_{recur} and *x-ray on-time* t_{on} can be related to the critical envelope mass

$$M_{env}^{crit} = M_{acc} - M_{env}^{min} = \dot{M}_{stable} \times t_{on}\left(1 - \frac{t_{on}}{t_{recur}}\right) \qquad (2)$$

with the minimum envelope mass for which nuclear burning can be sustained M_{env}^{min}. In this approach large primary masses $(1.0 - 1.4 \, M_\odot)$ and high mass transfer rates $(\sim 8 \times 10^{-8} - 4 \times 10^{-7} \, M_\odot \, \mathrm{yr}^{-1})$ have been derived for the transients (cf. Table 3). In Figure 2 the distributions of white dwarf masses and mass accretion rates deduced with the approach of recurrent nuclear burning are shown. From hydrodynamic calculations Prialnik & Kovetz, 1995 derived e.g. the recurrence period of outbursts for a grid of white dwarf masses, temperatures and mass accretion rates. The result is given in Fig. 3. Applying to RX J0513.9-6951 with a recurrence period of ~100-200 days derived from the optical light curve with the *MACHO* project (Alcock et al. 1995) and a very large mass accretion rate close to $\sim 10^{-6} \, M_\odot \mathrm{yr}^{-1}$ (Southwell et al. 1996) a white dwarf mass of $\sim 1.3 \, M_\odot$ is obtained. This may mean the source experiences a weak flash during the outburst which results in an envelope expansion without ejection. A strong wind mass loss phase may follow the envelope expansion phase. After the envelope has been consumed by wind mass loss and in part by nuclear burning the hot nuclear burning white dwarf is reveiled. This pattern repeats.

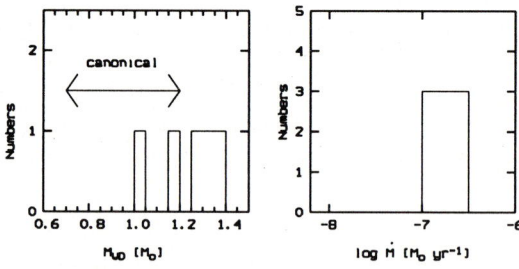

Fig. 2. Distribution of white dwarf masses and effective mass accretion rates for transient supersoft sources in the model of recurrent shell flashes (cf. Kahabka 1995). The values are taken from Tab. 3 and are in part highly uncertain.

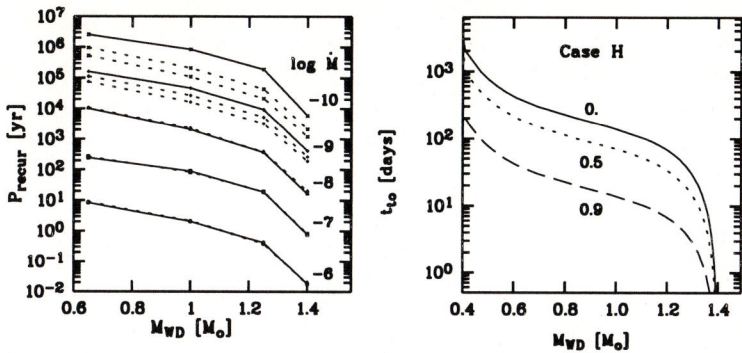

Fig. 3. Left panel: Recurrence period of outbursts (from Prialnik and Kovetz, 1995) as a function of the white dwarf mass M_{WD} and the mass accretion rate $\dot{M}$. Results are given for different temperatures of the white dwarf: 1.10^7 K (solid line), 3. and 5.10^7 K (dotted line). Right panel: X-ray turn-on time t_{to} for supersoft sources as a function of the white dwarf mass M_{WD} for case H (high mass accretion rate 10^{-7} $M_\odot$ yr^{-1}). A contracting envelope model is used. The solid, dotted and dashed curves refer to f_{loss} = 0, 0.5 and 0.9 respectively, the fraction of the envelope mass either ejected during the outburst or lost by a wind during the envelope contraction phase.

3.2 Cyclic Accretion at Rates Above the Stability Band

Another model has been proposed to account for the observed time variability on time scales of days to months, variable mass transfer onto massive white dwarfs (Pakull 1993, Southwell et al. 1996). In case cyclic (time variable) accretion is the correct model for a source (e.g. for RX J0513.9–6951), considerably larger (maximum) accretion rates are deduced than for systems accreting at rates below the stability band.

3.3 X-Ray Turn-Ons

An additional source of information are the observed X-ray turn-ons in several of the supersoft sources. These have been modeled as envelope contraction of the blown-up white dwarf following a nuclear flash (Nariai 1974, Kahabka 1996b). The observed turn-on time in X-rays t_{contr}^{obs} corresponds to a temperature change of the WD envelope from ~ 1 to 5×10^5 K and is in this approach given by Equ. 3 assuming that a fraction f_{loss} of the mass in the envelope is either ejected during the outburst or lost during the contraction phase by a wind.

$$t_{contr}^{obs} \approx 1200 \left(\frac{M_{env}(1 - f_{loss})}{10^{-4} M_\odot} \right) \left(\frac{0.01 R_\odot}{R_{core}} \right) \text{ days.} \tag{3}$$

From Fig. 3 it becomes obvious that even if $\sim$ 90% of the envelope mass is lost (ejected) X-ray turn-on (contraction) time scales of $\approx$10 days afford very massive white dwarfs ($M_{WD} \gtrsim 1.2$ $M_\odot$). For a discussion of the effects of (optically thick) winds on nova outbursts and supersoft sources we refer to Kato & Hachisu, 1994

Table 3. Parameters of transient and recurrent supersoft sources which can be considered as candidate progenitors of SNe Ia (and of AIC). The mass of the white dwarf M_{WD} [$M_\odot$], mass accretion rate $\dot{M}_7$ [10^{-7} $M_\odot$ yr^{-1}], wind velocity vw [km/sec], bolometric luminosity L[$L_\odot$] of the white dwarf, orbital period P_{orb} [d] and mass of the donor star M_d [$M_\odot$] is given. Values for the white dwarf masses enclosed in brackets are deduced from the X-ray turn-on times.

Galaxy	Source	WD				donor		Ref.
		M_{WD}	$\dot{M}_7$	vw	log(L)	P_{orb}	M_d	
LMC	RX J0513.9-6951	(1.33-1.36)		~4000	3.5-4.1	0.76		1-4
	RX J0527.8-6954	≤ 1.18	≤ 1.6		2.0-3.9			1-2
Milky	RX J0925.7-4758	?	?			3.8		
Way	RX J0019.8+2156	0.97-1.10	1.2-2.6	~1000	~3.0	0.66	~1	6
M31	RX J0045.4+415	1.34-1.38	0.09-1.9					2,7
		(1.37-1.38)						4

Ref.: (1) Kahabka & Trümper 1996 (from van Teeseling et al. 1994; (2) Kahabka unpublished results (cf. Kahabka 1995); (3) Crampton et al. 1996; (4) Kahabka 1996b; (5) Jordan et al. 1996; (6) Beuermann et al. 1995; (7) White et al. 1995.

and Hachisu, Kato & Nomoto, 1996. Using for R_{core} a mass-radius relation (e.g. Nauenberg 1972) the dependence of t_{contr}^{obs} on the mass of the white dwarf M_{WD} is obtained. The result is shown in Fig. 3. This model allows to deduce independent estimates for the white dwarf masses, which turn out to be consistent with the masses deduced from the analysis of the recurrence behavior of the sources. A compilation of the system parameters deduced from this and other work is given in Tab. 3.

4 The Case for SNe Ia

The homogeneity of SNe Ia in terms of spectra, light curves and absolute peak luminosities (cf. Livio et al., 1995 and references therein) has been considered as indicator for a rather well confined class of progenitors, i.e. thermonuclear explosions of accreting C-O WDs. This comprises accreting WDs and WD mergers (with one component finally accreting from the other component disrupted into a He accretion disk). Several scenarios have been considered to lead to an explosion of the WD. The main two cases are (1) ignition of carbon at the center of the WD after it reaches the Chandrasekhar mass and (2) off-center detonation after conversion of $\lesssim 0.2$ $M_\odot$ of hydrogen accreted on top of a C-O WD into helium. The initial mass of the WD before the start of the accretion event should be below ~ 1.2 $M_\odot$ (excluding ONeMg WDs) in order to avoid a collapse. Case (1) has been considered to be most promising in explaining the observed characteristics of SNe Ia (cf. Branch et al., 1995). Case (2) may account for part of the smaller fraction of optically fainter SNe Ia events but specific (and highly uncertain) explosion scenarios have to be set up in order to get rough consistency within the observations (cf. Nomoto et al., 1984). For a discussion of sub-Chandrasekhar mass models for SNe Ia it is referred to Woosley and Weaver, 1994 and Canal, Ruiz-Lapuente and Burkert, 1996. Rappaport et al., 1994 derive from population synthesis calculations of steady supersoft binary sources SN Ia rates due to Chandrasekhar core collapse as high as $3 - 6 \times 10^{-3}$ yr^{-1} which are at the upper bound close to the upper limit of the observationally inferred galactic SN Ia rate of $\sim 3 - 7 \times 10^{-3}$ yr^{-1}. Yungelson et al., 1996 find from their population synthesis calculations a SN Ia rate for case (1) of $\sim 3.5 \times 10^{-5}$ yr^{-1}. This is well below the observed rate. For case (2) they derive a rate of $\sim 3 \times 10^{-4}$ yr^{-1}. The supersoft sources given in Table 3 can be considered to be good candidate progenitors of SNe Ia, the symbiotic novae may contain lower mass white dwarfs and may explode via channel (2).

5 The Case for AIC

WDs with masses in excess of ~ 1.2 $M_\odot$ before the onset of accretion (mostly ONeMg) may grow in mass for sufficiently high mass accretion rates and experience an Accretion Induced Collapse (AIC) if they reach the Chandrasekhar mass (cf. Isern & Hernanz, 1995). Yungelson et al., 1996 derive for the Milky Way a rate of permanent supersoft sources experiencing the AIC of $(3-4) \times 10^{-5}$ yr^{-1}. As ~ 0.2 $M_\odot$ of the mass of the WD is removed by neutrino emission, neutron stars formed by this channel are expected to have masses of ~ 1.2 $M_\odot$. RX J0045.4+4154, a supersoft source in M31, can be considered to be a strong candidate for the AIC (Kahabka, 1995). Finally it should be noted that an independent way to derive an estimate of the SN Ia rate of a galaxy is by means of the nucleosynthesis products, the heavy element content (Tsujimoto et al. 1995).

Acknowledgement: P. Kahabka is a Human Capital and Mobility fellow.

References

Alcock C., et al., 1995, Phys. Rev. Lett., 74, 2867
Beuermann K., Reinsch K., Barwig H., et al., 1995, A&A 294, L1
Branch D., Livio M., Yungelson L.R., et al., 1995, PASP 107, 1019
Canal R., Ruiz-Lapuente P., Burkert A., 1996, ApJ 456, L101
Cowley A.P., Schmidtke P.C., Hutchings J.P., 1993, ApJ 418, L63
Cowley A.P., Schmidtke P.C., Crampton D., 1996, in: IAU Symposium 165, Compact Stars in Binaries, ed. J. van Paradijs, E.P.J. van den Heuvel & E. Kuulkers, p. 439
Crampton D., Hutchings J.B., Cowley A.P., et al., 1996 ApJ 456, 320
Di Stefano R., Rappaport S., 1994, ApJ 437, 733
Gänsicke B.T., Beuermann K., de Martino D., 1996, this volume p. 107
Greiner J., Hasinger G., Kahabka P., 1991, A&A 246, L17
Greiner J., Wenzel W., 1995, A&A 294, L5

Greiner J., Bickert K., Luthardt R., Viotti R., Altamore A., González-Riestra R., 1996, this volume p. 267
Hachisu I., Kato M., Nomoto K., 1996, this volume p. 205
Hasinger G., 1994, Rev. Mod. Astr. 7, 129
Isern J., Hernanz M., 1995, Mem.S.A.It. Vol.65-1, 339
Jordan S., Mürset U., Werner K., 1994, A&A 283,
Jordan S., Schmutz W., Wolff B., et al., 1996, A&A (subm.)
Kahabka P., Pietsch W., Hasinger G., 1994, A&A 288, 538
Kahabka P., 1995, A&A 304, 227
Kahabka P., 1996a, A&A 306, 795
Kahabka P., 1996b, in: Röntgenstrahlung from the Universe, ed. H.U. Zimmermann, J.E. Trümper and H. Yorke, MPE Report 263, p. 151
Kahabka P., Trümper J., 1996, in: IAU Symposium 165, Compact Stars in Binaries, ed. J. van Paradijs, E.P.J. van den Heuvel & E. Kuulkers, p. 425
Kato M., Hachisu I., 1994, ApJ 437, 802
Krautter J., Ögelman H., Wichmann R., et al., 1995, ApJ (subm.)
Livio M., Branch D., Yungelson L.R., et al., 1995, in: Cataclysmic variables and related objects, IAU Colloquium 158, (Kluwer)
Long K.S., Helfand D.J., Grabelsky D.A., 1981, ApJ 248, 925
Mikolajewska J., Kenyon S.J., Mikolajewski M., et al., 1995, AJ 109, 1289
Motch C., Hasinger G., Pietsch W., 1994, A&A 284, 827
Motch C., 1996, this volume p. 83
Nariai K., 1974, PASJ 26, 57
Nauenberg M., 1972, ApJ 175, 417
Nomoto K., Thielemann F., Yokoi K., 1984, ApJ 286, 644
Ögelman H., Orio M., Krautter J., Starrfield S., 1993, Nat. 361, 331
Pakull M.W., Beuermann K., van der Klis M., 1988, A&A 203, L27
Pakull MW., Motch C., Bianchi L., et al., 1993, A&A 278, L39
Prialnik D., Kovetz A., 1995, ApJ 445, 789
Rappaport S.A., Di Stephano R., Smith M., 1994, ApJ 426, 692
Reinsch K., van Teeseling A., Beuermann K., et al., 1996 A&A (in press)
Schaeidt S., Hasinger G., Trümper J., 1993, A&A 270, L9
Schaeidt S., 1996, this volume p. 159
Schmidtke P.C., McGrath T.K., Cowley A.P., 1993, PASP 105, 863
Schmidtke P.C., Cowley A.P., McGrath T.K., 1996a, AJ 111, 788
Schmidtke P.C., Cowley A.P., 1996b, this volume p. 123
Smale A.P., Corbet R.H.D., Charles P.A., et al., 1988, MNRAS 233, 51
Southwell K.A., Livio M., Charles P.A., et al., 1996, this volume p. 165
Tsujimoto T., Nomoto K., Yoshii Y., et al., 1995, MNRAS 277, 945
van den Heuvel E.P.J., Bhattacharya D., Nomoto K., Rappaport S.A., 1992, A&A 262, 97
van Teeseling A., Heise J., Kahabka P., 1994, preprint
White N.E., Giommi P., Heise J., 1995, ApJ 445, L125
Woosley S.E., Weaver T.A., 1994, ApJ 423, 371
Yungelson L., Livio M., Truran J.W., et al., 1996, ApJ (in press)

Part VI

Symbiotics and Other Related Systems

ROSAT Observations of Symbiotic Binaries and Related Objects

K.F. Bickert[1], J. Greiner[1], R.E. Stencel[2]

[1] Max-Planck-Institut für extraterrestrische Physik, Garching FRG
[2] Department of Physics and Astronomy, University of Denver, Denver CO

Abstract. We present X-ray observations of 217 confirmed or suspected symbiotic binaries with the ROSAT PSPC during the all-sky survey and 129 archived pointings. A table lists 3σ upper limits or detected count rates for survey and pointing observations (available for 46 objects) plus a collection of UBVJHK brightnesses, stellar types, outburst status, and structure (jets, ejecta, winds, clouds). The detection of 28 symbiotics (21 in survey) will be handled in greater detail in a forthcoming paper.

1 Introduction

Symbiotic stars form a heterogeneous class of long period, interacting binaries hosting a cool red giant (K- or M-class or Mira) and a hot compact subdwarf, white dwarf, or neutron star (Mikolajewska and Kenyon 1992). They are strong UV emitters, but *Einstein* and *EXOSAT* observatories detected only a few systems (Allen 1981) of a small subset (ca. 25) observed, presumably due to circumbinary and interstellar reddening. The ROSAT all-sky survey provides the first statistically complete sampling of the more than 200 objects. Long exposure pointings added to detections, improved statistsics and show sometimes drastic time variation of the X-ray emission.

A key question in symbiotics is the source of the high energy radiation, whether it is a) thermal radiation from the compact companion surface, or b) compact's photosphere, or c) from a possible accretion disk or d) its corona, e) from a heated outburst shell, or f) collisional excitation from e.g. stellar wind shock regions, or g) from the giant's corona. X-ray observations reveal the most energetic regions in the system (10–30 eV typ.), spectra and variability should help assigning models to individual systems.

2 Observations

The German X-ray satellite ROSAT (cf. Trümper 1984) hosts two imaging telescopes: the UK EUV wide field camera, and the German soft X-ray telescope (XRT). The XRT mirrors feed MPE's Position Sensitive Proportional Counters (PSPC-C and -B, 4 color energy resolution) and the US SAO's High Resolution Imager (HRI, b/w). Here we used only PSPC data because of higher sensitivity and better background rejection at low energies. EUV detected EX Hya (in a pointing) and (marginally) AG Dra (in the all-sky survey).

We have examined the survey data for detections from a list of 212 symbiotic and symbiotic-like objects compiled by Allen (1984), Kenyon (1986), Vaidis (1988) and others. For completeness we included all sources that may be or might have been symbiotics. Some of the objects might need reobservations to confirm their status, some might be mis-classified planetary nebulae, contain super-giants or their assignment may simply be in error. Some may have changed type due to outbursts or evolution. We include a flag denoting the confirmation status.

The ROSAT all-sky survey covered 96% of the sky. For 208 objects all-sky survey data were available (exposure > 32 sec). We used EXSAS V4.0 standard survey procedures on 40 x 40 arcmin fields around the candidate source positions (involving a local detect algorithm for channels 11–235, using a 5 arcmin cutradius; sources with a likelihood > 8 were cut out the data before calculating the local background map; reading supplied exposure map for dead-time and vignetting corrected time). Two maximum likelihood upper limit runs with fixed and free position fits at 3σ confidence level and a detection threshold of ML > 6 delivered upper limit or source counts plus error. For detected sources, the distance between the optical and free fit X-ray position was calculated. The analysis of bright X-ray sources suggests that PSPC positions can be accurately determined to within 30 arcsec, and to less than 1 arcmin for soft and weak sources. Source confusion may disturb our results for LMC S63 (HV 12671).

We performed a number of follow-up pointed observations on symbiotic stars which were detected in the all-sky survey above at least 2σ. There were also pointings by other observers on stars from our list, either on purpose or coincidentally within a search radius of 30 arcmin. Beyond 35 arcmin, weak soft sources are difficult to detect due to blurring and the upper limit determination gets unreliable due to possible source confusion and problems in the background determination. A total of 129 such datasets are available (now in the ROSAT archive) covering 46 stars. Due to off-center pointings, 4 objects fall outside of our 30 arcmin detection area and thus have no pointing entries in the table.

3 Results List

The results were compacted to a table with 21 columns (described next) and 217 double line rows, sorted by right ascension.
1: sequential number
2: One or two most popular names (some objects have up to 14), variable star notation was preferred.
3: Flags:
    c - symbiotic confirmation status (t -true, s - suspected, blank - unknown)
    i - mid-infrared excess = $25\mu/12\mu$ IRAS class (D dusty or S stellar)
    j - jet flag set if any extended structure known (j jet, o optical, i infrared, r radio)
    bur - burst type (rn - recurrent nova, sn - slow nova, bu - some outburst detected)

4-6: Coordinates:
   right ascension + declination epoch 2000.0 fk5,
   galactic (system II) longitude l and latitude b
   ecliptic latitude (estimates survey visibility)
   survey total coverage [sec] = upper limit for survey exposure
7-13: Count rates (upper line pointing results, lower line survey):
   detection flag: u upper limit, d 3σ detection (ml >= 6.)
   maximum likelihood for point source existence (from source detection)
      (ml >= $14.3 = 5\sigma, 9.65 = 4\sigma, 5.91 = 3\sigma, 3.08 = 2\sigma$)
   source count rate in [1/sec] if detected, else upper limit rate
      (vignetting and dead-time corrected)
   rate error scaled to [1/1000sec]. only for detections
   actual exposure time (vignetting and dead-time corrected)
   positon offset determined with parameter free (in compute/upper) [arcsec]
   observation dates [yymmddxmmdd]
      (x = - if dates span < 1 year, + if <2, or a digit)
14 top: offcenter, distance between pointing position and symbiotic in [arcmin]
14 bot: structure flags: denote if star has any known extended structures
   (j-jet, e-ejecta, w-wind, c-clouds, x-complex, b-eclipse)
15 top: hydrogen column density nH in [$10^{20} / cm^2$]
15 bot: jet, optical, infrared, or radio diameter in [arcsec]
16: stellar types of red and blue component
17-19: brightness in UBV and JHK bands in [magnitudes] (c is color mag)
20 top: year or Julian date of outburst (slow or recurrent nova or other)
20 bot: estimates of stellar distance [kpc]
21: References: 1-Allen 1980, 2-Paresce 1990, 3-Kenny et al. 1991, 4-Burgarella & Paresce 1991, 5-Leedyarv 1989, 6-Whitelock 1988, 7-Taylor 1988, 8-Schulte-Ladbeck 1988, 9-Kenyon 1988, 10-Anandaro et al. 1988, 11-Solf 1984, 12-Kohoutek 1987, 13-Viotti 1988, 14-Downes & Shara 1993, 15-Kholopov et al. 1985, 16-Kenyon & Truran 1983, 17-Kenyon 1986, 18-Vaidis 1988, 19-Luthardt 1992, 20-Argelander, 21-Perek & Kohoutek 1967, 22-Kholopov et al. 1987, 23-Acker et al. 1982, 24-Kholopov 1982, 25-Henize 1976, 26-Cannon & Pickering, 27-Sanduleak & Stephenson 1973, 28-SIMBAD catalogue, 29-Whitelock 1988, 30-Gezari et al. 1993, 31-Inca consortium 1992.

All structural, burst, and brightness information was taken from the literature. The data are not completely up to date (literature search stopped mid 1994), especially as the stars are often highly variable and may undergo substantial changes in stellar type and brightness within few months or years. Reported are changes of appearance from symbiotic to planetary nebulae and back within a few months. If dates of acquired information was given, the year was appended (value'year). If data was unsecure, a question mark was added. The data is highly incomplete, mostly due to a real lack of observations of the stars, but also in part due to lack of time devoted to searching data. Search has been resumed for the detected stars.

4 Analysis

We tried correlations of the X-ray count rate of all sources with an ML above 6 (3σ, 28 objects), 3 (2σ, 49) or 2 (1.5σ, 71) with any of the data collected from the literature. We only summarize here the first results, a deeper study of the detections is in preparation, where we also discuss previous observations. A preliminary version is available.

Too few data were cited on the stellar type of the compact object, on U magnitude or UV brightness. On B mags, information was too often imprecise of whether absorption and reddening was included or not, completely upsetting results. No significant correlation to the stellar class of the giant and the V, J, H, K magnitudes was found. For IRAS infrared classification, 6/8/8 of 24 dusty systems were detected versus 10/22/37 of 107 stellars. When corrected for distance and N_H, there also seems to be no correlation.

Information on the spatial structure (jets, ejecta, winds, clouds) is scarce (stars must be well within 1 kpc distance), but 8/9/9 out of 11 objects were detected. 7/8/10 of 17 bursters are seen, those nearby or with low N_H and in later outburst phases (AG Dra is absorbed in beginning, no others seen in early phase). A few sources found by Einstein/EXOSAT were now below detection level, even in deep pointing.

Symbiotics seem not to be standard candles, neither within nova (4 comparable) or steady burner group (5). A rough logN/logS shows strong deviation from powerlaw (but bad statistic).

Their high known variability in the radio, infrared, optical, and UV is also seen in soft X-rays: up to a factor of 300 (EX Hya), mean factor of 4 of 23 compared (without EX Hya) on time scale of 1..3 years (survey to pointing). Only 7 keep their level.

Most stars detected are within 1–2 kpc stellar distance, and the 3 which are further away have a hard spectrum. Typical spectra are very soft (10–25 eV) (the only really hard one, GX 1+4, harbours a neutron star). Neither blackbody nor powerlaw models fit the X-ray spectra well. Better fits are sometimes achieved by detailed white dwarf atmosphere model with or without winds (e.g. Mürset et al. 1996).

No straightforward support of any of the possible emission mechanisms is possible. Too few detections are comparable without in-depth model studies. Symbiotics are too heterogeneous.

Symbiotics cluster in the Galactic plane/center and are strongly affected by galactic and (in some systems high) intrinsic reddening and absorption. Accounting for distance and absorbing column, one would expect to detect X-ray emission only from a minor fraction of our sample of symbiotics. The fairly high detection rate among this subsample in the all-sky survey and pointed observations suggests, that all symbiotic systems emit X-ray emission at some level.

Acknowledgement: JG is supported by the Deutsche Agentur für Raumfahrtangelegenheiten (DARA) GmbH under contract FKZ 50 OR 9201. The ROSAT project is supported by the German Bundesministerium für Bildung, Forschung, Wissenschaft und Technik (BMBW/DARA) and the Max-Planck-Society. We made use of the SIMBAD database, operated at CDS, Strasbourg France.

References

Acker A., et al. ., 1982, Publ. Spec. CDS. No. 3
Allen D.A., 1980, MNRAS 192, 521
Allen D., 1981, MNRAS 197, 739 (erratum MNRAS 201, 1199)
Allen D., 1984, Proc. Astron. Soc. Austral. 5, 369
Anandaro B.G., Taylor A.R., Pottasch S.R., 1988, A&A 203, 361
Anderson C., Cassinelli J., Sanders W., 1981, ApJ 247, L127
Argelander F.W.A., Bonner Durchmusterung
Burgarella D., Paresce F., 1991, ApJ 370, 590
Cannon A.J., Pickering E.C., Henry Draper Memorial HD catal.
Cassatella A., et al. 1984, ApSS 131, 763
Cordova F., Mason K., 1984, MNRAS 206, 879
Downes R.A., Shara M.M., 1993, PASP 105, 127
Gezari D.Y., et.al. 1993, NASA Ref. Pub. 1294, Cat. of Infrared Observations
Henize K.G., 1976, ApJ Suppl. 30, 491
inca consortium: 1992, ESA SP-1136, Hipparcos input catalogue
Jura M., Helfand D., 1984, ApJ 287, 785
Kenny H.T., Taylor A.R. Seaquist E.R., 1991, ApJ 366, 549
Kenyon S.J., 1986, The Symbiotic Stars (Cambridge)
Kenyon S.J., 1988, AJ 96, 337
Kenyon S.J., Truran J.W., 1983, ApJ 273, 280
Kenyon S., Fernandez-Castro T., Stencel R., 1988, AJ 95, 1817
Kholopov P.N. (ed.), 1982 NSV New catal o Suspected Variable Stars
Kholopov P.N. (ed.), 1985, General Catalogue of Variable Stars (GCVS)
Kholopov P.N. (ed.), 1987, General Catalogue of Variable Stars II
Kohoutek L., 1987, ApSS 131, 781
Leahy D., Taylor A., 1987, A&A 176, 302
Leedyarv L.E., 1989, Ap 31, 598
Luthardt R., 1992, Measurements from the Sonneberg Plate Archive (priv. comm.)
Mikolajewska J., Kenyon S., 1992, MNRAS 256, 177
Mürset U., Jordan S., Wolff B., 1996, this volume p. 251
Nussbaumer H., Vogel M., 1990, A&A 236, 117
Paresce F., 1990, ApJ 357, 231
Perek L., Kohoutek L., 1967, Catal. Galactic Planetary Nebulae, Czechosl.
Sanduleak N., Stephenson C.B., 1973, ApJ 185, 899
Schulte-Ladbeck R., 1988, AAp 189, 97
Solf J., 1984, A&A 139, 296
Taylor A.R., 1988, in The Symbiotic Phenomenon, ed. J. Mikolajewska, p. 77
Trümper J., 1984, Phys. Scripta T7, 209
Vaidis J.P., 1988, Bull. Astron. Franc. Obs. Etoiles Var. No.45/3+ff
Viotti R., 1988, in The Symbiotic Phenomenon, ed. J. Mikolajewska, p. 269
Viotti R., Piro L., Friedjung M., Cassatella A., 1987, ApJ 319, L7
Whitelock P., 1988, in The Symbiotic Phenomenon, ed. J. Mikolajewska, p. 47

Table 1. ROSAT Observations of Symbiotic Stars

seq	name..........	cij	ra.[e2000]	......dec	....ecl	P	mlik	.prate	perr	pexpo	pxoff
	name-2........	bur	..gal.lII.	.gal.bII.	surexpo	S	mlik	.srate	serr	sexpo	sxoff
1	V641 Cas	t	00h09m26s4	+63d57m14	54.797						
			118.341	1.457	582.0	u	0.9	0.0066		430.4	
2	V410 Cas	t	00h23m27s8	+61d46m28	52.108						
			119.630	−0.918	673.9	u	0.0	0.0158		492.6	
3	SMC1	t	00h29m10s8	−74d57s40	−64.048						
			304.877	−42.083							
4	SMC2	t	00h42m48s2	−74d42m01	−64.750						
			303.704	−42.415							
5	EG And	tS	00h44m37s2	+40d40m46	32.662	u	3.6	0.0063		8037.7	
	HD 4174		121.537	−22.174	510.0	u	3.1	0.0089		492.5	
6	SMC3	t	00h48m20s8	−73d31m53	−64.615						
			303.237	−43.595							
7	SMC-S 18		00h54m09s4	−72d41m51	−64.643	u	0.4	0.0009		69799	
	LHA 115-S 18		302.650	−44.429	483.3	u	0.9	0.0037		471.9	
8	SMC-N 60	tS	00h57m06s0	−74d13m18	−65.448						
	LHA 115-S 60		302.407	−42.901	647.1	u	4.8	0.0131		616.5	
9	SMC Ln 358	tS	00h59m12s2	−75d05m19	−65.886						
	Lin 358		302.260	−42.029	661.7	u	4.5	0.0173		627.6	
10	SMC-N 73	tS	01h04m39s1	−75d48m24	−66.438						
	LHA 115-N 73		301.855	−41.294	582.0	u	4.2	0.0134		561.5	
11	AX Per	tS	01h36m16s9	+54d15m35	40.536						
	MWC 411		129.519	−8.038	548.7	u	0.7	0.0116		372.9	
12	V741 Per	tD	01h58m49s7	+52d53m48	37.794						
	PK 133-08 1		133.119	−8.637	44.4	u	1.1	0.0623		37.0	
13	TT Ari	t	02h06m53s1	+15d17m42	2.309	d	>10k	0.3770	2.3	102833	10.8
			148.528	−43.795	392.5	d	275	0.3943	38.8	296.5	6.1
14	o Cet	t	02h19m20s7	−02d58m27	−15.932	d	145	0.0079	1.0	37132	10.1
			167.747	−57.982	385.4	u	2.2	0.0079		285.6	
15	LW Cas	s	02h57m20s6	+60d37m15	41.537	u	4.4	0.0021		25274	
	SVS 1144		137.780	1.431	558.6	u	3.0	0.0120		536.2	
16	s 32	s	04h37m45s3	−01d19m09	−23.170						
			197.484	−30.038	730.5	d	996	0.7551	39.0	549.1	9.1
17	S147	t	04h54m04s8	−70d59m34	−82.606						
			282.662	−34.971							
18	UV Aur	tS	05h21m49s0	+32d30m44	9.343						
	HD 34842		174.219	−2.349	460.9	u	0.7	0.0029		444.2	
19	LMC1	t	05h25m01s1	−62d28m47	−84.459						
			271.817	−33.766							
20	TV Col	t	05h29m25s4	−32d49m07	−55.952	u	2.1	0.0029		16676	
			236.791	−30.606	687.4	d	372	0.3510	28.7	511.6	7.0
21	N67	t	05h36m02s9	−64d43m24	−86.923						
			274.356	−32.351							
22	Sanduleak's *	t	05h45m19s7	−71d16m10	−85.112	u	1.1	0.0010		75807	
			281.914	−30.873	2474.0	u	2.1	0.0032		2365.6	

ROSAT Observations of Symbiotic Binaries and Related Objects

point.dates surveydates	offc stru	...nH .diam	..red.type ..blue.type	........mU ....mJ	........mB ....mH	........mV ....mK	burst.date .dist.[kpc]	refs....... ....
910121-0808		71.74			2.2c			30,
910121-0806		71.33				15.5-18		14,
		3.95						
		4.46						
930102-0102 910107-0118	25.4	6.58	M2IIIepv		8.9 2.79c	7.08-7.8 2.59c	0.34	15,30, 5,
		5.16						
930509-0512 901021-1101	6.1	5.16	B0		13.96	13.55 10.69m		28,30,
901017-1027		4.77	C			13.		17,
901015-1025		4.59	C			15.8 11.5		18,17,
901015-1023		4.41	C		17.0	11.5		18,17,
900730-0813		18.56	M5IIIe	6.67c'88	9.5-13.0 5.73m'88	9.0-12.9 5.45m'88	2.70	19,30,18,5
900730-0802		16.75	G2e	10.9m'88	10.56m'88	13.3 9.83m'88		19,30,17,
910730-0802 910118-0123	0.3	6.19						
930713-0802 910111-0121	0.4	2.44	M5:e	-1.7'80	-2.3	3.6-9.2 -2.5	0.12	31,30, 6,
920310-0311 900808-0817	4.0	92.38	AoIII		15.4-17.1 10.3	16.9 9.1		18,30,
900815-0218		5.34						
		8.61						
900830-0907		64.56	Ne(C8ep)	4.32c'88	9.8-11.1 3.06c	7.4-10.6 2.28,2.08	1.00	17,30,10,
		4.60						
920927-0930 900825-0905	18.5	2.02						
		5.17						
910218-0219 900909-1013	22.6	7.49		15.15'83	14.58	12.87		30,

seq	name......... name-2.......	cij bur	ra.[e2000] ..gal.lII.	......dec .gal.bII.	....ecl surexpo	P S	mlik mlik	.prate .srate	perr serr	pexpo sexpo	pxoff sxoff
23	MaC 2-4	s	05h47m36s0 204.912	+00d39m00 −13.858	−22.753 650.8	u	0.0	0.0129		492.4	
24	LMC S63 HV 12671	tS	05h48m44s2 277.629	−67d36m13 −30.878	−88.487 6207.0	d	14.5	0.0078		5899.0	99.4
25	WY Gem	t	06h11m56s2 187.914	+23d12m25 2.257	−0.204 533.7	u	0.0	0.0107		410.7	
26	SS73 1		06h48m45s3 219.236	−07d24m44 −3.965	−30.260 323.3	u	0.9	0.0042		297.5	
27	GH Gem CSV 950	s	07h03m35s5 203.515	+12d02m19 8.120	−10.532 459.1	u	1.3	0.0065		437.6	
28	NSV 3539 BD-19 1821		07h20m04s6 233.496	−19d30m45 −2.737	−41.203 416.1	u	2.1	0.0056		396.1	
29	BX Mon HV 10446	tS	07h25m23s9 220.040	−03d35m59 5.884	−25.300 392.7	u	0.5	0.0032		375.2	
30	MWC 560	s	07h25m51s2 223.760	−07d44m05 4.046	−29.361 534.7	u u	0.9 2.6	0.0010 0.0194		18895 406.5	
31	NQ Gem HD 59643		07h31m55s0 194.635	+24d30m11 19.354	2.710 452.1	u	1.4	0.0066		431.0	
32	Wray 15-157 IRAS 08045-282	tD	08h06m35s1 246.606	−28d31m59 1.948	−47.371 288.5	u	0.3	0.0060		260.7	
33	RX Pup HV 3372	tDr bu	08h14m12s4 258.510	−41d42m58 −3.939	−59.090 539.5	d d	504 18.0	0.0448 0.0346	3.8 10.2	13171 524.6	23.2 37.1
34	He 3-160 Wray 15-208	tS	08h24m52s8 267.678	−51d28m35 −7.871	−66.587 809.4	u	0.7	0.0032		767.5	
35	AS 201 PK 249+06 1	tD	08h31m42s8 249.078	−27d45m32 6.969	−44.827 416.1	u	0.8	0.0044		401.9	
36	WY Vel HV 3655		09h21m59s2 274.142	−52d33m51 −1.816	−61.924 508.6	u	0.7	0.0047		488.5	
37	KM Vel PK 274+02 1		09h41m13s4 274.188	−49d22m42 2.577	−57.544 514.2	u	0.3	0.0049		505.3	
38	He 2-38 V366 Car	tD	09h54m43s3 280.808	−57d18m53 −2.241	−62.111 33.1	u u	0.6 0.0	0.0037 0.0000		22319 25.0	
39	NSV 4775 BD-57 2768		10h11m03s1 282.857	−57d48m13 −1.324	−60.856 197.9	d d	533 37.7	0.0253 0.1123	1.9 27.6	29693 195.0	18.1 1.3
40	He 3-404 Wray 15-549		10h21m33s9 284.177	−58d05m45 −0.789	−60.046 309.3	u	0.0	0.0199		227.1	
41	He 3-461	s	10h39m08s5 282.905	−51d24m12 6.250	−53.399 101.5	u	0.1	0.0037		96.9	
42	RT Car HV 107	s	10h44m47s1 287.442	−59d24m59 −0.408	−58.769 328.4	d u	10.5 1.4	0.0018 0.0193	0.5	78721 246.7	122.7
43	SS73 29	tS	11h08m27s5 292.632	−65d47m17 −4.997	−60.827 453.4	u	0.4	0.0022		435.3	
44	He 3-653 Wray 15-807	s	11h25m32s5 292.365	−59d56m32 1.156	−55.481 264.4	u	0.4	0.0035		255.3	
45	SY Mus HV 3376	tS	11h32m10s5 294.808	−65d25m10 −3.805	−58.691 425.1	u	4.3	0.0093		406.9	

ROSAT Observations of Symbiotic Binaries and Related Objects

point.dates surveydates	offc stru	...nH .diam	..red.type .blue.type	........mU ....mJ	........mB ....mH	........mV ....mK	burst.date .dist.[kpc]	refs.......
900903-0911		22.60						
900907-1217		5.67	C	12.48'83	11.60	11.3		30,
900911-0918		59.49	M2Iabpe	2.98c'83	2.09c	7.66'92 1.83		31,30,
900922-0930		57.13	M2ep		13.5	12.0		18,27,
900924-0930		11.77			12.4-14.6	14.6 9.7'74		19,30,
901005-1011		69.71	M3ep		9.5-11.5	9.5		18,
901002-1009		17.97	M4ep FOII-III	6.91c'88	9.5-13.4 5.97c	12. 5.89,5.58		18,30,
920429-0430 901002-1011	0.4	33.35						
900930-1007		5.29	C6,2/R9ev	4.05	10.0 3.33	7.4-8.0		31,30,22,
901023-1029		45.34	G			13.1 9.4		18,17,
930412-0608	0.4 w	78.43 3.70o	Mep	4.05-5.92	11.1-14.1 2.82-4.70	9.3-12.3 1.94-3.49	1972 1.00- 1.30	18,30, 7,15 2,10,
901115-1126		38.90	M7			12.5 7.5	20.00	25,17, 1,
901029-1107		14.22	G2III			11.4 9.0'74	1.26	18,30,17,12
901203-1213		128.3	M3epIb A2		8.8-10.2	13.5		22,
901205-1213		37.59		9.92'90011	7.93	15.6 6.10	3.00	18,30,6
911211-1212 901219-0117	25.0	158.6	M	6.5'83	5.29	3.7V	2.80	30, 6
920724-0728 901226-0117	0.3	189.4	M3ep OB		11.0	10.5		18,
900714-0117		178.5		8.64'80	7.64	12.5 7.00		25,30,
901226-0117		33.56				9.8 3.9		25,17,
920610-0615 901230-0118	16.5	126.3	M2Ia O	3.1'880623	10.6 3.16c	8.2-9.9 1.82		31,30,
910113-0125		42.15	G Be(pec)		13.0	13. 9.4'75		18,30,
910109-0120		83.31	Be	6.97'75	5.79	12.5 5.40		25,30,
910120-0125		64.63	M2	5.9'82	11.3-12.3 4.95	10.2-12.7 4.62	1.00	18,30, 5,

seq	name.........	cij	ra.[e2000]	......dec	....ecl	P	mlik	.prate	perr	pexpo	pxoff
	name-2........	bur	..gal.lII.	.gal.bII.	surexpo	S	mlik	.srate	serr	sexpo	sxoff
46	BI Cru	tD	12h23m26s4	−62d38m16	−52.766						
	SVS 1855		299.722	0.059	49.7	u	0.0	0.0093		43.8	
47	CQ Dra	t	12h30m07s1	+69d12m07	61.188	d	149	0.0084	0.9	61032	28.8
	4 Dra		125.748	47.809	951.7	d	14.4	0.0211	7.4	717.6	33.7
48	TX CVn	tS	12h44m42s2	+36d45m49	37.654						
	BD+37 2318		130.929	80.260	557.5	u	0.6	0.0032		528.9	
49	He 2-87	tS	12h45m47s1	−63d00m32	−51.418						
	PK 302-00 1		302.292	−0.144	193.3	u	0.0	0.0349		149.5	
50	He 3-828	tS	12h50m58s0	−57d50m46	−46.892						
	Wray 15-1022		302.870	5.026	105.1	u	0.2	0.0084		90.5	
51	SS73 38	tD	12h51m26s3	−64d59m58	−52.571						
	CD-64 665		302.933	−2.128	14.3		0.0	0.0000		20.6	
52	EX Hya	t	12h52m24s4	−29d14m56	−21.692	d	> 10k	5.8116	4.0	63794	5.2
	4U1249-28		303.185	33.622	433.1	u	0.2	0.0166		318.2	
53	He 3-863	tS	13h07m43s8	−48d00m12	−37.184						
			305.750	14.777	316.3	u	3.3	0.0175		302.7	
54	St 3-22	tS	13h14m30s6	−58d51m52	−46.023						
	PK 305+03 1		305.918	3.874	193.4	u	0.0	0.0349		149.5	
55	He 3-886	s	13h16m01s8	−37d00m07	−26.661						
	NSV 6160		308.375	25.610	295.5	u	3.8	0.0219		277.7	
56	V840 Cen		13h20m49s3	−55d50m13							
57	He 3-905	tS	13h30m37s2	−57d58m17	−44.152						
	Wray 15-1108		308.126	4.505	55.3		0.0	0.0000		38.5	
58	RW Hya	tS	13h34m18s2	−25d22m49	−14.449	u	1.2	0.0039		4096.4	
	MWC 412		314.994	36.486	338.2	u	0.5	0.0040		330.7	
59	He 3-916	tS	13h35m28s9	−64d45m44	−49.576						
	Wray 15-1123		307.608	−2.291	235.4	u	3.6	0.0200		225.8	
60	V704 Cen	s	13h54m56s1	−58d27m18	−42.989						
	PK 311+03 1		311.175	3.397	381.6	u	2.5	0.0217		293.9	
61	He 2-104	s	14h11m52s0	−51d26m23	−35.627						
	V852 Cen		315.480	9.462	281.5	u	0.0	0.0018		279.9	
62	V835 Cen	tD	14h14m09s0	−63d25m44	−46.243						
	NSV 6587		312.027	−2.027	444.8	u	1.6	0.0082		433.2	
63	BD-21 3873	tS	14h16m34s3	−21d46m51	−7.650						
			327.874	36.934	345.3	u	1.7	0.0077		326.2	
64	V748 Cen	s	14h59m34s6	−33d25m27	−15.727						
	CD-32 10517		331.507	22.242	311.1	u	1.5	0.0061		301.6	
65	He 2-127	tD	15h24m49s8	−51d49m48	−31.986						
	PK 325+04 2		325.542	4.179	398.7	u	0.4	0.0038		381.2	
66	He 3-1092	tS	15h47m10s8	−66d29m16	−45.115						
	PK 319-09 1		319.225	−9.349	326.7	u	1.8	0.0045		306.8	
67	He 3-1103	tS	15h48m28s4	−44d19m00	−23.701						
	Wray 15-1359		333.206	7.903	427.3	u	1.3	0.0091		409.8	
68	HD 330036	sD	15h51m15s8	−48d44m54	−27.892						
	PK 330+04 1		330.780	4.155	125.8	u	4.1	0.0314		116.1	

point.dates surveydates	offc stru	...nH .diam	..red.type .blue.type	........mU ....mJ	........mB ....mH	........mV ....mK	burst.date .dist.[kpc]	refs....... ....
		183.6	M		11.0-14.0	11.0-14.0		18,30, 6,
910123-0125				7.38'88111	6.17	4.7V	2.00	
910404-0405	0.2	1.69	M3IIIa			4.958'92		31,30,
901025-1105				1.36'84				
		1.34	KOIII-M4		9.2-11.8	9.1-10.0		18,30,
901203-1209			B1-B9Veq	7.43'88	6.61c	6.26		
		159.6	M7.5					30,18, 1,
900730-0813						6.06	13.00	
		30.70	M6		13.5	12.5-13.5		18,17,1
910121-0125						7.1	10.00	
		108.6				12.5-13.8		18,29, 6,
900730+0813						6.1	4.20	
910716-0717	0.1	6.10				11		32,
910103-0118			CVint.pola					
		12.80	K4			11.0		18,
910120-0125						8.5		
		37.99	M					17,
900730-0813						8.5		
		4.42			12.6	10.		25,30,
910113-0121					5.96c'72	5.67		
		40.95	K4			12.5		18,17, 1,
900730+0813							2.30	
920130-0201	0.2	5.29	M2IIIep		10.0-11.2	8.1-9.6		18,30, 1,
910111-0120				5.73c'88	4.90c	4.70	0.60- 1.00	
		115.5	M6			12.5		18,17, 1,
900730-0806						7.8	30.00	
		57.23			14.3-<16	13.		18,30,
900730-0813					10.1'74	8.39		
		29.84				14.6		23,30, 6,
900730-0804				10.7'89	8.7	6.9	4.30	
		134.8	M		5.7	4.4		18,30, 6,
900804-0812				10.1'73		4.3V	2.80	
		7.04	G			9.6		20, 9,
910121-0125						7.2V		
		8.84	M			11.6-13.7		18,30,
900730-0806			Fep			8.12		
		45.76	M7			15.6		23,30,18,6
900810-0817						9.5'74	14.00	
		15.97	K5		13.5	14.		18,30,17,1
900819-0827				8.98'74	7.96	7.67	2.10	
		20.56	M0			12.2		17, 1,
900812-0819						8.4	3.50	
		57.99			11.8	11.040'92		31,17,
900815-0821						7.6		

seq	name..........	cij	ra.[e2000]	......dec	....ecl	P	mlik	.prate	perr	pexpo	pxoff
	name-2........	bur	..gal.lII.	.gal.bII.	surexpo	S	mlik	.srate	serr	sexpo	sxoff
69	He 2-139	tD	15h54m44s7	−55d29m36	−34.301	u	0.1	0.0045		11622	
	PK 326-01 1		326.914	−1.400	380.7	u	0.5	0.0039		363.4	
70	T CorB	tS	15h59m30s2	+25d55m12	45.275						
	MWC 413	rn	42.375	48.164	557.8	u	0.7	0.0040		551.0	
71	AG Dra	tS	16h01m41s0	+66d48m09	78.399	d	>1k	0.8808	20.2	9100.9	43.2
	BD+67 922		100.288	40.971	1492.0	d	>1k	1.1489	29.4	1425.8	7.1
72	V345 Nor		16h06m44s7	−52d03m03	−30.540						
	HV 8827	bu	330.507	0.043	376.0	u	1.4	0.0044		354.8	
73	Wray 16-202	tS	16h06m57s1	−49d26m41	−27.988						
	PK 332+01 1		332.279	1.955	350.5	u	0.5	0.0034		342.3	
74	SZ 134		16h09m12s1	−41d40m22	−20.300						
	Wray 15-1423		337.842	7.417	288.3	u	2.8	0.0098		283.2	
75	He 2-147	tD	16h14m00s8	−56d59m31	−35.132						
	V347 Nor		327.919	−4.298	370.3	u	1.5	0.0092		350.5	
76	UKS Ce-1	tS	16h15m29s3	−22d12m16	−0.922						
	UKS 1612-22.0		353.022	20.248	292.3	u	0.0	0.0013		294.8	
77	QS Nor		16h21m07s9	−42d23m54	−20.598						
	PK 338+05 2		338.936	5.356	333.1	u	0.3	0.0033		321.6	
78	Wray 15-1470	tS	16h23m21s6	−27d40m10	−6.005						
			350.062	15.238	376.8	u	2.0	0.0111		361.9	
79	He 2-171	tD	16h34m04s2	−35d05m23	−12.976						
	PK 346+08 1		346.028	8.551	324.3	u	1.9	0.0116		300.3	
80	He 3-1213	tS	16h35m15s2	−51d42m24	−29.357						
	Wray 15-1511		333.869	−2.815	342.6	u	2.2	0.0118		331.6	
81	He 2-173	tS	16h36m24s6	−39d51m42	−17.626						
	PK 342+05 1		342.772	5.011	347.9	u	0.0	0.0155		330.6	
82	HD 149407	t	16h37m20s8	−53d46m35	−31.346						
	CPD-53 8091		332.223	−3.936	471.8	u	0.8	0.0146		350.6	
83	He 2-176	tD	16h41m31s2	−45d13m05	−22.786						
	PK 339+00 1		339.394	0.745	334.0	u	1.0	0.0060		320.2	
84	He 3-1242	tS	16h44m35s5	−62d37m13	−39.910						
	KX TrA		326.415	−10.939	166.5	u	1.9	0.0161		164.1	
85	AS 210	tD	16h51m20s5	−26d00m27	−3.482						
	Wray 16-237		355.516	11.553	339.4	u	1.2	0.0054		326.9	
86	HK Sco	tS	16h54m40s8	−30d23m04	−7.745						
	HV 4493		352.486	8.267	333.7	u	0.0	0.0102		319.8	
87	CL Sco	tS	16h54m51s9	−30d37m17	−7.976						
	HV 4035		352.323	8.089	331.3	u	4.8	0.0199		318.9	
88	MaC 1-3	s	17h01m30s0	−47d46m00	−24.873						
			339.618	−3.528	384.1	u	1.0	0.0147		288.8	
89	V455 Sco	sS	17h07m21s8	−34d05m14	−11.153						
	AS 217		351.175	3.886	313.8	u	2.7	0.0132		304.1	
90	V2051 Oph	t	17h08m19s1	−25d48m28	−2.889	d	28.1	0.0141	4.1	3887.3	4.2
			358.010	8.624	421.4	d	7.6	0.0278		327.0	23.8
91	He 3-1341	tS	17h08m36s7	−17d26m29	5.451						
	NSV 8226		5.016	13.393	363.2	u	2.2	0.0110		351.6	

ROSAT Observations of Symbiotic Binaries and Related Objects 237

| point.dates | offc | ...nH | ..red.type | ....mU | ....mB | ....mV | burst.date | refs....... |
surveydates	stru	.diam	.blue.type	........mJ	........mH	........mK	.dist.[kpc]	....
930309-0313	26.3	156.8	M			16.8		18,30, 6,
900817-0823				8.10'74	6.40	5.3V	5.00	
		4.67	gM3III			2.0-10.8	1946,1866	31,30, 5,
900730-0804			sdBe	5.83'88	4.97	4,72	1.24- 1.35	
920416-0416	4.7	3.15	K1IIpevar		8.9-11.8	7.6-10.5		31,30, 1,
901118-1207				7.09c'88	6.34	6.18	1.00- 1.20	
		251.1			13-<18		1985	18,14,
900817-0825								
		105.4	M					17,
900817-0823						6.8		
910313-0313		25.99	Mep		14.0	12.5		18,30,
900817-0823					9.98'83	9.62		
		39.68	M8			15.		28,30,17,6
900821-0827				7.06'74	5.73	4.92	3.40	
		10.86	C			15.0		17,
900812-0821						11.3		
		39.73				13.5-15.5		22,30,18,
900819-0825			Be	9.5'74	8.46	8.06		
		15.14	M4e			12.0		17,18, 1,
900817-0823						7.8	1.70	
		22.19	M					30, 6,
900819-0827				9.52'74	7.64	6.2V	4.70	
		64.72	M2e		13.5	10.4		18,17, 1,
900823-0830						6.7	1.00	
		37.14	M		13.5			18,30,17,
900821-0827				8.48'74	7.10	6.78		
		40.02	B8			8		26,
900823-0901								
		124.2	M7			15.		28,17,
900823-0830						5.7		
		14.10	M6		13.1	12.4		18,30,17, 1
900827-0903						6.20	5.00	
		15.98	G		14.0	11.5		18,30,17,6
900823-0830				9.87c'88	8.08c	6.52,6.35	4.90	
		17.49	M		13.1-15.8	15.		18,17, 1,
900823-0830						7.9	2.80	
		19.18	M		11.2-13.9	12.		18,17, 1,
900823-0901						7.9	2.30	
		57.05						
900827-0903								
		29.98	M		12.8-<16.5	15.		18,17, 1,
900827-0903						5.9	5.00	
920920-0921	0.1	21.66						
900825-0903								
		16.90	M2		13.5-16.9	12.0		18, 9, 1,
900825-0901			B			7.60	2.40	

seq	name.......... name-2........	cij bur	ra.[e2000] ..gal.lII.	......dec .gal.bII.	....ecl surexpo	P S	mlik mlik	.prate .srate	perr serr	pexpo sexpo	pxoff sxoff
92	He 3-1342	tS	17h08m55s1 0.079	$-23d23m37$ 9.920	-0.472 331.4	u	1.1	0.0059		319.5	
93	AS 221 Wray 15-1637	tS	17h12m12s3 352.952	$-32d37m48$ 3.934	-9.612 162.9	u	0.3	0.0064		152.8	
94	PN H 2-5 PK 354+04 2	tS	17h15m18s6 354.202	$-31d34m03$ 4.020	-8.500 131.3	u	1.4	0.0330		110.8	
95	PN Sa 3-43 PK 355+04 1	s	17h17m55s7 355.787	$-30d01m44$ 4.447	-6.923 191.8	u	1.2	0.0063		192.6	
96	Draco C-1	t	17h19m57s6 86.273	$+57d50m05$ 34.759	80.131 1678.0	d d	397 57.6	0.0437 0.0491	3.1 7.0	21882 1607.0	14.5 15.6
97	He 3-1383 Wray 15-1680		17h20m31s7 353.529	$-33d09m48$ 2.201	-10.010 460.5	u	1.0	0.0153		329.7	
98	PN Th 3-7 PK 356+04 3	tS	17h21m02s6 356.708	$-29d22m51$ 4.261	-6.229 330.1	u	0.1	0.0041		317.5	
99	PN Th 3-9	s	17h24m00s1 355.703	$-31d02m00$ 2.801	-7.837 400.8	u	0.2	0.0088		310.2	
100	PN Th 3-17 PK 357+03 3	tS	17h27m31s8 357.764	$-29d03m56$ 3.266	-5.827 295.4	u	0.8	0.0093		285.8	
101	PN Th 3-18 PK 358+03 5	tS	17h28m26s9 358.229	$-28d38m33$ 3.332	-5.393 298.1	u	1.5	0.0075		286.6	
102	He 3-1410 NSV 8805	tS	17h29m06s4 357.406	$-29d43m22$ 2.616	-6.464 312.5	u	0.0	0.0126		299.2	
103	GX 1+4 V2116 Oph	tS	17h32m02s2 1.934	$-24d45m00$ 4.792	-1.465 301.5	d d	20.7 12.4	0.0052 0.0321	1.6 13.0	9489.5 290.8	19.0 1.1
104	PN Th 3-29 PK 358+02 3		17h32m27s6 358.341	$-29d05m07$ 2.355	-5.790 378.4	u	1.6	0.0130		297.4	
105	PN Th 3-30 PK 359+02 1	tS	17h33m43s4 359.300	$-28d07m21$ 2.646	-4.816 310.8	u	0.0	0.0161		297.6	
106	PN Th 3-31 PK 358+01 2	tS	17h34m16s8 358.220	$-29d29m11$ 1.804	-6.173 321.1	u	1.1	0.0071		307.1	
107	PN M 1-21 PK 006+07 1	tS	17h34m17s3 6.962	$-19d09m21$ 7.358	4.148 309.2	u	0.5	0.0032		297.9	
108	He 2-251 PK 358+01 3		17h35m21s7 358.119	$-29d45m25$ 1.460)	-6.433	u	4.8	0.0144		293.4	
109	V503 Her	s	17h36m46s1 47.001	$+23d18m14$ 26.232	46.587 669.2	u	0.8	0.0021		637.5	
110	Pt-1 PK 003+03 3	tS	17h38m49s7 3.485	$-23d54m03$ 3.942	-0.550 332.7	u	0.0	0.0116		318.0	
111	LuSt-1	t	17h39m18s1 359.854	$-28d15m00$ 1.541	-4.893 423.8	u u	1.7 1.1	0.0061 0.0199		9223.3 315.0	
112	RT Ser MWC 265	tS sn	17h39m51s9 13.896	$-11d56m37$ 9.971	11.408 354.7	u	3.1	0.0138		343.7	
113	AE Ara HV 5491	tS	17h41m04s9 343.999	$-47d03m20$ -8.659	-23.674 175.6	u	2.6	0.0173		169.6	
114	SS73 96	tSr	17h41m28s3 352.847	$-36d47m42$ -3.377	-13.417 319.4	u	0.8	0.0034		313.2	

ROSAT Observations of Symbiotic Binaries and Related Objects   239

point.dates	offc	...nH	..red.type	........mU	........mB	........mV	burst.date	refs........
surveydates	stru	.diam	.blue.type	........mJ	........mH	........mK	.dist.[kpc]	....
		18.39	M2		14.0	12.0		18,30, 1,
900827-0903			B	8.9c'88	7.94	7.60	5.00	
		35.90	M4		17.0	11.0-13.1		18,30,17, 1
900827-0903				8.07'74	7.49		5.00	
		35.31	M					30,17,
900830-0905				7.13'83	5.98	5.55		
		40.68						17,
900830-0905						7.9		
920401-0401	0.3	2.54	C1,2			17.0		17,
900730-0817								
		71.46	Mep			12.5		
900830-0907								
		42.18	M			14.0		21,30,17, 1
900830-0905					8.55'73	8.05	10.00	
		58.84						30,
900830-0907					9.0'75	8.75		
		49.09	M3			14.		21,30,17, 1
900901-0907					8.38'75	8.18	5.00	
		49.09	M2			12.6		21,30,17, 1
900901-0907				9.6'74	8.38	8.03	3.50	
		51.22	M			12.5-13.1		18,30, 1,
900901-0907			B	9.8'74	9.08	8.41	5.00	
930918-0918	0.0	32.09	M6III			18.7-19.4		18,17, 1,
900901-0907						8.1	13.00	
		64.40				13.4		21,
900901-0909								
		42.98	K5			13.1		21,30,17,
900903-0909						9.2'74		
		64.40	M5			13.6		21,30,17, 1
900903-0909				9.23'74	8.03	7.57	8.00	
		23.92	M2			14.		18,30,17, 1
900901-0909				8.77'74	7.63	7.21	2.40	
						16.5		28,30,
					10.7'75	7.3		
		6.97	M2p		11.1-14.4	12.5		18,
900830-0907								
		31.42	M1			15.		17,
900903-0909						8.6		
920229-0301	17.0	53.30						
900903-0911								
		19.62	M		10.6-17.0	13.	1909	18,30,16, 1
900903-0909			A8(pec)	8.36'88	7.34	6.94	9.00	
		22.05	M2		11.5-13.8	11.5-13.8		18, 1,
900905-0911			Be			6.3	1.60	
		47.93	M2		14.0	12.		18,30, 7,17
	w	0.17r	Be	8.11'75	6.85	6.39	1.70	1,

seq	name.........	cij	ra.[e2000]	......dec	....ecl	P	mlik	.prate	perr	pexpo	pxoff
	name-2........	bur	..gal.lII.	.gal.bII.	surexpo	S	mlik	.srate	serr	sexpo	sxoff
115	UU Ser	tS	17h42m38s4	−15d24m40	7.965						
	HV 8771		11.232	7.616	326.2	u	0.8	0.0071		317.0	
116	V2110 Oph	tD	17h43m32s5	−22d45m34	0.626						
	AS 239	rn	5.029	3.623	343.7	u	0.3	0.0038		323.0	
117	SSM 1	tS	17h43m54s8	−36d03m24	−12.664						
	V916 Sco	bu	353.735	−3.408	315.4	u	0.1	0.0024		302.4	
118	NSV 9601		17h44m26s2	−02d04m47	21.303						
	HD 161224		23.230	13.873	528.6	u	0.8	0.0055		405.0	
119	PN H 2-19		17h45m14s3	−38d17m25	−14.889						
	AS 241		351.959	−4.794	266.0	u	1.3	0.0045		262.7	
120	HD 161213		17h45m33s1	−38d39m47	−15.260						
	PK 351-05 1		351.670	−5.039	251.5	u	0.1	0.0027		247.2	
121	PN Bl 3-2	s	17h46m00s1	−26d58m00	−3.565						
			1.726	0.953	400.3	u	1.9	0.0166		309.7	
122	V917 Sco	tS	17h48m03s9	−36d08m18	−12.725						
			354.100	−4.168	273.4	u	1.1	0.0073		262.2	
123	PN Bl 3-11	s	17h48m12s1	−28d01m00	−4.604	u	0.0	0.0037		5895.6	
			1.081	−0.007	390.2	u	0.7	0.0083		303.6	
124	PN H 1-36	tDr	17h49m48s2	−37d01m27	−13.603						
	PK 353-04 1		353.515	−4.921	243.4	d	13.0	0.0745	24.4	236.7	54.5
125	RS Oph	tSr	17h50m13s2	−06d42m27	16.710	d	97.8	0.0106	1.5	21709	16.2
	MWC 414	rn	19.801	10.372	355.9	u	4.6	0.0155		336.2	
126	Wray 16-312	s	17h50m16s8	−30d57m35	−7.538						
	PK 358-01 2		358.790	−1.910	324.5	u	0.5	0.0045		306.3	
127	PN Th 4-4	s	17h50m23s8	−19d53m45	3.525						
	V4141 Sgr		8.312	3.732	320.6	u	1.2	0.0056		305.7	
128	AS 245	s	17h50m58s0	−22d19m24	1.099						
			6.289	2.379	458.0	u	1.2	0.0108		335.2	
129	PN H 2-28	tS	17h51m01s2	−22d19m34	1.097						
	AS 245		6.293	2.368	336.2	u	0.4	0.0024		320.7	
130	He 2-294	tS	17h51m45s4	−32d54m53	−9.488						
	PK 357-03 1		357.266	−3.176	292.5	u	1.9	0.0099		279.9	
131	KW Sgr		17h52m00s3	−28d01m30	−4.598						
	HD 316496		1.505	−0.733	322.3	u	2.5	0.0073		315.0	
132	PN Bl 3-14	tS	17h52m26s0	−29d45m59	−6.338						
	PK 000-01 4		0.054	−1.700	323.0	u	2.9	0.0130		309.9	
133	PN Bl 3-6	t	17h52m56s4	−31d19m20	−7.892						
	PK 358-02 2		358.768	−2.584	304.1	u	2.8	0.0133		292.5	
134	PN Bl L	tS	17h53m13s5	−30d18m01	−6.870						
	PK 359-02 1		359.680	−2.119	310.2	u	0.6	0.0046		302.7	
135	MaC 1-9	s	17h55m54s1	−14d07m00	9.319						
			13.986	5.503	423.0	u	0.3	0.0172		304.3	
136	AS 255	tD	17h57m08s8	−35d15m34	−11.822						
	MHA 363-45		355.794	−5.318	201.8	u	0.0	0.0216		192.0	
137	V2416 Sgr	tS	17h57m15s9	−21d41m28	1.747						
	PK 007+01 2		7.573	1.440	282.6	u	1.6	0.0078		273.6	

ROSAT Observations of Symbiotic Binaries and Related Objects 241

point.dates surveydates	offc stru	...nH .diam	..red.type .blue.type	........mU ....mJ	........mB ....mH	........mV ....mK	burst.date .dist.[kpc]	refs........ ....
900903-0909		20.91	M		14.6-16.0	16. 9.1		18,17,
900905-0911		32.54	M8	9.92'80	8.41	12.-19.0 7.5V	<1980 20.00	17,30,16, 1
900905-0911		44.14	M			8.3	1967	17,14,
900903-0911		13.86	K0		10.4	9.4		18,
900905-0911		27.95		9.42'75	12. 8.28	7.77		18,30,
900905-0911		27.95	K0		10.6	9.9 9.3'74		18,30,26,
900905-0911		83.90						
900905-0911		32.84	M7			13.0 8.0		17,
930316-0316 900905-0911	26.0	149.0						
	xw	29.31 5.00r	M8		7.73 10.3'75	7.4'87	7.60	28,30, 7, 6
910302-0305 920304-0304	0.4 e	24.79 0.20r	M2ep Oep	7.74'88	5.3 6.98	4.3-12.5 6.73,6.59,	1967 0.60- 1.60	17,30, 7,15 1,
900905-0913		63.81				7.6	7.80	29, 6,
900905-0911		26.46				14.1 7.9		21,17,
900905-0913		45.00				7.5	12.00	29, 6,
900905-0913		45.00	M6	8.40		11.0 10.3?'74	9.00	17,30, 1,
900907-0913		39.50	M3			9.6'74		30,17,
920303-0303 900907-0913		97.62	M0-M4	2.72'73	11.0-13.2 1.81	11.0-13.2 5.2		31,30,
900907-0913		52.26	M6			14.4 8.7		17,
900907-0913		43.02						
900907-0913		52.26	M6	9.55'75	8.19	16.0 7.81	16.00	17,30, 1,
900907-0913		33.03						
900907-0913		26.01	K3		13.5	12.0 8.4		18,17,
900907-0913		96.70	M	6.41c'88	14.6 5.09c	13. 4.50	2.30	18,30,17, 1

seq	name.......... name-2........	cij bur	ra.[e2000] ..gal.lII.	......dec .gal.bII.	....ecl surexpo	P S	mlik mlik	.prate .srate	perr serr	pexpo sexpo	pxoff sxoff
138	PN H 2-34 PK 001-02 1		17h58m28s1 1.765	−28d33m14 −2.228	−5.115 274.6	u	1.1	0.0061		267.2	
139	He 2-325 PK 003-01 1		18h01m05s9 3.963	−26d21m26 −1.644	−2.918 277.2	u	0.0	0.0146		272.1	
140	SS73 117	tS	18h02m23s1 359.193	−31d59m10 −4.658	−8.548 198.3	u	0.3	0.0078		186.8	
141	AS 269	s	18h03m24s1 358.672	−32d42m00 −5.192	−9.262 242.4	u	0.0	0.0238		178.7	
142	PN Ap 1-8 PK 002-03 1	tS	18h04m29s6 2.588	−28d21m36 −3.285	−4.925 262.1	u u	0.0 1.3	0.0020 0.0084		29469 252.2	
143	SS73 122 IRAS 18015-270	tS	18h04m41s1 3.663	−27d09m12 −2.732	−3.718 268.4	u	0.8	0.0072		260.1	
144	NSV 10142 WR 107		18h04m45s6 8.291	−21d51m28 −0.154	1.577 270.4	u u	2.2 3.3	0.0022 0.0095		21730 264.7	
145	AS 270	tS	18h05m33s7 9.703	−20d20m34 0.425	3.090 275.4	u	2.4	0.0081		261.2	
146	PN H 2-38 PK 002-03 4	tD	18h06m01s2 2.818	−28d17m03 −3.541	−4.852 259.3	d d	7.5 6.6	0.0018 0.0496	0.6 20.0	41079 240.5	46.8 124.3
147	SS73 129 PK 001-04 3	tS	18h07m05s7 1.772	−29d36m25 −4.387	−6.177 228.8	u	0.0	0.0194		218.7	
148	He 3-1591 NSV 10219	tS	18h07m32s0 5.073	−25d53m46 −2.677	−2.468 274.4	u	0.4	0.0067		262.7	
149	V615 Sgr HV 7199	tS	18h07m40s0 356.088	−36d06m20 −7.597	−12.677 139.6	u	1.0	0.0175		135.2	
150	Ve 2-57 NSV 10241	s	18h08m22s2 6.333	−24d33m43 −2.196	−1.136 285.1	u	0.5	0.0032		275.5	
151	V1148 Sgr NOVA Sgr 1943	s bu	18h09m06s2 5.161	−25d59m28 −3.032	−2.568 323.1	u	1.4	0.0078		252.1	
152	AS 276 MHA 363-7	tD	18h09m09s5 351.639	−41d13m25 −10.236	−17.799 137.3	u	3.3	0.0276		131.7	
153	PN Ap 1-9 Wray 16-377	tS	18h10m28s9 3.430	−28d07m41 −4.327	−4.710 225.9	u	1.0	0.0101		211.4	
154	AS 281 Wray 16-378	tS	18h10m43s8 3.601	−27d57m50 −4.297	−4.547 221.1	u u	2.8 0.4	0.0020 0.0103		40730 211.4	
155	V2506 Sgr AS 282	tS bu	18h11m01s7 3.121	−28d32m40 −4.632	−5.128 213.3	u	2.3	0.0154		203.6	
156	SS73 141 PK 359-07 2	tS	18h12m11s3 359.134	−33d10m42 −7.042	−9.766 167.0	u	2.0	0.0253		158.0	
157	AS 289 F1-11	tS	18h12m22s2 18.091	−11d40m07 3.189	11.737 293.1	u	2.8	0.0104		288.4	
158	Y CrA HV 169	tS	18h14m22s9 350.611	−42d50m30 −11.830	−19.437 153.9	u	1.7	0.0087		149.9	
159	YY Her AS 297	tS	18h14m34s1 48.144	+20d59m18 17.243	44.367 434.0	u	0.5	0.0063		422.5	
160	V2756 Sgr AS 293	tS	18h14m34s5 2.357	−29d49m23 −5.919	−6.424 159.7	u	0.7	0.0081		155.1	

ROSAT Observations of Symbiotic Binaries and Related Objects 243

| point.dates | offc | ...nH | ..red.type | ....mU | ....mB | ....mV | burst.date | refs....... |
surveydates	stru	.diam	.blue.type	........mJ	........mH	........mK	.dist.[kpc]	....
900907-0913		45.83				9.9'75		30,
900909-0916		51.87	Mep B		12.5 9.6'74	11.5 9.1		18,30,
900909-0916		29.94	M6	8.68'75	14.0 7.57	12.5 7.07	10.00	18,30,17, 1
900909-0916		29.94		10.77'85	9.88	8.88		30,
930331-0401 900909-0916	23.0	35.58	M0 Be	9.6'75	8.3	7.9	2.60	30,18, 1,
900909-0916		34.38	M7.5 Be		12.0 6.6			17,29,18,
921003-1007 900909-0916	19.9	153.5	Be(pec		9.0-13.5 9.9'75	12. 9.0		24,30,
900909-0916		181.5	M1	7.16c'88	11.0 6.00c	12.5 5.57	0.70	18,30,17, 1
930331-0401 900909-0916	0.5	29.52	M8	8.40'75	7.57	6.67	8.10	30,17, 6,
900909-0916		30.33	K			12.5 8.0		17,
900909-0916		36.60	G Be		13.0 9.5'75	12.-13. 9.0		18,30,17,
900909-0916		13.92	M1 Be	8.94'74	13.2-14.8 7.94	12.5 7.69		18,30,
900909-0918		53.75			8.37'75	12.5 7.87		18,30,
900909-0918		36.60			8.0-<16.	<16	1943	18,15,
900909-0918		9.38	M4 B			11.5-12.5 8.1		18,
900911-0918		29.52	K4 Be		9.1'74	12.5 8.8	2.60	17,30,18, 1
930328-0406 900911-0918	0.3	27.20	M5	8.28'74	7.26	12.0 6.95	7.00	17,30, 1,
900911-0918		21.46	M		8.9'74	12. 8.42	1946	17,30,14,
900911-0918		17.90	M Be		13.0	12.5 10.5?'75		18,30,
900911-0918		66.83	M3.4 B	6.77c'88	13.0 5.57c	10.5 5.10		18,30,
900911-0918		10.38	M		12.0-13.8	12. 6.6	5.00	18,17, 1,
900914-0920		10.46	M2epvar	9.02c'88	11.1-<14 8.14c	12.5-13.5 7.90		18,30,
900911-0918		17.61	M	9.01'74	13.2-15.2 8.03	11.5 7.76	3.00	18,30,17, 1

seq	name..........	cij	ra.[e2000]	......dec	....ecl	P	mlik	.prate	perr	pexpo	pxoff
	name-2.........	bur	..gal.lII.	.gal.bII.	surexpo	S	mlik	.srate	serr	sexpo	sxoff
161	AS 296	tS	18h15m06s4	−00d18m59	23.069	u	2.3	0.0013		34268	
	JP11 5251		28.480	7.931	316.4	u	0.5	0.0028		305.9	
162	HD 319167	tS	18h15m24s6	−30d31m55	−7.137						
	PK 001-06 1		1.812	−6.410	168.7	u	0.6	0.0119		158.1	
163	He 2-374	tS	18h15m31s0	−21d35m24	1.801						
	PK 009-02 1		9.732	−2.213	236.5	u	2.1	0.0189		226.9	
164	AS 295B	t	18h16m05s5	−30d51m12	−7.463						
	V4074 Sgr		1.594	−6.689	166.8	u	1.1	0.0174		155.9	
165	V2905 Sgr		18h17m20s5	−28d09m45	−4.781						
	AS 299		4.117	−5.677	177.8	u	4.1	0.0322		141.9	
166	He 3-1674	tS	18h20m19s0	−26d22m44	−3.020						
	Wray 15-1864		6.012	−5.432	206.6	u	0.7	0.0091		186.4	
167	AR Pav	tS	18h20m28s6	−66d04m48	−42.691						
	MWC 600		328.535	−21.606	121.7	u	0.9	0.0181		118.0	
168	He 2-390	tD	18h20m58s7	−26d48m27	−3.454						
	PK 005-05 2		5.699	−5.761	183.7	u	0.8	0.0121		172.9	
169	V3804 Sgr	tS	18h21m28s4	−31d32m04	−8.181						
	AS 302		1.517	−8.024	173.9	u	0.0	0.0196		163.6	
170	V443 Her	tS	18h22m07s7	+23d27m20	46.752						
	MWC 603		51.232	16.596	570.5	u	0.0	0.0049		547.5	
171	FR Sct		18h23m22s8	−12d40m54	10.640	d	6.1	0.0016	0.6	30754	102.0
	SVS 588		18.472	0.340	317.2	u	0.3	0.0032		307.5	
172	V3811 Sgr	tS	18h23m29s0	−21d53m09	1.442						
	PK 010-03 1		10.341	−3.985	235.9	u	0.9	0.0043		226.5	
173	AS 304	tS	18h25m26s7	−28d35m57	−5.284						
	CD-28 14567		4.551	−7.460	161.9	u	0.6	0.0127		150.7	
174	RY Sct		18h25m31s8	−12d41m27	10.608						
	MWC 295		18.710	−0.128	313.7	u	0.0	0.0059		300.3	
175	V1988 Sgr	s	18h27m58s5	−27d37m29	−4.336						
			5.682	−7.518	140.7	u	0.0	0.0013		139.8	
176	V1017 Sgr	tS	18h32m04s1	−29d23m53	−6.154						
	HV 3519	bu	4.481	−9.113	138.4	u	2.2	0.0107		135.8	
177	V2601 Sgr	tS	18h38m02s1	−22d41m50	0.453						
	AS 313		11.152	−7.352	318.8	u	0.5	0.0050		309.6	
178	PN Sa 3-142		18h40m25s9	−08d44m10	14.342						
	PK 023-01 1		23.908	−1.549	332.7	u	1.8	0.0098		322.1	
179	AS 316	tS	18h42m32s8	−21d17m47	1.778						
	PK 012-07 1		12.886	−7.669	292.6	u	2.2	0.0091		284.5	
180	K 3-12	s	18h44m42s1	+06d07m00	29.063						
			37.621	4.278	562.1	u	1.2	0.0053		424.1	
181	MWC 960	tS	18h47m55s8	−20d05m51	2.875	u	3.1	0.0016		40511	
	MHA 204-22		14.532	−8.273	283.9	u	0.4	0.0086		269.1	
182	AS 323		18h48m35s5	−06d41m35	16.218	u	2.1	0.0014		36969	
	PK 026-02 2		26.649	−2.417	346.4	u	3.9	0.0129		334.2	
183	V603 Aql	t	18h48m54s5	+00d35m02	23.458	d	>10k	0.3090	6.2	34284	2.5
			33.163	0.830	484.2	d	388	0.3838	34.3	350.0	3.4

ROSAT Observations of Symbiotic Binaries and Related Objects 245

point.dates surveydates	offc stru	...nH .diam	..red.type .blue.type	........mU ....mJ	........mB ....mH	........mV ....mK	burst.date .dist.[kpc]	refs....... ....
920307-0323 900911-0920	0.4	25.95	M5	5.79'89	13.0 4.72	10.5 4.46	1.80	18,30,17, 1
900911-0918		17.61	M3	8.92'74	13.0 7.78	12.5 7.50	3.50	18,30,17, 1
900911-0918		63.12	M	8.05c'83	14.0 6.96c	12.0 6.55		18,30,17,
900911-0918		15.83	M		8.6	11.5		18,17,
900911-0920		17.74	pec Be(pec)	8.58'75	7.32	10 7.07		22,30,18,
900914-0920		21.22	M5		12.5	12.5 7.7	8.00	18,17, 1,
900911-0918		8.06	M3III+.. Of	8.24'73	8.5-13.62 7.38	6.8-13.0 7.10	3.80-10.00	31,30, 5,
900914-0920		16.60	M		13.0 10.8'74	12 7.0V	5.20	18,30, 6,
900914-0920		14.10	M6			11.5-12.2 7.3	10.00	18, 1,
900916-0924		10.00	M3ep O	6.57c'88	5.66c	11.4-11.7 5.38		18,30,
921010-1013 900914-0920	19.5	182.3	M2.5Iabpe+ OB	3.94'73	11.7-12.5 2.67	10.46'92 2.15		31,30,
900914-0920		33.05	M			14. 8.5		17,
900914-0920		14.82	M4			11.5 7.6	5.00	17, 1,
900916-0922		202.5	O8:e.. B0ep		9.7-10.3 5.80c'73	9.1-9.7 5.48		31,30,
900916-0922		14.09	M7		12.7-15.9	13.0 2.47'69		18,30,
900916-0922		13.36	G5IIIp		6.2-14.4	14 10.58'74	1919	18,30,14,
900918-0924		17.62	M5e			15 8.0	2.30	17,18, 1,
900920-0926		55.45				9.1?'74		30,
900918-0926		17.50	M			12.0 7.8		17,
900920-0928		59.07				9.2'74		30,
930407-0414 900920-0926	10.4	19.06	M0		13.5	12.0 7.8	2.50	18,17, 1,
921013-1014 900922-0928	9.0	57.17		9.21'74	8.30	12. 7.91		18,30,
910406-0415 900922-0930	0.3	149.6						

seq	name.......... name-2........	cij bur	ra.[e2000] ..gal.lII.	......dec .gal.bII.	....ecl surexpo	P S	mlik mlik	.prate .srate	perr serr	pexpo sexpo	pxoff sxoff
184	AS 327 PK 011-11 1	tS	18h53m16s7 11.144	−24d22m57 −11.230	−1.502 308.5	u	2.7	0.0164		298.8	
185	FN Sgr AS 329	tS	18h53m54s4 16.157	−18d59m42 −9.061	3.849 315.3	u	0.6	0.0034		302.9	
186	PN Pe 2-16 PK 029-02 1	tS	18h54m10s0 29.104	−04d38m52 −2.723	18.126 219.4	u	0.8	0.0111		201.6	
187	CM Aql	tS	19h03m35s1 31.595	−03d03m14 −4.093	19.460 384.8	u	1.1	0.0039		368.1	
188	V919 Sgr AS 337	tS	19h03m45s1 19.008	−16d59m54 −10.312	5.598 290.8	u	4.0	0.0224		285.8	
189	AS 338 PK 048+04 1	tS	19h03m46s8 48.975	+16d26m19 4.766	38.791 583.1	u	0.5	0.0024		554.0	
190	BL Tel HD 177300	s	19h06m38s2 345.526	−51d25m03 −23.155	−28.656 135.2	u	0.6	0.0143		129.2	
191	PN Ap 3-1 PK 037-02 1	tS	19h10m36s0 37.637	+02d49m31 −2.967	25.076 425.7	u	0.4	0.0043		412.5	
192	MaC 1-17	s	19h13m00s1 30.602	−05d21m00 −7.224	16.892 519.8	u	0.7	0.0109		375.1	
193	V352 Aql	s	19h13m30s1 37.503	+02d18m00 −3.851	24.457 560.4	u	0.0	0.0081		426.8	
194	BF Cyg MWC 315	tSj	19h23m53s4 62.928	+29d40m28 6.697	50.981 628.7	u	0.6	0.0032		599.1	
195	CH Cyg HD 182917	tSj rn	19h24m33s1 81.859	+50d14m30 15.580	70.492 833.3	d d	>10k 60.0	0.3323 0.0595	6.5 10.4	33547 800.1	8.0 18.5
196	SS Sge SVS 81	s bu	19h39m07s6 53.205	+16d42m38 −2.578	37.568 492.4	u u	0.2 0.4	0.0066 0.0019		6203.0 470.3	
197	HM Sge PK 053-03 2	tDo sn	19h41m57s0 53.567	+16d44m39 −3.150	37.457 553.3	d d	956 20.1	0.0456 0.0280	2.7 8.8	27416 533.8	6.0 1.8
198	He 3-1761 NSV 12264	tS	19h42m25s9 327.675	−68d07m41 −29.757	−45.874 126.5	u	0.0	0.0284		128.8	
199	AS 360 QW Sge	tS	19h45m49s3 55.650	+18d36m45 −3.023	39.074 333.2	u	1.0	0.0087		305.4	
200	V1290 Aql ZI 1783	s	19h46m31s4 48.910	+10d47m07 −7.053	31.398 513.4	u u	0.2 0.0	0.0007 0.0073		27310 488.7	
201	CI Cyg MWC 415	tS	19h50m12s2 70.898	+35d41m03 4.741	55.249 567.9	u	3.8	0.0107		543.9	
202	OX Cyg AN 108.1928		19h54m38s9 74.433	+39d14m57 5.786	58.296 688.4	u	0.5	0.0033		652.1	
203	V1016 Cyg AS 373	tDo sn	19h57m05s0 75.173	+39d49m36 5.678	58.660 708.5	d d	386 21.8	0.0140 0.0336	1.0	65272 685.0	3.4 0.0
204	RR Tel HV 3181	tD	20h04m18s5 342.163	−55d43m34 −32.242	−34.493 195.1	d d	>1k 86.8	0.2242 0.2706	10.3 41.2	9199.7 189.7	10.6 14.1
205	FG Sge PK 060-07 1		20h11m56s0 60.334	+20d20m07 −7.392	39.176 558.9	u	5.4	0.0121		536.3	
206	PU Vul NOVA Vul 1979	tS rn	20h21m13s3 62.576	+21d34m18 −8.532	39.733 566.1	d u	137 0.1	0.0042 0.0009	0.5	110241 547.6	6.2

ROSAT Observations of Symbiotic Binaries and Related Objects 247

point.dates surveydates	offc stru	...nH .diam	..red.type .blue.type	........mU ....mJ	........mB ....mH	........mV ....mK	burst.date .dist.[kpc]	refs........
900920-0928		16.75	M		8.68c'72	11.5-12.0 8.52		18,30,17,
900922-0928		13.49	Mep	9.14'73	9.0-13.9 8.17	9.0-13.9 7.92'79	5.00	18,30, 1,
900924-0930		41.22	M5 OB		8.59'74	16.0 8.09	10.00	17,30,18, 1
900926-1002		39.13	M4	9.14c'88	13.0-16.5 8.08c	13.0-16.5 7.70		15,30,17,
900924-0930		14.73	Me		12.0-<14.2	10.5-12.5 7.2	2.10	18,17, 1,
900928-1007		35.20	M3	8.76'74	11.5 7.89	11.5 7.54	7.00	18,30, 1,
900920-0928		5.37	F8Iab		7.72-9.82	7.1-9.4 4.77		31,30,
900928-1005		48.60			9.07'74	8.50		30,
900928-1005		25.60						
900928-1007		35.67	ea			13.3-18.5 9.6?'74		15,30,
	j	21.49	M5III Bep	7.58c'90	9.3-13.4 6.65	11. 6.23	4.20	18,30,11, 5
920326-0329	0.1 ej	7.23 1.50r	M7IIIvar Be	.83'90	6.0-9.1 -.17	5.6-8.5 -0.68	0.20- 0.40	31,30, 7,10
921105-1109 901009-1017	25.2	42.61				15.	1916	19,15,
921101-1103	0.3 ew	42.61 1.50o	M	7.98'89	11-<17 6.11	11-17 4.25,4.16,	1975 0.40- 2.10	18,30, 7,16 10,
900922-0930		6.59	M3			10.4 5.6	1.40	17, 1,
901013-1019		38.90	M6	8.33c'88	11.0 7.39c	11.0 7.03	10.00	18,30,17, 1
910506-0507 901011-1017	11.4	21.16			15.-16.5	15-<16.5		15,
	b	32.83	Bep+.. Bep	5.76c'88	9.9-13.1 4.79c	9.6-12.0 4.48	1.70	31,30, 8, 5
901023-1103		25.58	M2..4		14.8-16.2	15.5		18,22,
921118-1119	0.8 we	25.77 0.40o	M6	6.98'88	10.5-17.5 5.57	12.5 4.33,5.20,	1964 2.20- 3.30	18,30, 7,16 10,
930327-0330 900930-1007	0.6	4.39	M F5epv		6.5-16.5	14 3.6V	1944 2.50- 2.60	18,30,16,10
901021-1027		25.67	B4-K2pec	6.99'90	9.45-13.7 6.65	8.1-12.3 6.49		31,30,
921109-1112 901023-1101	0.4	16.00	M4-5 F8	7.71'82	8.5-16 7.2	11 5.93	1978	18,30,13,

seq	name...	cij	ra.[e2000]	...dec	....ecl	P	mlik	.prate	perr	pexpo	pxoff
	name-2...	bur	..gal.lII.	.gal.bII.	surexpo	S	mlik	.srate	serr	sexpo	sxoff
207	He 2-467	tS	20h35m57s3	+20d11m33	37.363						
	LT Del		63.399	−12.150	563.7	u		2.1	0.0079	539.2	
208	He 2-468	tS	20h41m18s9	+34d44m52	50.543						
	PK 075-04 1		75.943	−4.441	692.1	u		0.9	0.0043	665.3	
209	ER Del	t	20h42m46s7	+08d40m58	25.890						
			54.458	−19.994	650.1	u		1.3	0.0059	492.7	
210	V1329 Cyg	tS	20h51m01s2	+35d34m54	50.479						
	HBV 475	bu	77.836	−5.477	684.9	u		1.4	0.0070	659.5	
211	Hen 1924	tS	21h00m06s3	−42d38m50	−24.506	d	694	0.0584	3.2	27225	8.1
	CD-43 14304		358.654	−41.104	379.4	d	44.7	0.1057	20.6	364.8	14.9
212	V407 Cyg	t	21h02m11s7	+45d46m28	58.404						
	AS 453		86.986	−0.487	724.6	u		0.7	0.0030	695.5	
213	S 190	s	21h41m45s0	+02d43m52	15.611						
			58.417	−35.433	486.5	u		0.6	0.0140	364.2	
214	AG Peg	tSr	21h51m02s0	+12d37m31	24.055	d	>1k	0.0694	3.5	24568	6.2
	MWC 379	sn	69.278	−30.887	322.7	d	24.3	0.0564	16.5	305.6	14.7
215	W Cep		22h36m27s8	+58d25m35	58.861						
	HD 214369		106.024	0.057	538.0	u		0.0	0.0056	512.2	
216	Z And	tS	23h33m40s0	+48d49m06	46.100	d	17.9	0.0019	0.5	40305	10.1
	MWC 416		109.980	−12.088	432.6	d	6.3	0.0187	9.5	411.5	43.7
217	R Aqr	tSj	23h43m49s5	−15d17m02	−12.403						
	MWC 400		66.519	−70.325	328.7	d	57.1	0.1284	23.2	309.5	6.1

ROSAT Observations of Symbiotic Binaries and Related Objects

point.dates	offc	...nH	..red.type	........mU	........mB	........mV	burst.date	refs.......
surveydates	stru	.diam	.blue.type	........mJ	........mH	........mK	.dist.[kpc]	....
		9.94	G5		13.1-14.1	13		18,30,
901027-1105					9.874	9.36		
		48.54	M					30,17,
901105-1113					8.25	7.96		
		9.32						
901025-1103								
		30.57	M5II		12.1-18	12.6-14.0	1969	18,30,16,
901109-1117				8.88'88	7.27c	6.87		
930408-0408	0.3	3.82	K3			10.0		1,
901015-1021						7.6	1.20	
		100.7	Mep		13.3-<17.0	11.5-15		18,30,17,
901118-1129				4.74'90	3.64	2.90		
		5.12						
901107-1115								
930608-0610	0.3	6.40	M1-3,II-II			6.0-9.4	1857	31,30, 7,16
	e	20.0o	WN6	5.03-5.05	4.15-4.18	3.84-3.89	0.60- 1.00	3,
		79.82	K0ep-M2epI			6.9-8.9		31,30,
901228-0118			O			2.3		
930104-0109	0.2	14.74	M2III		8.0-12.4	7.8-11.3		15,30, 5,
901230-0118			B1eq	6.13'88	5.24c	4.95	1.20	
		1.70	M5-M8.5III			5.8-12.4		15,30, 7, 4
	e?j	15.5o	pec	0.18-1.23	-0.80-0.14	-1.32–0.5	0.24- 0.33	

X-Ray Properties of Symbiotic Stars:
I. The Supersoft Symbiotic Novae RR Tel and SMC3 (= RX J0048.4−7332)

U. Mürset[1], S. Jordan[2], B. Wolff[2]

[1] Institut für Astronomie, ETH-Zentrum, CH-8092 Zürich, Switzerland
[2] Institut für Astronomie und Astrophysik, Universität Kiel, 24098 Kiel, Germany

Abstract. We searched the ROSAT archive for pointed PSPC observations covering positions of symbiotic stars. 16 systems are detected. We find three distinct types of energy distributions, one of which is supersoft. This class consists of seven objects, among them two symbiotic novae, RR Tel and SMC3, which are discussed in more detail in this paper. The supersoft emission is produced by photospheric emission from the hot star. For RR Tel the ROSAT and IUE observations are simultaneously reproduced with a white dwarf type atmosphere with $T_{\text{eff}} = 142\,000$ K. SMC3 can only be fitted with a WR-type atmosphere with considerable mass loss ($\dot{M} \sim 10^{-5} M_\odot/\text{yr}$), and a temperature in excess of $T_{\text{eff}} \gtrsim 260\,000$ K.

1 Introduction

Some supersoft sources have been identified as symbiotic stars. Symbiotics are a class of interacting binaries with about 150 confirmed members. The stellar components are usually a RGB or AGB star and a very hot ($T_{\text{eff}} \sim 10^5$ K) white dwarf. The hot star radiatively ionizes the surrounding gas in the system, which possibly originates from the wind of the cool star. The optical spectrum is composite, with emission lines from the ionization nebula superimposed on the continuum of the red giant. The UV is dominated by nebular emission, but towards short wavelengths the contribution from the hot star increases and becomes dominant at the short wavelength end of the IUE window. For the X-ray range either supersoft photospheric emission from the hot star can be expected, or harder radiation from shocks in the gas streams.

With EINSTEIN and EXOSAT, a total of 6 symbiotic objects were detected (Allen 1981, Anderson et al. 1981, Willson et al. 1984, Kwok & Leahy 1984, Piro et al. 1985, Viotti et al. 1987, Leahy & Taylor 1987). Bickert et al. (1993) reported 14 objects to be detected in the ROSAT all-sky-survey. Since the completion of that survey, several objects were recorded during pointed observations and hidden in the ROSAT archive. In many cases the symbiotic stars were not the very target of the observation, but accidently covered by the 2° PSPC field centered on another object. In Jordan et al. (1994), Mürset et al. (1995), and

Jordan et al. (1996) we explored the interpretation possibilities of such pointed observations.

In this Paper we summarize work on the two supersoft symbiotic novae RR Tel and SMC3, and we give first results from a search for further supersoft symbiotic sources.

2 Symbiotic Stars in the ROSAT Archive

We systematically searched in the ROSAT archive for symbiotic stars. We found 30 objects situated in fields covered by pointed PSPC observations; 16 of them were in fact detected with a significant count rate (see Table 1). Details will be given in Paper II (Mürset et al. 1996b).

Table 1. Symbiotic stars detected on pointed PSPC observations.

Star	Date	Count rate [10^{-2} s^{-1}]	Type
EG And	July 1991	0.8 ± 0.3	β
SMC3	April 1992	20.4 ± 0.4	α
Ln358	Oct – Nov 1992	2.9 ± 0.2	α
RX Pup	April – June 1993	6.0 ± 0.6	β
AG Dra	1992/1993	$92 \ldots 114 \pm 3.0$	α
Draco C-1	April 1992	5.3 ± 0.6	α
GX 1+4	Sept 1993	0.6 ± 0.4	γ
Hen1591	Sept 1993	1.4 ± 0.6	γ
CH Cyg	March 1992	37.2 ± 0.9	β
HM Sge	Nov 1992	4.5 ± 0.4	β
V1016 Cyg	Nov 1992	1.2 ± 0.3	β
RR Tel	April 1992	18.3 ± 0.8	α
PU Vul	Nov 1992	0.4 ± 0.1	β
CD−43.14304	April – Oct 1993	9.1 ± 0.4	α
AG Peg	June 1993	6.7 ± 0.5	β
R Aqr	Dec 1992	4.1 ± 0.3	α

not detected: SMC2, RAW1691, BE191, N67, Sanduleak's star, MWC 560, He2-38, RW Hya, Ap1-8, H2-38, Ap1-9, AS281, MWC 960, Z And

The extracted pulse height distributions can be divided into three distinct groups (cf. Fig. 1):

α: supersoft spectra. The two symbiotic novae RR Tel and SMC3 are typical members of this group (Jordan et al. 1994, 1996).

β: objects with a harder pulse height distribution that typically peaks at about 0.8 keV. AG Peg (Mürset et al. 1995) may serve as a prototype.

γ: relatively hard X-ray sources like GX1+4. In fact, GX1+4 has the hardest spectrum of all known X-ray binaries (Predehl et al. 1995).

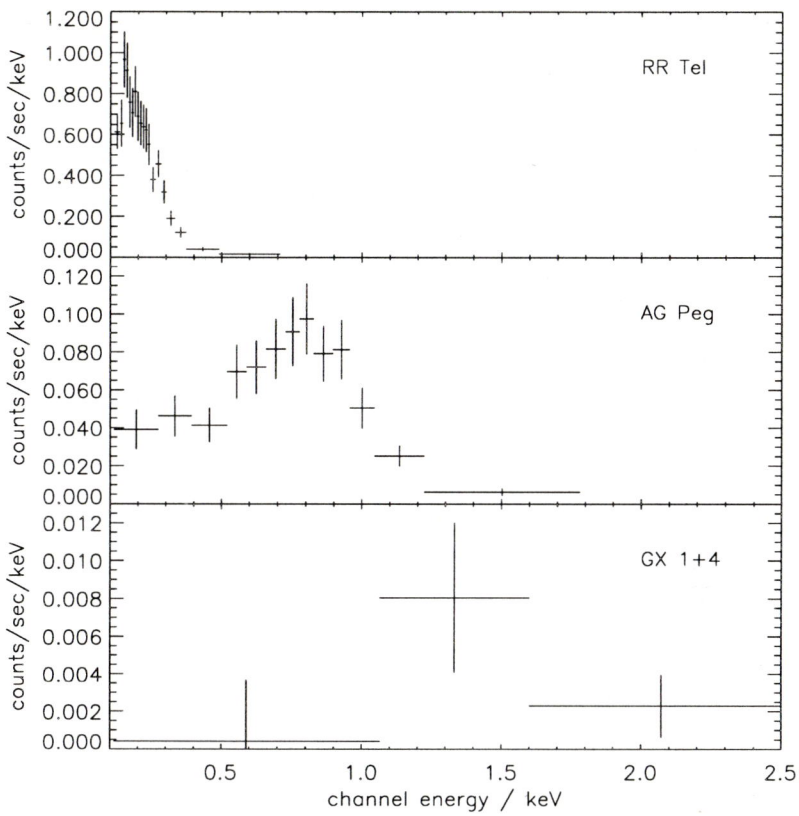

Fig. 1. Observed PSPC pulse height distributions for RR Tel (type α, Jordan et al. 1994), AG Peg (type β, Mürset et al. 1995), and GX1+4 (type γ, Predehl et al. 1995).

So far, the three X-ray types of symbiotic stars are a mere observational discovery. They become a meaningful classification scheme if they are related to physical parameters of the system. Indeed, a physical interpretation of the $\alpha/\beta/\gamma$-system appears well possible:

α: a supersoft spectrum is expected from those symbiotic systems which contain particularly hot white dwarfs. In these cases the photospheres produce photons hard enough ($h\nu \approx 0.2$ keV) to leak into the ROSAT window.

β: The classical model of a symbiotic nebula consists of the ionized section of the wind that flows radially away from the cool star. However, in several cases there is now clear evidence that the hot star loses mass as well. In AG Peg, for instance, N v λ1240 displays P Cyg profiles (Nussbaumer et al. 1995) which can be modelled with a wind of $v_\infty = 950$ km/s and $\dot{M} = 10^{-6.7}$ $M_\odot$/yr (Schmutz 1996). If the hot star sheds a substantial wind, the shock region where both winds collide is expected to have a temperature of several million

K. Mürset et al. (1995) have shown that the β–type X-ray emission of AG Peg indeed matches the theoretical predictions for the colliding wind scenario. β–type plasma emission will be analyzed in Paper II.

γ: GX1+4, which is one of the two γ–type objects, hosts a pulsar as the hot component (e.g. Davidsen et al. 1977). We may speculate that this is also the case for the other γ–type object, Hen1591.

The $\alpha/\beta/\gamma$–classification, thus, mirrors basic properties (temperature, mass-loss, white dwarf/neutron star) of the hot star. It reminds us of the infrared classes "s" and "d" (Webster & Allen 1975) which mainly correspond to basic parameters of the cool star.

A weak X-ray component of type β was detected in RR Tel which is an object of type α. From this we learn that the different emission sources do not necessarily exclude each other mutually. The classification of an object simply mirrors the relative strengths of the components.

Out of the 16 detections, 7 are supersoft (type α). For the galaxy the ratio known : observed : detected : supersoft is 150:22:13:4, whereas it is 13:8:3:3 for the extragalactic symbiotics. The surprisingly high fraction of detections of extragalactic supersoft symbiotics is partly due to the comparably long exposure times on SMC fields, but it also shows the influence of the interstellar extinction: The galactic symbiotic stars are concentrated towards the center of the Milky Way (Kenyon 1986), and many of them are strongly reddened.

3 A Model for the Supersoft Symbiotic Stars

3.1 What Can We Learn from the Supersoft X-Rays?

Supersoft X-ray data have the potential to crucially improve our understanding of the hot stars in symbiotic systems:

- It is noteworthy that the X-rays directly originate from the hot star. In other regimes of the spectrum the nebula or the cool star are much brighter than the hot star. The hot star can then only be investigated indirectly, e.g. by modelling the emission nebula with the properties of the hot star as model parameters. Indirect methods can certainly yield correct results, but at a lower level of confidence than direct observations. In the case of RR Tel Jordan et al. (1994) find similar parameters as Mürset & Nussbaumer (1994) guessed from the UV alone. On the other hand, the ROSAT based results of Jordan et al. (1996) for SMC3 strongly differ from the UV investigation by Vogel & Morgan (1994).
- The ROSAT window is even that part of the stellar spectrum which, due to the presence of ionization edges, depends most sensitively on the temperature and the chemical composition. The determination of the chemical composition of the hot star is certainly beyond the possibilities of IUE data alone.

Table 2. Solutions for RR Tel and SMC3

Parameter	RR Tel	SMC3
Type of atmosphere	hydrostatic	wind model
Elemental composition	solar	$C/He > 2 \cdot 10^{-4}$
T_{eff} [K]	142 000	> 260 000
L [$L_\odot$]	3 500	12 000
R [$R_\odot$]	0.1	< 0.05
$\log g$ [cgs]	6.5	—
$\dfrac{\dot{M}}{v_\infty} \left[\dfrac{M_\odot/\text{yr}}{\text{km/s}}\right]$	—	$\sim 10^{-8}$
N_H [cm^{-2}]	$2 \cdot 10^{20}$	$9 \cdot 10^{20}$

It would, however, be wrong to base an analysis solely on X-ray data. These show an extreme end of the hot star's spectrum, and in addition the spectral resolution of the PSPC is not satisfactory. A reliable result is obtained by *simultaneously* fitting the ROSAT observations, the UV continuum (stellar + nebular) and crucial emission lines.

This was done by Jordan et al. (1994) for RR Tel and by Jordan et al. (1996) for SMC3. First, they computed a grid of model spectra appropriate for very hot stars. Then the nebular emission was calculated with a photo-ionization code. The resulting stellar + nebular emission was requested to match the ROSAT and IUE data. The stellar radius, the size of the nebula and the interstellar absorption column enter the models as additional fit parameters. Table 3.1 summarizes the solutions for the hot stars.

3.2 RR Tel

During the first decades of this century, in RR Tel the Mira type variations of the cool component were observed (Mayall 1949), and no signatures of a compact companion showed up. In 1944, the outburst started with a 7 mag brightness rise to $M_V = -5$. In the next few years, the object was wrapped into an expanded photosphere ($R \approx 100$ $R_\odot$, Mürset & Nussbaumer 1994). Then, the object shrank and turned hotter, exciting low ionization emission lines in the surrounding gas. During the transition phase, strong WN-type wind features were observed by Thackeray & Webster (1974). After that, RR Tel has been smoothly evolving to higher and higher temperatures, as could be seen from the increasing ionization (e.g. Thackeray 1977). Presently, the largest observed ionization potential is that of [Mg VI]. The light curve and ionization records can be found in Mürset & Nussbaumer (1994).

A hydrostatic, white dwarf-type model atmosphere with solar elemental composition and a temperature of $T_{\text{eff}} = 142\,000$ K is a possible solution to the present data (Jordan et al. 1994). Jordan et al. did not really exploit the phase space of possible chemical compositions of the atmosphere. However, we belief that the temperatures and luminosities which could possibly be derived with other abundances would not differ too much.

3.3 SMC3

SMC3 (= RX J0048.4−7332) has been in outburst since 1981 (Morgan 1992). Few is, however, known about its further history. The overall spectrum of SMC3 is remarkable. The optical spectrum (Morgan 1992, Mürset et al. 1996a) reveals nebular ionization that is unusually high for a symbiotic (Fe^{+9}). It proves that the hot star emits a significant rate of photons harder than $h\nu > 235$ eV. On the other hand, the UV spectrum (Vogel & Morgan 1994) is typical for a system with rather moderate excitation. Finally, the PSPC count rate (Kahabka et al. 1994) is higher than the count rate of RR Tel, despite of the 20 times larger distance! In summary, the spectrum looks confusing at first sight and provides a challenge to models of symbiotic stars.

Using plane-parallel, hydrostatic model atmospheres, Jordan et al. (1996) could not find a solution fitting all data consistently. It turned out that every atmosphere that is hot and luminous enough to reproduce the X-ray observations is at the same time too bright in the EUV range between the He II ionization edge and the ROSAT range leading to a brighter emission nebula than observed. The situation is different with WR-type atmosphere models. If the mass loss is sufficiently strong the wind is opaque to the emission in this spectral region. The mass loss can be adjusted to yield the observed nebular He II emission. Figure 2 shows WR-type models that differ slightly in mass loss. In this manner Jordan et al. derived a mass loss of several times 10^{-6} $M_\odot$/yr. Figure 3 displays fits to the ROSAT pulse height distribution. In order to reproduce the X-rays $T_{\text{eff}} \gtrsim 260\,000$ K is required (Fig. 4 of Jordan et al. 1996). No upper limit can be specified, because with increasing temperature the energy distribution degenerates to a common shape.

4 Concluding Remarks

- The use of multi-wavelength data (optical+UV+X-ray) is extremely important for the analysis of the hot components in symbiotic systems.
- We further emphasize that sophisticated non-LTE atmospheres have to be used. Tests with black body models clearly led to wrong results.
- Accretion onto the hot components of symbiotic systems should not be adopted as self-evident; some objects lose mass at least episodically. As SMC3 shows, the mass-loss rates can be extremely high. May be, simultaneous accretion and mass-loss is possible.

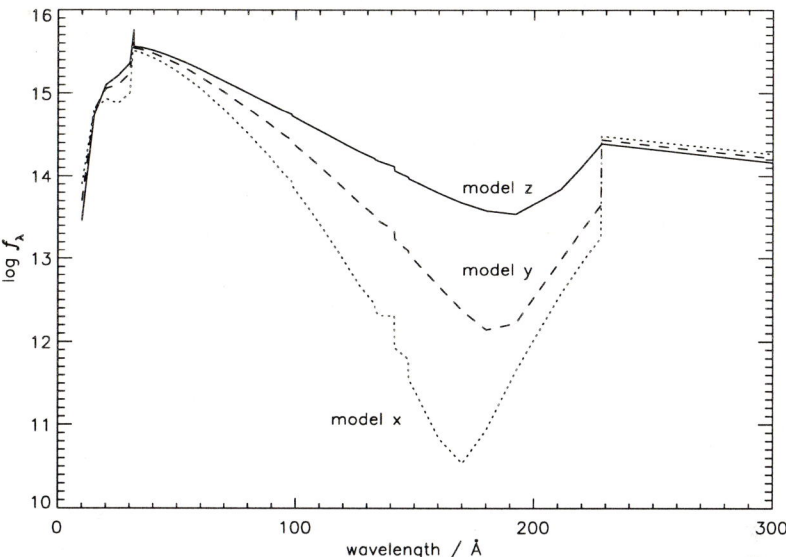

Fig. 2. Relative energy distribution of three WR-type models with approximately the same effective temperature of $T_{\text{eff}} = 300$ kK, but differing in the mass loss rate (models $\underline{x}$, $\underline{y}$, $\underline{z}$: $\dot{M}_{\underline{x}} = 10^{-5.1}, 10^{-5.2}, 10^{-5.3}$ M$_\odot$/yr.).

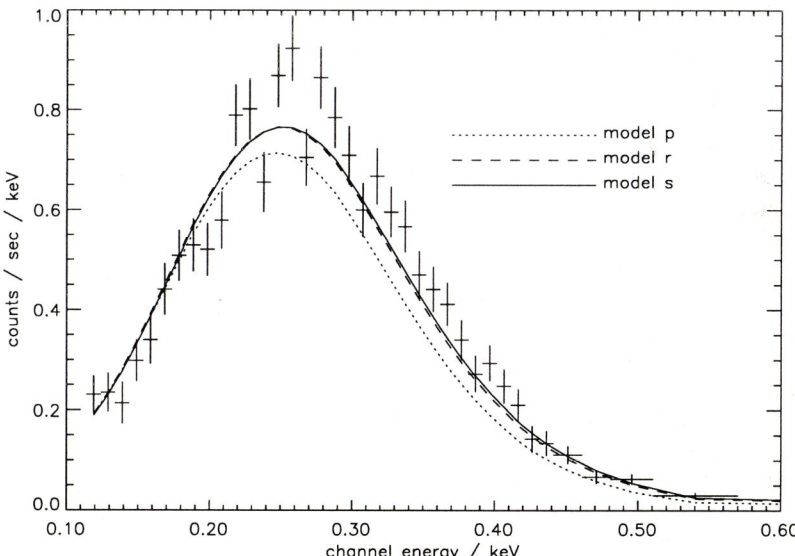

Fig. 3. Comparison of the observed PHD of SMC3 with three Wolf-Rayet type models with $T_{\text{eff}} = 220$ kK (model $\underline{p}$), 309 kK ($\underline{r}$), and 354 kK ($\underline{s}$). The predicted flux from model $\underline{p}$ is not hard enough to reproduce the observation, while the two hotter models ($\underline{r}$) and ($\underline{s}$) lead to relatively good fit to the data. The specific parameters of the models are listed in Table 2 of Jordan et al. (1996)

- The luminosities are well below the Eddington limit.
- The wind of SMC3 contains some carbon. The matter is therefore not completely CNO processed, or carbon is admixed from outside. In this respect, SMC3 differs from the galactic symbiotic nova AG Peg which exhibits a strong carbon deficiency (Schmutz 1996).

Acknowledgement: We thank Werner Schmutz for helpful comments. Work on ROSAT data in Kiel was supported by DARA grant 50 OR 94091.

References

Allen D.A., 1981, MNRAS 197, 739
Anderson C.M., Cassinelli J.P., Sanders W.T., 1981, ApJ 247, L127
Bickert K.F., Stencel R.E., Luthardt R., 1993, in: Planetary Nebulae, IAU Symp. 155, eds. R. Weinberger & A. Acker, p. 405
Davidsen A., Malina R., Bowyer S., 1977, ApJ 211, 866
Jordan S., Mürset U., Werner K., 1994, A&A 283, 475
Jordan S., Schmutz W., Wolff B., Werner K., Mürset U., 1996, A&A (in press)
Kahabka P., Pietsch W., Hasinger G., 1994, A&A 288, 538
Kenyon S.J., 1986, The symbiotic stars, Cambridge University Press
Kwok S., Leahy D.A., 1984, ApJ 283, 675
Leahy D.A., Taylor A.R., 1987, A&A 176, 262
Mayall M.W., 1949, Harv. Bull. 919, 15
Morgan D.H., 1992, MNRAS 258, 639
Mürset U., Nussbaumer H., 1994, A&A 282, 586
Mürset U., Jordan S., Walder R., 1995, A&A 297, L87
Mürset U., Schild H., Vogel M., 1996a, A&A (in press)
Mürset U., Wolff B., Jordan S., 1996b (in prep.) (Paper II)
Nussbaumer H., Schmutz W., Vogel M., 1995, A&A 293, L13
Predehl P., Friedrich S., Staubert R., 1995, A&A 294, L33
Piro L., Cassatella A., Spinoglio L., Viotti R., Altamore A., 1985, IAU Circ. 4082
Schmutz W., 1996, in: Science with the Hubble Space Telescope II, eds. P. Benvenuti, F.D. Macchetto, E.J. Schreier (in press)
Thackeray A.D., 1977, Mem. R. Soc. 83, 1
Thackeray A.D., Webster B.L., 1974, MNRAS 168, 101
Viotti R., Piro L., Friedjung M., Cassatella A., 1987, ApJ 319, L7
Vogel M., Morgan D.H., 1994, A&A 288, 842
Webster B.L., Allen D.A., 1975, MNRAS 171, 171
Willson L.A., Wallerstein G., Brugel E.W., Stencel R.E., 1984, A&A 133, 154

Multiwavelength Observations of the Symbiotic Star AG Dra During 1979–1995

R. Viotti[1], R. González-Riestra[2], F. Montagni[3], J. Mattei[4], M. Maesano[3], J. Greiner[5], M. Friedjung[6], A. Altamore[7]

[1] Istituto di Astrofisica Spaziale, CNR, Via Fermi 21, I-00044 Frascati (Roma)
[2] IUE Observatory, ESA, P.O. Box 50727, SP-28080 Madrid
[3] Istituto Astronomico, Università La Sapienza, Via Lancisi 29, I-00161 Roma
[4] American Association of Variable Star Observers, 25 Birch Street, Cambridge, MA 02138
[5] Max-Planck-Institut für Extraterrestrische Physik, D-85740 Garching
[6] Institut d'Astrophysique, CNRS, 98bis Boulevard Arago, F-75014 Paris
[7] Dip. Fisica, Università Roma III, Via della Vasca Navale 84, I-00146 Roma

Abstract. Since 1980 the high velocity symbiotic system AG Dra has undergone three phases of activity (or "outbursts") which were extensively observed in optical, ultraviolet and X-ray bands. We describe the historical light curve of the star, and summarize 17 years of UV monitoring with the IUE satellite.

1 Introduction

The category of symbiotic stars includes some two hundred objects whose main features are: (i) the light curve, characterized by a number of "outbursts", and (ii) the composite spectrum. The latter one generally arises from the emission of a binary system composed by a late-type giant or bright giant star, and an evolved hot dwarf companion, whose radiation ionizes an extended circumstellar nebula filled in by the cool star's wind, and by some kinds of "ejecta" from the dwarf companion. The typical time scale of the variations is up to many years, which makes possible the extensive multiwavelength monitoring of a number of "representative" objects. In this regard AG Dra has been in the last years the best studied symbiotic object (and one of the best studied cataclysmic-type variables) from the optical to the ultraviolet and to X-rays, and the amount of observational data so far collected might give important clues for understanding the physics of the *cataclysmic phenomenon*.

AG Dra is a high-velocity (-148 km s^{-1}), high-galactic latitude ($bII=+41°$) symbiotic system, known since the earlier HEAO-2 observations for being an intense soft-X ray source (Anderson et al. 1981). The higher sensitivity at lower energies allowed the EXOSAT and ROSAT satellites to disclose the supersoft nature of the source (Piro et al. 1985, Bickert et al. 1996, Greiner et al. 1996). AG Dra is a photometric and spectroscopic binary (T=$\sim 554^d$, Meinunger 1979, Mikolajewska et al. 1995), with a luminous K-star primary and a hot ($\sim 10^5$ K) dwarf companion which is a source of an intense UV radiation producing a rich emission line spectrum, and a strong UV continuum. The larger amplitude of the variation in the U–band, and the broadness of the minimum as described by Meinunger (1979) imply that the occulted region is an extended H II region in between the two stars, most probably the ionized cool star's wind. The system distance is so far unknown mainly because of the uncertainty on the luminosity class of the red component – e.g. III according to Viotti et al. (1983), Ib according to Huang et al. (1994). The classification is difficult because of the lower metallicity of the star, and of the non-negligible nebular contribution to the stellar continuum also in the longer wavelengths (as discussed below), which both should significantly affect the photospheric line ratio. Of special importance is the low interstellar absorption towards AG Dra (E_{B-V}=0.06, Viotti et al. 1983), which makes it an ideal target for far–UV and soft X-ray observations.

The most spectacular feature of AG Dra is the fact that during the last 16 years the star underwent many different phases of activity, which has given us the opportunity to study for the first time, and repeatedly, the outburst of a symbiotic system at UV and X-ray wavelengths.

The first "major" active phase of 1980–1983 was monitored with IUE, and the results were described by Viotti et al. (1984). No X-ray observation was made during the 1980 outburst because of a failure of the power supply of the HEAO-2 IPC detector. The second "minor" active phase of 1985–1986 was observed with IUE and EXOSAT, and was described by Viotti et al. (1995) and Mikolajewska et al. (1995). Viotti et al. found a marked anticorrelation of the UV and X-ray light curves, with a large fading of the X-ray source during the two light maxima, while the UV flux largely increased. We also noted that no eclipse of the X-ray source occurred on 4 November 1985, at orbital phase 0.49 according to Meinunger's (1979) ephemeris.

For the most recent 1994–1995 "major" active phase, a campaign of coordinated optical, ultraviolet, and X-ray observations was organized, which was based on numerous target of opportunity observations with ROSAT and IUE. The results of the campaign are for the first time presented in this Workshop. In the following we shall describe the IUE observations, and the optical light curve from the AAVSO databank and from BVRI photometry. We also recall that coordinated optical spectroscopy was obtained at the Observatoire de Haute Provence, and at the Padova-Asiago and Bologna-Loiano Observatories. The ROSAT X-ray data and their model implication are discussed by Greiner et al. (1996).

2 The Visual Light Curve

Photometrically, AG Dra is a variable star which is normally at minimum with V_{min}=9.8–10.0 (star's *quiescent phase*). Erratically, AG Dra undergoes 1–2 mag light maxima, or "outbursts" at time intervals from one to many years (*active phases*). Robinson (1966) recorded 11 outbursts between 1890 and 1966, and found that in many cases a second maximum occurred after 250–500 days. Iijima

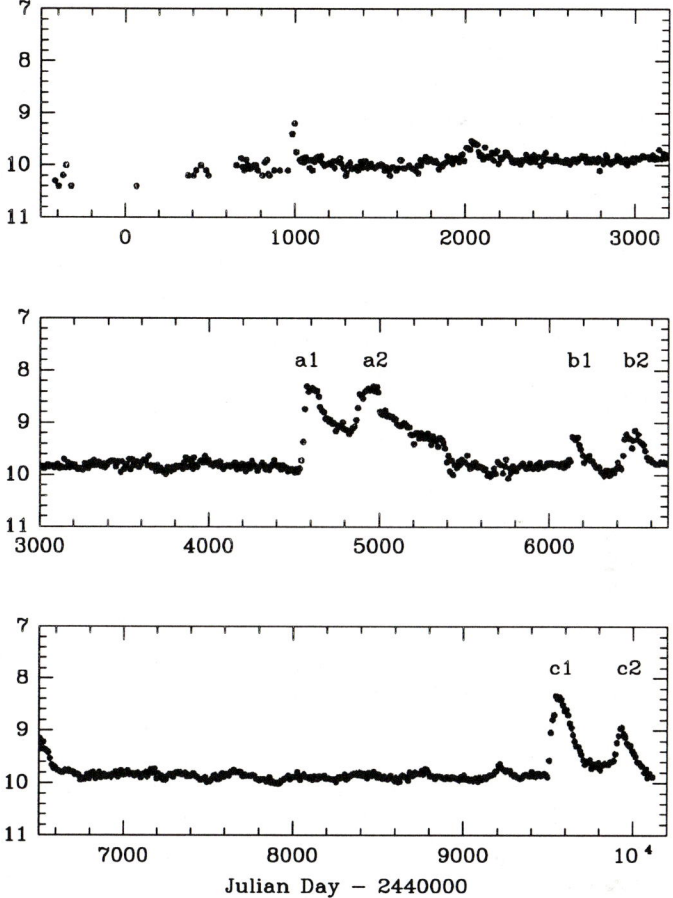

Fig. 1. The visual light curve of AG Dra from 1967 to 1995 (from AAVSO). The three most recent active phases are marked.

et al. (1987) noted that since 1935 the main outbursts repeated with an apparent periodicity of about 15 years, corresponding to nearly exactly 10 orbital periods.

Figure 1 gives the AAVSO light curve of AG Dra during 1967–1995. As shown in the figure, during the last three decades, the star has been in quiescence during 1967–1980, 1983–1985, and 1986–1994. In this period six light maxima have been recorded which can be grouped into three 1 year–splitted twin peaks ("primary" and "secondary" outbursts): November 1980–August 1981 ($a1$ and $a2$), March 1985–January 1986 ($b1$ and $b2$), and June 1994–July 1995 ($c1$ and $c2$). As concerns the two "major" active phases (1980–82 and 1994–95), the timelag was of about 4961^d, corresponding to 8.96 554^d–periods (or to 9.00 551.2^d–periods), which seems to confirm the existence of a relation between the recurrence time of the active phases and the orbital period.

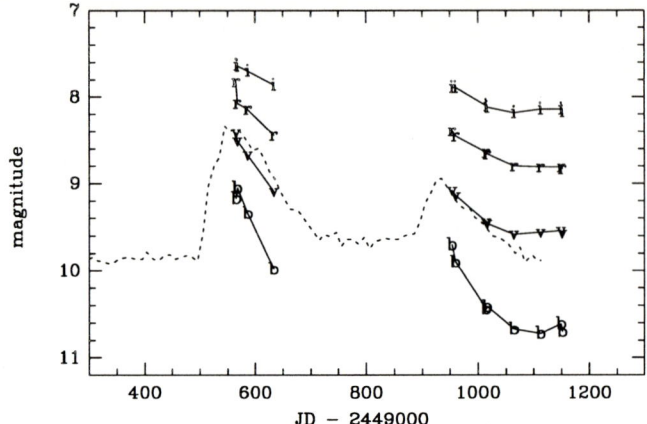

Fig. 2. The four-colour photometry of AG Dra during 1994-1995 (BVRI, from bottom to top). The dotted line is the AAVSO visual light curve.

The four–colour photometry obtained at the Greve (Firenze, Italy) and Vallinfreda (Roma) is shown in Figure 2. It appears that AG Dra largely varied in all the colours with an amplitude which is larger at the shorter wavelengths. This behaviour resembles that of AG Dra in quiescence during the orbital motion as described by Meinunger (1979). In both cases the effect is due to the larger contribution of the (variable) nebular emission towards shorter wavelengths. This shows that the *optical* outburst is the direct result of the increased volume of the ionized nebula.

3 Spectral Variation in the Ultraviolet

AG Dra has been monitored in the UV by the IUE satellite since 1979. This has given the unique opportunity to follow all the recent six "outbursts", and to analyse the behaviour of the nebular and hot dwarf's emission during different activity phases (e.g. Viotti et al. 1984, Lutz et al. 1987, Kafatos et al. 1993). It should also allow to enlighten the main features of the activity phases, which is fundamental for a correct modeling of the phenomenon. The variation of the UV continuum and emission line fluxes in the three phases of *activity* is shown in Figure 3. For the continuum we have chosen the far–UV (at 1340 Å) and near–UV (at 2860 Å) regions, which are less affected by the line emission, and where the dominant contribution is from the hot star (Rayleigh-Jeans black-body tail), and from the ionized nebula (Balmer continuum), respectively.

Fig. 3 shows that the "primary" outbursts were accompanied by an increase of the UV continuum and line flux, with an amplitude tyically much larger than that in the optical region. The large increase in the N V and He II lines is an indication of a marked increase of the flux of 54–77 eV photons at the time of the outburst. The He II$_{1640A}$/F$_{1340A}$ flux ratio, which is a measure of the hot

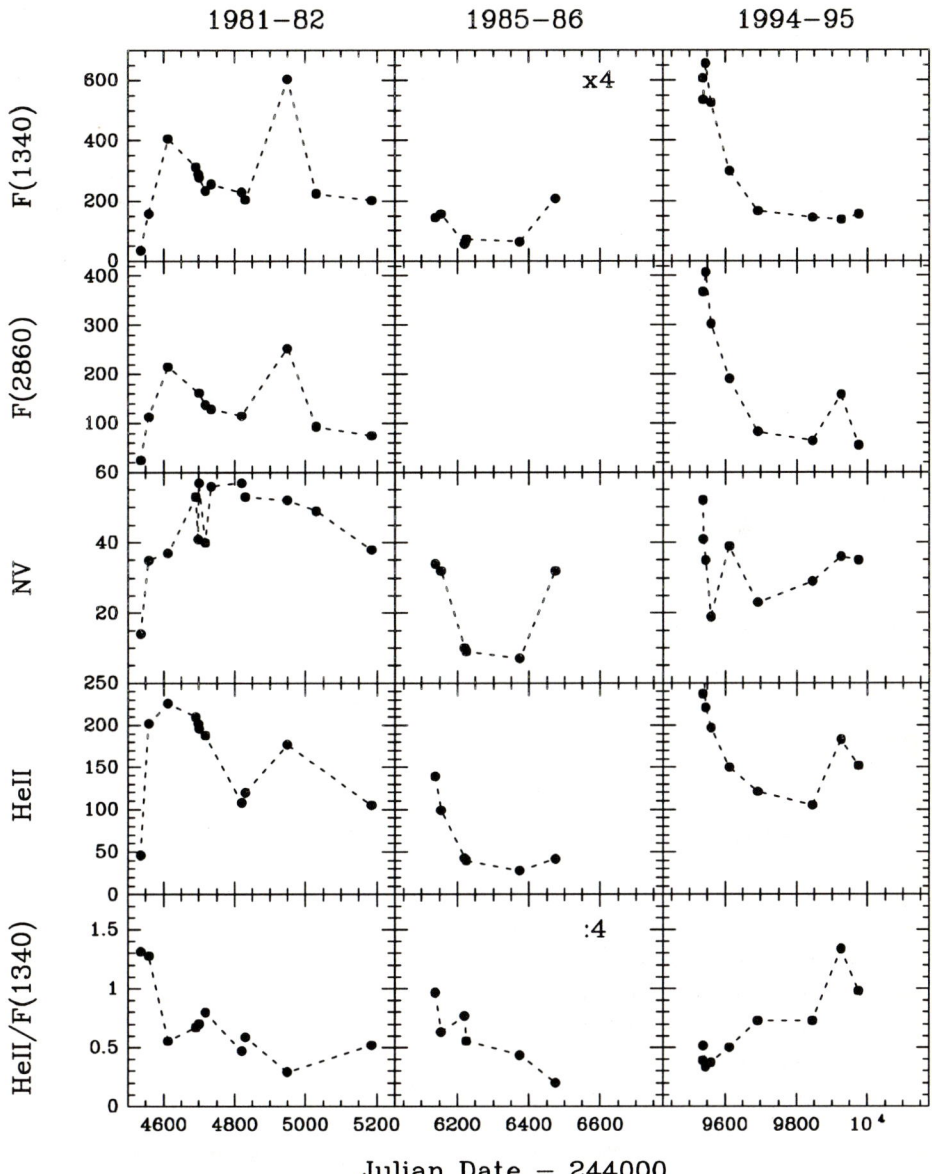

Fig. 3. The UV variation of AG Dra during the three active phases observed with the IUE satellite. From top to bottom: the far–UV continuum at 1340 Å, the near–UV continuum at 2860 Å, N v λ1240, He II λ1640, and He II/$F_{1340 A}$. All the fluxes are corrected for an interstellar extinction of $E_{B-V}=0.06$.

star's Zanstra temperature, did not change during the November 1980 rising phase, but largely decreased later on. We also note that the behaviour of the continuum and emission line flux during the "secondary" maximum and in the declining phases did not follow a common trend during the three active phases.

The high resolution IUE observations have disclosed the presence of a hot (N V), intermediate velocity ($-200/300$ km s^{-1}) wind which is present at least during the major outbursts (in 1981 and 1994). According to Viotti et al. (1983) this wind should arise from the hot WD, and we argue that it should originate from a lower gravity envelope of the WD during outburst.

Figure 4 shows the UV–to–optical energy distribution of AG Dra one month after the first light maximum, and in the following minimum phase. During outburst the flux increase has been very large in the far–UV, which is dominated by the tail of the hot star radiation. There is also an indication of a large Balmer jump. The flux increase during outburst was significant also at longer wavelengths. In particular we remark from the figure a 40% variation in the I-band, which implies that at maximum the "nebular" emission largely contributed also to the near–IR radiation of AG Dra. We cannot discard the possibility of a non-negligible contribution of the nebular continuum (f-f and f-b) to the longwave energy distribution also at minimum. In any case, both the VRI colours and the photospheric line depths might be affected by the nebular continuum, and they should therefore be used with some care in the determination of the stellar temperature and surface gravity.

AG Dra is surrounded by an extended nebula, ionized by the far–UV radiation of the hot dwarf, which is partly, or mostly formed from the K–star wind. The UV spectroscopy suggests that the CNO composition of the nebula is peculiar, with low C/N and (C+N)/O ratios (0.63 and 0.43, respectively), which is typical of metal poor cool giant atmospheres after CN cycle burning and a first dredge up phase (maybe plus mass transfer effects) (Schmid & Nussbaumer 1993). In this regard, it would be interesting to determine the chemical composition of the cool star atmosphere which was not yet done so far.

In quiescence, during the orbital motion the ionized nebula is partially occulted by the red star which explains the very broad minimum in U. EXOSAT observations at phase 0.50 gave no indication of an eclipse, which could be due either to the orbit inclination smaller than 90°, or to a very short duration of the eclipse combined with a slight phase shift. During the active phases, the increased flux of far–UV photons ionizes a much more extended region, which is masking the occultation effects. The major active phases seem to begin at the same orbital phase (0.1–0.2) which might suggest a link between orbital motion and outburst. This and the nearly regular recurrence time of the major outbursts could be understood if the orbit is eccentric, but the radial velocity curve gives no indication of a measurable orbital eccentricity (Kenyon & Garcia 1986, Mikolajewska et al. 1995). Alternatively, if the cool star is filling, or nearly filling its Roche lobe, an even small eccentricity or an increase of the surface activity of the red star, would substantially increase the accretion rate producing the outburst.

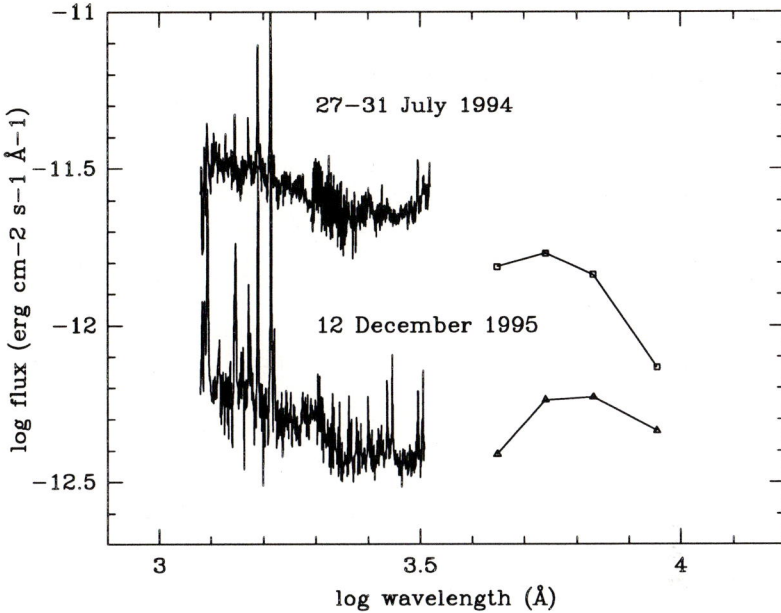

Fig. 4. Ultraviolet (IUE)–optical (BVRI) energy distribution of AG Dra during the 1994 outburst (27-31 July 1994), and at minimum (11 December 1995).

Acknowledgements: This work is based on data obtained from the IUE Data Bank, and on observations by IUE collected at the IUE Observatory of ESA at VILSPA within approved IUE programmes, or as targets–of–opportunity. We are grateful to the many amateurs who have continuously monitored AG Dra during the different phases of activity, and whose data are providing a precious database for this and future investigations. R.V. was partially supported by funds of the Agenzia Spaziale Italiana (ASI) under contract ASI 94-RS-59. JG is supported by the Deutsche Agentur für Raumfahrtangelegenheiten (DARA) GmbH under contract FKZ 50 OR 9201.

References

Anderson C.M., Cassinelli J.P., Sanders W.T., 1981, ApJ 247, L127
Bickert K., Greiner J., Stencel R.E., 1996, this volume p. 225
Greiner J., Bickert K., Luthardt R., Viotti R., Altamore A., Gonzalez-Riestra G., 1996, this volume p. 267
Huang C.-C., Friedjung M., Zhou Z.X., 1994, A&AS 106, 413
Iijima T., Vittone A., Chochol D., 1987, A&A 178, 203
Kafatos M., Meyer S.R., Martin I., 1993, ApJS 84, 201
Kenyon S.J., Garcia M.R., 1986, AJ 91, 125
Lutz J.H., Lutz T.E., Dull J.D., Kolb D.D., 1987, AJ 94, 463

Meinunger L., 1979, IBVS No. 1611
Mikolajewska J., Kenyon S.J., Mikolajewski M., et al., 1995, AJ 109, 1289
Mürset U., Nussbaumer H., Schmid H.M., Vogel M., 1991, A&A 248, 458
Piro L., Cassatella A., Spinoglio L., Viotti R., Altamore A., 1985, IAU Circ. 4082
Robinson L., 1969, Peren. Sviosdi 16, 587
Schmid H.M., Nussbaumer H., 1993, A&A 268, 159
Viotti R., Ricciardi O., Ponz D., et al., 1983, A&A 119, 285
Viotti R., Altamore A., Baratta G.B., et al., 1984, ApJ 283, 226
Viotti R., Giommi P., Friedjung M., Altamore A., 1995, Proc. Abano-Padova Conference on Cataclysmic Variables, A. Bianchini et al. (eds.), ASSL 205, 195

UV and X-Ray Monitoring of AG Draconis During the 1994/1995 Outbursts*

J. Greiner[1], K. Bickert[1], R. Luthardt[2], R. Viotti[3], A. Altamore[4],
R. González-Riestra[5]**

[1] Max-Planck-Institut für Extraterrestrische Physik, 85740 Garching, Germany
[2] Sternwarte Sonneberg, 96515 Sonneberg, Germany
[3] Istituto di Astrofisica Spaziale, CNR, Via Enrico Fermi 21, 00044 Frascati, Italy
[4] Dipartimento di Fisica E. Amaldi, Università Roma III, 00146 Roma, Italy
[5] IUE Observatory, ESA, Villafranca del Castillo, 28080 Madrid, Spain

Abstract. The recent 1994-1995 active phase of AG Draconis has given us for the first time the opportunity to follow the full X-ray behaviour of a symbiotic star during two successive outbursts and to compare with its quiescence X-ray emission. With *ROSAT* observations we have discovered a remarkable decrease of the X-ray flux during both optical maxima, followed by a gradual recovering to the pre-outburst flux. In the UV the events were characterized by a large increase of the emission line and continuum fluxes, comparable to the behaviour of AG Dra during the 1980-81 active phase. The anticorrelation of X-ray/UV flux and optical brightness evolution is shown to very likely be due to a temperature decrease of the hot component. Such a temperature decrease could be produced by an increased mass transfer to the burning compact object, causing it to slowly expand to about twice its original size.

1 Introduction

The symbiotic star AG Draconis (BD +67°922) plays an outstanding role inside the group of symbiotic stars (binary systems consisting of a cool, luminous visual primary and a hot compact object (white dwarf, subdwarf) as secondary component) because of its high galactic latitude (b^{II}=+41°), its large radial velocity of v_r=–148 km/s and its relatively early spectral type (K). AG Dra is probably a metal poor symbiotic binary in the galactic halo.

The historical light curve of AG Dra is characterized by a sequence of active and quiescent phases (e.g. Robinson 1969) The activity is represented by 1–2 mag light maxima (currently called *outbursts* or *eruptions*) frequently followed by one or more secondary maxima. Between the active phases AG Dra is spending long periods (few years to decades) at minimum light with small (0.1 mag) semiregular photometric variations in B and V with pseudo-periods of 300-400 days (Luthardt 1990). However, in the U band regular variations with amplitudes of 1 mag and a period of 554 days have been discovered by Meinunger (1979).

---

*Based on observations made with the *International Ultraviolet Explorer* collected at the Villafranca Satellite Tracking Station of the European Space Agency, Spain.
**Also Astronomy Division, Space Science Department, ESTEC.

This periodicity is associated with the orbital motion of the system, as confirmed by the radial velocity observations of Kenyon and Garcia (1986).

The optical spectrum of AG Dra is typical of a symbiotic star, with a probably stable cool component which dominates the yellow-red region, and a largely variable "nebular" component with a strong blue-ultraviolet continuum and a rich emission line spectrum (e.g. Boyarchuk 1966). According to most authors the cool component is a K3 giant, which together with its large radial velocity and high galactic latitude, would place AG Dra in the halo population at a distance of about 1.2 kpc, 0.8 kpc above the galactic plane. More recently, Huang et al. (1994) suggested that the cool component may be of spectral type K0Ib, which would place it at a distance of about 10 kpc, at z = 6.6 kpc.

The UV continuum and line flux is largely variable with the star's activity. Viotti et al. (1984) studied the IUE spectra of AG Dra during the major 1980–1983 active phase, and found that the outburst was most energetic in the ultraviolet with an overall rise of about a factor 10 in the continuum, much larger than in the visual, and of a factor 2–5 in the emission line flux.

First X-ray observations of AG Dra during the quiescent phase with *Einstein* before the 1981-1985 series of eruptions revealed a soft spectrum (Anderson et al. 1981). The data are consistent with a blackbody source of kT=0.016 keV (Kenyon 1988) in addition to the bremsstrahlung source (kT=0.1 keV) suggested by Anderson et al. (1981). *EXOSAT* was pointed on AG Dra four times during the 1985–86 minor active phase, which was characterized by two light maxima in February 1985 and January 1986. These observations revealed a large X-ray fading with respect to quiescence (Piro 1986), the source being at least 5–6 times weaker in the EXOSAT thin Lexan filter in March 1985, and not detected in February 1986 (Viotti et al. 1995).

In June/July 1994 AG Dra went into a major outburst (Graslo et al. 1994), after which it gradually declined to the quiescent level in November 1994. Like the 1981–82 and 1985–86 episodes, AG Dra underwent a secondary outburst in July 1995. Here, we use all available *ROSAT* data to document the X-ray light curve of AG Dra over the past 5 years and report on the results of the coordinated *ROSAT /IUE* campaign during the 1994/1995 outbursts.

2 IUE Observations

AG Dra was observed by IUE as a Target of Opportunity starting on June 29, 1994. Observations have continued until February 1996, covering therefore the 1994 and 1995 outbursts as well as the period between them and the return to quiescence. In general, priority was given to low resolution spectra, but high resolution data were also obtained on most of the dates. For most of the low resolution images, both IUE apertures were used in order to have in the Large Aperture the continuum and the emission lines well exposed and in the Small Aperture (not photometric) the strongest emission lines (essentially HeII 1640 Å) not saturated. The fluxes from the Small Aperture spectra were corrected for

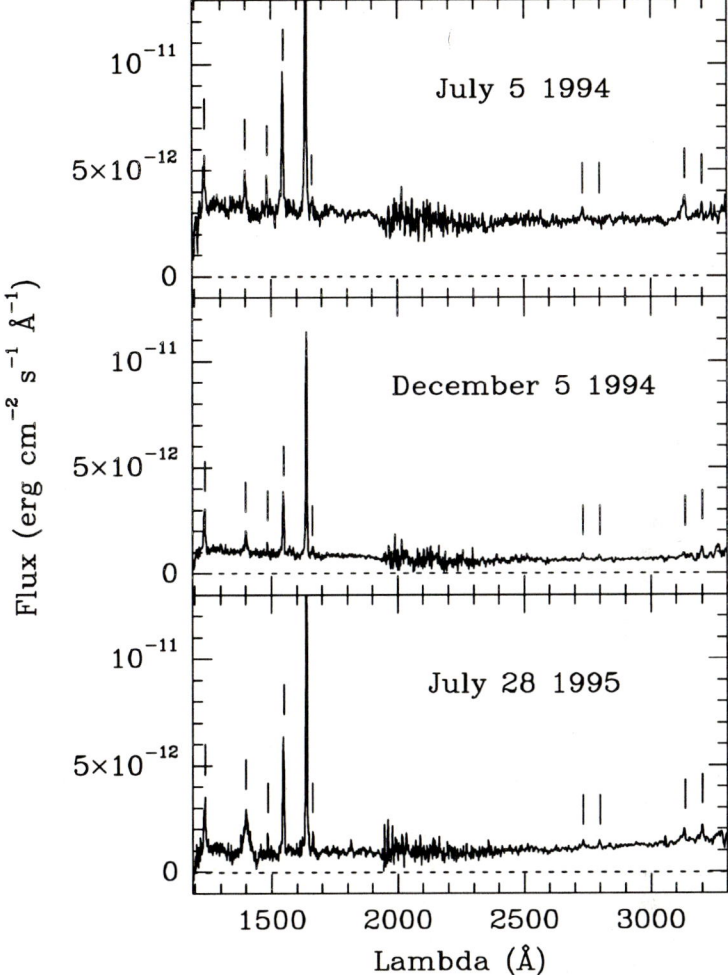

Fig. 1. IUE spectra of AG Dra at three different dates during the 1994-1995 activity phase: during the 1994 outburst (top), during the "quiescent" state between the outbursts (middle) and during the second outburst (bottom). The strongest line in the three spectra is HeII 1640 Å. The lines marked in the spectra are: NV 1240 Å, the blend SiIV + OIV] 1400 Å, NIV] 1486 Å, CIV 1550 Å, OIII] 1663 Å, HeII 2733 Å, MgII 2800 Å, OIII 3133 Å and HeII 3202 Å.

the smaller aperture transmission by comparison with the non-saturated parts of the Large Aperture data. Some representative IUE spectra are shown in Fig. 1.

The behaviour of the continuum and line intensities is shown in Fig. 2. All the fluxes are corrected for an interstellar absorption of $E(B-V)=0.06$ (Viotti et al. 1983). The feature at 1400 Å is a blend, unresolved at low resolution of the Si IV doublet at 1393.73 and 1402.73 Å, the OIV] multiplet (1399.77, 1401.16, 1404.81

270 Greiner et al.

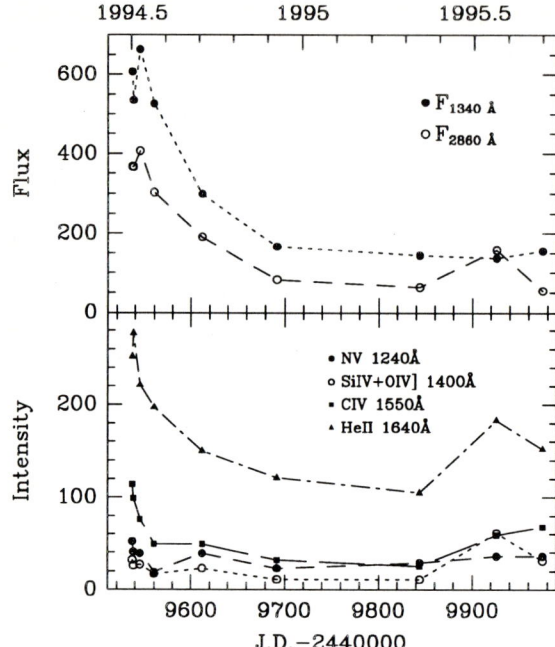

Fig. 2. Evolution of the UV continuum and emission lines of AG Dra during the 1994-1995 outburst. Units of continuum fluxes are 10^{-14} erg cm^{-2} s^{-1} Å^{-1}, and of line intensities 10^{-12} erg cm^{-2} s^{-1}. All the measurements are corrected for an interstellar reddening of E(B-V)=0.06.

and 1407.39 Å), and S IV] 1402.73 Å. O IV]1401.16 is the dominant contributor to the blend. Because of the variable intensity of the lines, at low resolution the 1400 Å feature shows extended wings with variable extension and strength (Fig. 1). The two continuum regions were chosen, as in previous works, for being less affected by the emission lines, and for being the 1340 Å respresentative of the continuum of the hot source (nearly the Rayleigh-Jeans tail of a 10^5 K blackbody), and the 2860 Å region representative of the HII Balmer continuum emission (see for instance Fernández-Castro et al. 1995).

3 ROSAT Observations

3.1 Observational Details

All-Sky-Survey AG Dra was scanned during the All-Sky-Survey over a time span of 10 days. The total observation time resulting from 95 individual scans adds up to 2.0 ksec.

Pointed Observations Several dedicated pointings on AG Dra have been performed in 1992 and 1993 with the ROSAT PSPC. All pointings were performed with the target on-axis. During the last *ROSAT* observation of AG Dra with the PSPC in the focal plane (already as a TOO) the Boron filter was erroneously left in front of the PSPC after a scheduled calibration observation.

When AG Dra was reported to go into outburst (Granslo et al. 1994) we immediately proposed for a target of opportunity observation (TOO) with ROSAT. AG Dra was scheduled to be observed during the last week of regular PSPC observations on July 7, 1994, but due to star tracker problems no photons were collected. For all the later *ROSAT* observations only the HRI could be used after the PSPC gas has been almost completely exhausted. Consequently, no spectral information is available for these observations. The first HRI observation took place on August 28, 1994, about 4 weeks after the optical maximum.

3.2 The X-Ray Lightcurve of AG Dra

The mean ROSAT PSPC countrate of AG Dra during the all-sky survey was determined to (0.99±0.15) cts/sec. Similar countrates were detected in several PSPC pointings during the quiescent time interval 1991–1993. The X-ray light curve (mean countrate over each pointing) of AG Dra as deduced from the All-Sky-Survey data taken in 1990, and 11 *ROSAT* PSPC pointings as well as 7 HRI pointings taken between 1991 and 1996 is shown in Fig. 3. The countrates of the HRI pointings have been converted with a factor of 7.5 and are also included in Fig. 3. This 5 yrs X-ray light curve displays several features:

1. The X-ray intensity has been more or less constant between 1990 and the last observation (May 1993) before the optical outburst. Using the mean best fit blackbody model with kT = 15 eV and the galactic column density $N_H = 3.15 \times 10^{20}$ cm^{-2} (see below), the unabsorbed intensity in the ROSAT band (0.1–2.4 keV) is 2.5×10^{-9} erg cm^{-2} s^{-1}.
2. During the times of the optical outbursts the observed X-ray flux drops substantially. The observed maximum amplitude of the intensity decrease is nearly a factor of 100. Due to the poor sampling we can not determine whether the amplitudes of the two observed X-ray intensity drops are similar. With the lowest intensity measurement being an upper limit, the true amplitude is certainly even larger.
3. Between the two X-ray minima the X-ray intensity nearly reached the pre-outburst level, i.e. the relaxation of whatever parameter caused these drops was nearly complete. We should, however, note that the relaxation to the pre-outburst level was faster at optical wavelength than at X-rays, and also was faster in the B band than in the U band. In December 1994 the optical V brightness ($\approx 9^m5$) was nearly back to the pre-outburst magnitude, while the X-ray intensity was still a factor of 2 lower than before the outburst.
4. The quiescent X-ray light curve shows two small, but significant intensity dips in May 1992 and April/May 1993. The latter and deeper dip coincides with the orbital minimum in the U lightcurve (Meinunger 1979, Skopal 1994).

3.3 The X-Ray Spectrum of AG Dra in Quiescence

For spectral fitting of the all-sky-survey data the photons in the amplitude channels 11–240 (though there are almost no photons above channel 50) were binned

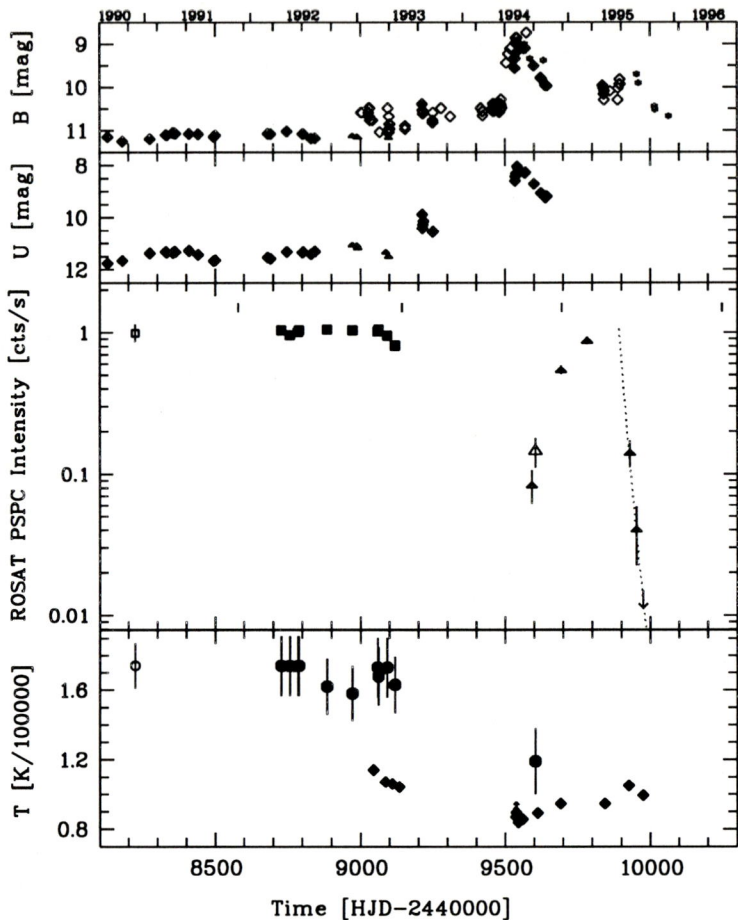

Fig. 3. The X-ray and optical light curve of AG Dra over the past 5 years. The two top panels show the U and B band variations as observed at Sonneberg Observatory (filled losanges = photoelectric photometry, open losanges = photographic sky patrol), Skalnate Pleso Observatory (Hric et al. 1994; filled triangles), and italian amateurs (Montagni et al. 1996; stars). The large middle panel shows the X-ray intensity as measured with the *ROSAT* satellite: filled squares denote PSPC observations, filled triangles are HRI observations with the countrate transformed to PSPC rates (see text), and the open triangle is the Boron filter observation corrected for the filter transmission. Statistical errors (1σ) are overplotted; those of the PSPC pointings are smaller than the symbol size. The vertical bars at the top indicate the minima of the U band lightcurve (Skopal 1994). The dotted line shows the fit of an expanding and cooling envelope to the X-ray decay light curve. The lower panel shows temperature estimates from blackbody fits to the *ROSAT* X-ray data (filled circles) and from the HeII(λ1640) flux to continuum flux at λ1340 Å as determined from the IUE spectra and assuming E(B–V)=0.06 (=Zanstra temperatures, filled losanges).

with a constant signal/noise ratio of 9σ. The fit of a blackbody model with all parameters free results in an effective temperature of $kT_{bb} = 11$ eV.

We investigated the possibility of X-ray spectral changes with time. First, we kept the absorbing column fixed at its galactic value and determined the temperature being the only fit parameter. We find no systematic trend of a temperature decrease (lower panel of Fig. 3). Second, we kept the temperature fixed (at 15 eV in the first run and at the best fit value of the two parameter fit in the second run) and checked for changes in N_H, again finding no correlation. Thus, no variations of the X-ray spectrum could be found along the orbit.

The independent estimate of the absorbing column towards AG Dra from the X-ray spectral fitting indicates that the detected AG Dra emission experiences the full galactic absorption. While fits with N_H as free parameter systematically give values slightly higher than the galactic value (which might lead to speculations of intrinsic absorption), we assess the difference to be not significant due to the strong interrelation of the fit parameters given the energy resolution of the PSPC and the softness of the X-ray spectrum. We will therefore use the galactic N_H value (3.15×10^{20} cm^{-2} according to Dickey and Lockman 1990) in the following discussion.

With fixed N_H the mean temperature during quiescence is about 14–15 eV, corresponding to 160000–175000 K. These best fit temperatures are plotted in a separate panel below the X-ray intensity (Fig. 3). The variations in temperature during the quiescent phase are consistent with a constant temperature of the hot component of AG Dra.

3.4 The X-Ray Spectrum of AG Dra in Outburst

As noted already earlier (e.g. Friedjung 1988), the observed fading of the X-ray emission during the optical outbursts of AG Dra can be caused by either a temperature decrease of the hot component or an increased absorbing layer between the X-ray source and the observer. In order to evaluate the effect of these possibilities, we have performed model calculations using the response of the ROSAT HRI. In a first step, we assume a 15 eV blackbody model and determine the increase of the absorbing column density necessary to reduce the ROSAT HRI countrate by a factor of hundred. The result is a factor of three increase. In a second step we start from the two parameter best fit and determine the temperature decrease which is necessary to reduce the ROSAT HRI countrate at a constant absorbing column (3.15×10^{20} cm^{-2}). We find that the temperature of the hot component has to decrease from 15 to 10 eV, or correspondingly from 175000 K to 115000 K.

The only ROSAT PSPC observation (i.e. with spectral resolution) during the optical outburst is the one with Boron filter. The three parameter fit as well as the two parameter fit give a consistently lower temperature. But since the Boron filter cuts away the high-end of the Wien tail of the blackbody, and we have only 19 photons to apply our model to, we do not regard this single measurement as evidence for a temperature decrease during the optical outburst.

What seems to be excluded, however, is any enhanced absorbing column during the Boron filter observation. The best fit absorbing column of the three parameter fit is 4.4×10^{20} cm^{-2}, consistent with the best fit absorbing column during quiescence. Since the low energy part of the spectrum in the PSPC is not affected by the Boron filter except a general reduction in efficiency by roughly a factor of 5, any increase of the absorbing column would still be easily detectable. For instance, an increase of the absorbing column by a factor of two (to 6.3×10^{20} cm^{-2}) would absorb all photons below 0.2 keV and would drop the countrate by a factor of 50 contrary to what is observed.

4 Discussion

4.1 Quiescent UV Emission

UV observations show that the hot components of symbiotic stars are located in the same quarters of the Hertzsprung-Russell diagram as the central stars of planetary nebula (Mürset et al. 1991). Due to the large binary separation in symbiotic systems the present hot component (or evolved component) should have evolved nearly undisturbed through the red giant phase. However, presently the outermost layers of the white dwarf might be enriched in hydrogen rich material accreted from the cool companion.

The far-UV radiation of the WD is ionizing a circumstellar nebula, mostly formed (or filled in) by the cool star wind. The UV spectrum in quiescence is the usual in symbiotic systems: a continuum increasing toward the shortest wavelengths with strong narrow emission lines superimposed. The most prominent lines in the quiescence UV spectrum of AG Dra are, in order of decreasing intensity: HeII 1640 Å, CIV 1550 Å, NV 1240 Å, the blend of SiV and OIV] at 1400 Å, NIV] 1486 Å and OIII] 1663 Å. The continuum becomes flatter longward approximately 2600 Å due to the contribution of the recombination continuum originated in the nebula. Although both continuum and emission lines have been found to be variable during quiescence (e.g. Mikolajewska et al. 1995), there is no clear relation with the orbital period of the system. The ratio intensity of the recombination He II 1640Å line to the far-UV continuum at 1340 Å suggests a Zanstra temperature of around $1.0\text{-}1.1\times10^5$ K during quiescence (see Fig. 3).

4.2 Quiescent X-Ray Emission

The X-Ray spectrum Previous analyses of X-ray emission of symbiotic stars have been interpreted either with blackbody emission from the hot component (Kenyon and Webbink 1984, Jordan et al. 1994) or with bremsstrahlung emission from a hot, gaseous nebula (Kwok and Leahy 1984). Our ROSAT PSPC spectra of AG Dra show no hint for any hard X-ray emission. The soft spectrum is well fitted with a blackbody model. A thermal bremsstrahlung fit is not acceptable to this soft energy distribution. There is also no need for a second component as proposed by Anderson et al. (1981).

With an adopted distance of 1 kpc and using the blackbody fit parameters while N_H was fixed at its galactic value, the unabsorbed bolometric luminosity of the hot star during quiescence (1990–1993) is 1.4×10^{36} (D/1 kpc)^2 erg/s (or equivalently 370 (D/1 kpc)^2 $L_\odot$) with an uncertainty of a factor of a few due to the errors in the absorbing column and the temperature. The blackbody radius is derived to be $R_{bb} = 1.33 \times 10^9$ cm (D/1 kpc).

Wind Mass Loss and Accretion Accepting the high luminosity during quiescence and consequently assuming that the hot component in the AG Dra system is in (or near to) the hydrogen burning regime, the companion has to supply the matter at the high rate of consumption by the hot component. The spectral classification of the companion in the AG Dra binary system, in particular its luminosity class, is very uncertain. The commonly used K3III classification would imply a mass loss rate of $(0.2-7) \times 10^{-10}$ $M_\odot$/yr according to the formula of Reimers (1975) or that of De Jager et al. (1988) for population I objects. In contrast, a K0 supergiant could have a mass loss rate of about 5×10^{-8} $M_\odot$/yr, more appropriate for the burning state of the compact object in AG Dra.

While a K3III giant of 1.1 $M_\odot$ would not fill its Roche lobe, a K0Ib supergiant would by far overfill its Roche lobe for any WD mass. Thus, in the case of a K3III companion the mass transfer to the WD would have to occur via Bondi-Hoyle accretion from the wind of the giant which by itself is already too low by orders of magnitude to sustain a burning compact object. Alternatively, a supergiant could either supply a high mass loss for wind accretion or could even feed an accretion disk around the WD by Roche lobe overflow.

We therefore think that in order to sustain the high mass comsumption of the compact object in AG Dra, the companion must be at least a bright giant to fill its Roche lobe. Adopting a bright giant solution, the distance to AG Dra would be of the order of 4 kpc, and thus the luminosity would be 2.2×10^{37} erg/s. The implied compact object mass would be 0.4 $M_\odot$ (assuming the core mass-luminosity relation $L/L_\odot \approx 4.6 \times 10^4$ ($M_{core}/M_\odot - 0.26$) according to Iben and Tutukov 1989), and the mass ratio 13.

4.3 X-Ray Emission During the Outbursts

Using our finding of an anticorrelation of the optical and X-ray intensity and the lack of considerable changes in the temperature of the hot component during the 1994/95 outburst of AG Dra, we propose the following rough scenario. (1) The white dwarf is already burning hydrogen stably before the outburst. (2) Increased mass transfer, possibly episodic, from the cool companion results in a slow expansion of the white dwarf. (3) The expansion is restricted either due to the finite excess mass accreted or by the wind driven mass loss from the expanding photosphere. This wind possibly also suppresses further accretion onto the white dwarf. The photosphere is expected to get cooler with increasing radius. (4) The white dwarf is contracting back to its original state once the accretion rate drops below the burning rate. Since the white dwarf is very sensitive it is not expected to return into a steady state immediately (Paczynski and Rudak 1980).

Instead, it might oscillate around the equilibrium state giving rise to secondary or even a sequence of smaller outbursts following the first one.

The scenario of an expanding white dwarf photosphere due to the increase in mass of the hydrogen-rich envelope has already been proposed by Sugimoto et al. (1979). The expansion velocity was shown to be rather low (Fujimoto 1982). If we assume that the luminosity remains constant during this expansion, and for the moment also assume that the change in the envelope mass is small during the early phase (justified by the result; see below) then we can determine the expansion velocity (or equivalently the excess accretion rate $\dot{M} - \dot{M}_{RG}$) simply by folding the corresponding temperature decrease by the response of the ROSAT detector. Fitting the countrate decrease of a factor of 3.5 within 23 days (from 0.14 cts/s on HJD 9929 to 0.04 cts/s on HJD 9952) we find that

$$\left(\frac{\dot{M} - \dot{M}_{RG}}{10^{-6} M_\odot yr^{-1}} \right) \approx 0.27$$

i.e. the necessary change in the accretion rate is as small as only 27%. This translates into an expansion rate of ≈ 2 m/s. The corresponding modelled countrate decrease is shown as the dotted line in Fig. 3, and the extrapolation to the quiescent countrate level gives an expected onset of the expansion at JD ≈ 9891.

Since we did not observe the X-ray intensity decline during the first outburst, and have no appropriate data yet for the X-ray intensity rise after the second outburst, the following estimates are only crude. Assuming similar durations of the first and the second outburst, we find that during each event the radius of the white dwarf roughly doubles (from 0.02 $R_\odot$ to 0.043 $R_\odot$) while the temperature decreases by about 35% (from 14.8 eV to 9.5 eV for the two parameter fit).

4.4 Relation to the Class of Supersoft X-Ray Sources

With its soft X-ray properties AG Dra is very similar to the supersoft X-ray sources (SSS) which are characterized by very soft X-ray radiation (kT$\approx$ 25–40 eV) of high luminosity (Greiner et al. 1991, Heise et al. 1994). Among the optically identified objects there are several different types of objects: close binaries like the prototype CAL 83 (Long et al. 1981), novae, planetary nebulae, and symbiotic systems. In addition to AG Dra, two other luminous symbiotic systems are known (RR Tel, SMC 3) which both are symbiotic novae.

The similarity in the quiescent X-ray properties of AG Dra to those of the symbiotic novae RR Tel and SMC 3 might support the speculation that AG Dra is a symbiotic nova in the post-outburst stage for which the turn-on is not documented. We have checked some early atlasses such as the Bonner Durchmusterung, but always find AG Dra at the 10–11^m intensity level. Thus, if AG Dra should be a symbiotic nova, its hypothetical turn on would have occurred before 1855. We note, however, that there are a number of observational differences between AG Dra and RR Tel which would be difficult to understand if AG Dra were a symbiotic nova.

If AG Dra is not a symbiotic nova, it would be the first wide binary supersoft source (as opposed to the classical close binaries) the existence of which has been predicted by DiStefano et al. (1996). These systems are believed to have donor companions more massive than the accreting white dwarf which makes the mass transfer unstable on a thermal timescale. Recent calculations have shown that there is a maximum mass ratio of about 7, above which no system survives. The mass ratio of 13 for AG Dra which results from adopting a luminosity class II companion is far above this upper limit and thus suggests that more work has to be done to improve our understanding of the system parameters in AG Dra and/or that of the mass transfer in high mass ratio binary systems.

5 Summary

The major results and conclusions can be summarised as follows:
- The X-ray spectrum in quiescence is very soft, with a blackbody temperature of about 15 eV.
- The quiescent bolometric luminosity of 1.4×10^{36} (D/1 kpc)^2 erg/s suggests stable hydrogen burning in quiescence.
- The monitoring at X-rays and UV wavelengths did not yield any hints for a predicted eclipse during the U light minima.
- In order to sustain the high luminosity the companion is required to fill its Roche lobe, and consequently is assumed to be of luminosity class II. With this assumption, the distance of AG Dra would be about 4 kpc, the X-ray luminosity would suggest a low-mass white dwarf, and the mass ratio would exceed 10.
- During the optical outburst in 1994 the UV continuum increased by a factor of 10, the UV line intensity by a factor of 2 and the X-ray intensity dropped by at least a factor of 100. There is no substantial time lag between the variations in the different energy bands.
- There is no hint for an increase of the absorbing column during one X-ray observation with spectral resolution performed during the decline of the optical outburst in 1994. Instead, a temperature decrease is consistent with the X-ray data and also supported by the *IUE* spectral results.
- Modelling the X-ray intensity drop by a slowly expanding white dwarf with concordant cooling, we find that the necessary excess accretion rate is only 27% of the quiescent one. Accordingly, the white dwarf expands to approximately its double size within the about three months rise of the optical outburst. The cooling is moderate: the temperature decreases by only ~35%.
- AG Dra could either be a symbiotic nova with a turn on before 1855, or the first example of the wide binary supersoft source class.

Acknowledgement: We are grateful to J. Trümper and W. Wamsteker for granting the numerous target of opportunity observations with *ROSAT* and *IUE*. JG is supported by the Deutsche Agentur für Raumfahrtangelegenheiten (DARA)

GmbH under contract FKZ 50 OR 9201. RV is partially supported by the Italian Space Agency under contract ASI 94-RS-59. The *ROSAT* project is supported by the German Bundesministerium für Bildung, Wissenschaft, Forschung und Technologie (BMBW/DARA) and the Max-Planck-Society.

References

Anderson C.M., Cassinelli J.P., Sanders W.T., 1981, ApJ 247, L127
Boyarchuk A.A., 1966, Astrofizika 2, 101
De Jager C., Nieuwenhuijzen H., van der Hucht K.A., 1988, A&A Suppl. 72, 259
Dickey J.M., Lockman F.J., 1990, Ann. Rev. Astron. Astrophys. 28, 215
Di Stefano R., et al. 1996, ApJ (in press)
Fernández-Castro T., González-Riestra R., Cassatella A., et al. 1995, ApJ 442, 366
Friedjung M., 1988, in "The symbiotic phenomenon", eds. J. Mikolajewska et al. , IAU Coll. 103, ASSL 145, p. 199
Fujimoto M.Y., 1982, ApJ 257, 767
Granslo B.H., Poyner G., Takenaka Y., Schmeer P., 1994, IAU Circ. 6009
Greiner J., Hasinger G., Kahabka P. 1991, A&A 246, L17
Heise J., van Teeseling A., Kahabka P., 1994, A&A 288, L45
Hric L., Skopal A., Chochol D., et al. 1994, Contrib. Astron. Obs. Skalnaté Pleso 24, 31
Huang C.C., Friedjung M., Zhou Z.X., 1994, A&AS 106, 413
Iben I., Tutukov A.V., 1989, ApJ 342, 430
Jordan S., Mürset U., Werner K., 1994, A&A 283, 475
Kenyon S.J., 1988, in "The symbiotic phenomenon", eds. J. Mikolajewska et al. , IAU Coll. 103, ASSL 145, p. 11
Kenyon S.J., Webbink R.F., 1984, ApJ 279, 252
Kenyon S.J., Garcia M.R., 1986, AJ 91, 125
Kwok S., Leahy D.A., 1984, ApJ 283, 675
Luthardt R., 1990, Contr. of the Astr. Obs. Skalnate Pleso, CSFR, Vol. 20, p. 129
Meinunger L., 1979, IVBS No. 1611
Mikolajewska J., Kenyon S.J., Mikolajewski M., et al. 1995, AJ 109, 1289
Montagni F., Maesano M., Altamore A., Viotti R., 1996, IBVS (subm.)
Mürset U., Nussbaumer H., Schmid H.M., Vogel M., 1991, A&A 248, 458
Paczynski B., Rudak B., 1980, Acta Astron. 82, 349
Piro L., 1986 (priv. comm.)
Reimers D., 1975, Mem. Soc. R. Sci. Liege 8, 369
Robinson L.J., 1969, Peremennye Zvezdy 16, 507
Skopal A., 1994, IBVS No. 4096
Sugimoto D., Fujimoto M.Y., Nariai K., Nomoto K., 1979, in White dwarfs and degenerate variable stars, IAU Coll. 53, eds. H.M. van Horn et al., Rochester: p. 280
Viotti R., Ricciardi O., Ponz D., et al. 1983, A&A 119, 285
Viotti R., Altamore A., Baratta G.B., Cassatella A., Friedjung M., 1984, ApJ 283, 226
Viotti R., Giommi P., Friedjung M., Altamore A., 1995, Proc. Abano-Padova Conference on Cataclysmic Variables: eds. A. Bianchini et al. , ASSL 205, 195

A Candidate Isolated Old Neutron Star

R. Neuhäuser[1], F.M. Walter[2], S.J. Wolk[2]

[1] Max-Planck-Institut für extraterrestrische Physik, D-85740 Garching, Germany
[2] Earth and Space Science Dept., State University of New York, Stony Brook NY 11794-2100, USA

Abstract. We report on the serendipitous detection of a bright, very soft ROSAT X-ray source, RX J1856.5–3754, near the R CrA molecular cloud. Walter, Wolk, & Neuhäuser (1996) very recently have argued that this is a nearby isolated neutron star. As the source has also been detected with the *Einstein Observatory* slew survey, the mean X-ray count rate (3.6 cts/sec, as observed with ROSAT) remained steady over a 14 year interval. The X-ray spectrum is close to that of a blackbody with kT=57 eV. The absorption column places it foreground to the R CrA molecular cloud, i.e. at a distance of less than 120 pc. There is no optical counterpart to V≈23, meaning that the X-ray to optical flux ratio exceeds 7000. The emitting area is 480 km^2 at a distance of 100 pc. From pulsar birthrates and the inferred number of supernova required to account for the heavy element abundance in the Galaxy, it is likely that the Galaxy is populated with about 10^8-10^9 neutron stars, of which only a small fraction has been discovered, yet. It has been proposed that a few thousand old, isolated (i.e., those not in binary systems) neutron stars may be detectable in the ROSAT All-Sky Survey as hot thermal sources, either from a cooling surface or from accretion of interstellar material. However, this is only the second promising candidate.

1 Introduction

It has been proposed that the Galaxy is populated with about 10^8 to 10^9 neutron stars. This follows either from an extrapolation of pulsar birthrates (Narayan & Ostriker 1990), or from the inferred number of supernovae required to account for the heavy element abundance in our Galaxy (Arnett, Schramm, & Truran 1989), if one assumes that a significant fraction of the supernovae produce neutron stars. Only about 700 neutron stars are known as rotation powered pulsars and about 1000 as pulsars powered by accretion (Taylor et al. 1995), so that there must be a very large number of undetected neutron stars in the Galaxy.

It has been proposed that old isolated (i.e., those not in binary systems) neutron stars (IONS) may be detectable as hot thermal sources, either from a cooling surface or from accretion of interstellar material (Helfand, Chanan, & Novick 1980). The expected emission peaks in the EUV/soft X-rays, with black body temperatures less than $\sim 10^6\,K$. Treves & Colpi (1991) suggested that about 5000 isolated neutron stars, accreting out of the interstellar medium, could be detectable in the ROSAT All Sky Survey (RASS). Blaes & Madau (1993) revised the estimate, concluding that for reasonable distributions of spatial velocities, some 2000 isolated neutron stars, accreting isotropically out of the interstellar medium, should be detectable in the RASS. This estimate has not been verified, and more recent estimates are significantly smaller (e.g., Madau & Blaes 1994, Blaes, Warren, & Madau 1995).

Despite the optimistic projection of thousands of nearby, detectable IONS, evidence for only two such object has been presented to date (Stocke et al. 1995, Walter, Wolk, & Neuhäuser 1996). The identification of the source presented by Stocke et al. (1995) is ambiguous; based on its ASCA X-ray spectrum, which shows a hard X-ray tail (J. Stocke, private communication), and proximity to the nucleus of the bright galaxy NGC 1313, it may well be an extragalactic massive accreting compact halo object. We report here about the soft X-ray source RX J1856.5−3754, which first appeared as very bright but unidentified source in the RASS, and was later observed in detail with deep ROSAT pointed observations. For details on the X-ray satellite ROSAT and its intruments, we refer to Trümper (1983).

2 The ROSAT Source RX J1856.5−3754

It has recently been claimed (Walter, Wolk, & Neuhäuser 1996) that the soft X-ray source RX J1856.5−3754 is an IONS; see also the discussion by Wang (1996) about this source. RX J1856.5−3754 is a bright source projected on the R CrA molecular cloud, a region of on-going low- and intermediate-mass star formation; the source's galactic latitude is $l_{II} = -17°$. The X-ray flux of RX J1856.5−3754 has been observed best with our 6.3 ksec deep ROSAT pointed observation with the Position Sensitive Proportional Counter (PSPC). In the PSPC energy range (from 0.11 to 2.4 keV) the flux is $f_X = (1.46 \pm 0.02) \cdot 10^{-11}\ erg\ cm^{-2}s^{-1}$.

The spectrum can be evaluated using the so-called X-ray hardness ratio, defined as follows. If Z_s, Z_{h1}, and Z_{h2} denote count rates for the ROSAT PSPC energy channels soft (0.1 to 0.4 keV), hard 1 (0.5 to 0.9 keV), and hard 2 (0.9 to 2.1 keV), respectively, we define hardness ratios HR 1 and HR 2 as follows:

$$\text{HR 1} = \frac{Z_{h1} + Z_{h2} - Z_s}{Z_{h1} + Z_{h2} + Z_s} \quad \text{and} \quad \text{HR 2} = \frac{Z_{h2} - Z_{h1}}{Z_{h2} + Z_{h1}} \quad (1)$$

The RASS hardness ratios of our source are as follows:

$$\text{HR 1} = -0.932 \pm 0.018 \quad \text{and} \quad \text{HR 2} = -0.679 \pm 0.071 \quad (2)$$

Hence, the spectrum is very soft. The X-ray spectrum as observed with our ROSAT PSPC pointed observation is shown in Walter et al. (1996).

We have performed spectral fits with several different models including two models on thermal plasma (Mewe and Raymond-Smith), see Tab. 1 for fit results. Additionally, it has to be noted that a power law fit gave $\alpha = 6.7 \pm 0.1$ and $n_H = (4.0 \pm 0.1) \cdot 10^{20}\ cm^{-2}$ with $\chi^2_\nu = 4.9$ only. We conclude that the source spectrum can be fit best by a 57 eV black body, with an absorption column of $(1.4 \pm 0.1) \cdot 10^{20}\ cm^{-2}$ corresponding to $A_V = 0.1$ mag.

Table 1. X-ray spectral fits

model	$k \cdot T$ [eV]	n_H [10^{20} cm^{-2}]	χ^2_ν
blackbody	57 ± 1	1.4 ± 0.1	2.3
Mewe	68 ± 6	6.1 ± 0.1	2.5
Bremsstrahlung	1000 ± 100	2.1 ± 0.5	2.8
Raymond-Smith	54 ± 2	5.0 ± 0.2	3.8

The source is *not* a soft transient. It was first detected in the *Einstein Observatory* slew survey (1980) with ten photons correponding to 0.86 ± 0.42 counts per second. It is visible in the RASS (1990), and in our ROSAT PSPC (1992) and ROSAT High Resolution Image (HRI, 1994) pointed observations. There is no evidence for source variability on any timescale (from milliseconds to years).

It is possible to study variability of a time scale of hours in detail using the RASS data. The PSPC on-board ROSAT performed the All-Sky Survey scanning the sky in great circles. With a 2° diameter field of view each object is observed once in every $\sim$ 90 minutes for up to $\sim$ 30 *sec* per scan. Vignetting corrected RASS exposure times vary roughly with $1/\cos\beta$ where β denotes ecliptic latitude. The CrA area with our soft source has been observed during twelve scans, so that the patrol time is $\sim 12 \cdot 90$ minutes, i.e. two thirds of a day. In Fig. 1 , the RASS light curve is shown.

This supersoft X-ray source has not been studied in detail before, e.g. by groups searching particulary for IONS, because the hardness ratio HR 1 as observed in the RASS and as obtained from the Standard Analysis Software System (SASS) does not fulfil the formal criterion for "supersoft" sources (namely that HR 1 plus its error is smaller than −0.9), among which such IONS are expected (Greiner 1996). However, the source's HR 1 as observed in the RASS and as obtained from the Extended Scientific Analysis Software System (EXSAS) does fulfil this criterion, see above equ. 2. While SASS is used to evaluate all the RASS data, EXSAS is superior in some details and is used only for parts of the RASS data (that are requested by investigators). A possible reason for different results from SASS and EXSAS is that (among other different details in these software packages) SASS evaluates RASS data in strips (see above), while EXSAS can be run on merged strips, so that a more realistic background can be estimated.

3 The Nature of the Source

At the position of the source, the integrated column density through the R CrA molecular cloud is about $5 \cdot 10^{20}$ cm^{-2} (Wang 1994); this requires that the source be no more distant than the 130 pc distance to the cloud (Marraco & Rydgren 1981). Walter, Wolk, & Neuhäuser (1996) took a recent optical CCD image of

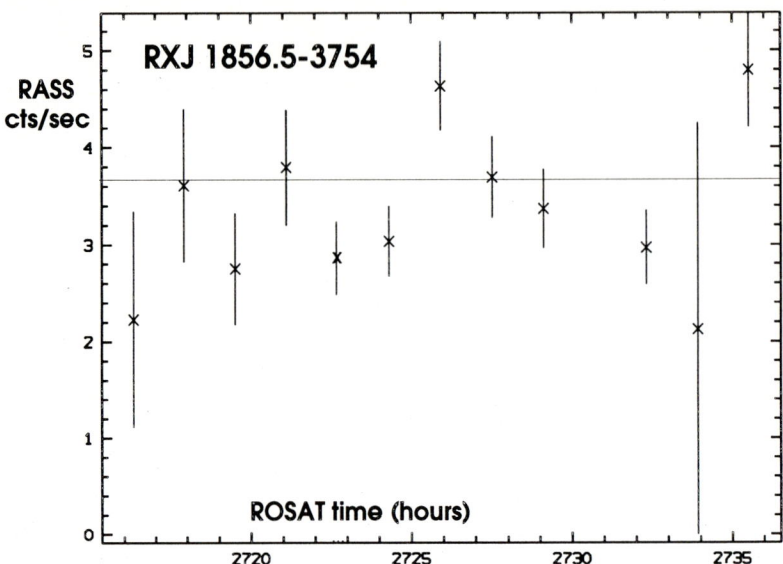

Fig. 1. RASS light curve of RX J1856.5–3754: RASS count rate in counts per second (corrected for vignetting and background) versus ROSAT observation time (in hours after the start of the ROSAT mission). The source has been observed in twelve consecutive RASS scans with exposure times between a few seconds and about 30 seconds. Error bars are one Gaussian sigma errors. No significant variability can be seen. The horizontal line indicates the mean RASS count rate of 3.67 ± 0.15 counts per second, consistent with the ROSAT PSPC pointed observation.

the area around the ROSAT HRI position. There is no optical counterpart down to $V = 23.1$ mag, yielding a lower limit to f_X/f_V of about 7000. The large f_X/f_V rules out *all* potential counterparts except isolated neutron stars. At a distance d the X-ray luminosity is $5 \cdot 10^{31} \left(\frac{d}{100 \text{ pc}}\right)^2 erg\ s^{-1}$, and the emitting area is $480 \left(\frac{d}{100 \text{ pc}}\right)^2 km^2$.

There is no known radio source or pulsar known at this position. Also, there is no ROSAT WFC, HEAO-1, EUVE, VLA, or IRAS source at or near this source (Walter, Wolk, & Neuhäuser 1996). The high X-ray to optical flux ratio f_X/f_V and the fact that the emission remains constant at time scales from milliseconds to 14 years rules out any accreting binary system or accretion powered system. The area suggests emission from the entire hot surface of a nearby IONS. The temperature is consistent with a cooling IONS with an age of $\sim 10^6$ years (Lattimer et al. 1994), or more, if the surfaces reheat due to internal friction (Reisennegger 1995). The luminosity and temperature are also consistent with the accretion heating model of Blaes & Madau (1993), for quasi-isotropic accretion by a neutron star moving at a velocity v of $\sim 50\ km\ s^{-1}$ (or little less) through a medium with a density of $n_H = 1\ cm^{-3}$. This object affords a unique opportunity to study the surface emission from a neutron star, unaffected by magnetospheric emission or by accretional processes in a binary system.

4 Future Observations

Our proposal (PI F.M. Walter) for observations of this IONS with the Hubble Space Telescope WFPC2 has been approved in order to obtain an optical identification (with expected $V = 27.4$ mag). Once we identify the counterpart, we will propose to obtain the spectral energy distribution in the optical and UV regime, the radius to within 25%, the parallax (to better than 20%), and the proper motion (if its transverse space velocity exceeds $\sim 10\ km/s$).

Another proposal (PI R. Neuhäuser) to obtain a broad band X-ray spectrum with the SAX LECS has been approved, too, in order to search for the 5 to 15 keV cyclotron resonance line emission expected from a strongly magnetized IONS and to determine the magnetic field from the line strength.

The study of IONS is useful for a number of fields in galactic astrophysics. Examination of non-pulsating IONS would permit study of the cooling of neutron star surfaces. If the IONS are heated by accretion from the interstellar medium, studies of the spatial structure of the interstellar medium on small spatial scales is possible. Since any detectable IONS must be nearby, it will be possible to study the kinematics of this population. Determination of the stellar radii would lead to improved constraints on the interior equation of state.

5 Summary

A very bright and soft, unidentified X-ray source has been observed with ROSAT. The X-ray spectrum can be fit best with a blackbody at $kT = 57\,eV$. The absorbing column density is less than the one observed towards the R CrA molecular cloud, that appears to be in the background of this soft X-ray source. Hence, the X-ray source cannot be more distant than $\sim 130\ pc$. There is no variability on time scales from milliseconds to several years, so that any accretion powered system can be excluded as counterpart. No optical counterpart can be identified down to a limiting magnitude of $V = 23.1$ mag. Thus, together with the observed X-ray flux, we estimate the X-ray to optical flux ratio to be at least 7000, so that we can exclude any possible counterpart exept isolated old neutron stars.

Acknowledgement: We would like to thank J.T. Stocke, J.C.L. Wang, W. Becker, J. Greiner, and J. Trümper for useful conversation. Werner Becker also looked in detail into our ROSAT PSPC pointed observations and also could not find any significant variability on time scales down to milliseconds. The ROSAT project is supported by the German Bundesministerium für Bildung, Wissenschaft, Forschung und Technologie (BMBF/DARA) and the Max-Planck-Society.

References

Arnett W.D., Schramm D.N., Truran J.W., 1989, ApJL 339, L25
Blaes O., Madau P., 1993, ApJ 403, 690
Blaes O., Warren O., Madau P., 1995, ApJ 454, 370

Greiner J., 1996 (private comm.)
Helfand D.J., Chanan G.A., Novick R., 1980, Nat. 283, 337
Lattimer J.M., van Riper K.A., Prakash M., Prakash M., 1994, ApJ 425, 802
Madau P., Blaes O., 1994, ApJ 423, 748
Marraco H.G., Rydgren A.E., 1981, AJ 86, 62
Narayan R., Ostriker J.P., 1990, ApJ 352, 222
Reisenegger A., 1995, ApJ 422, 749
Stocke J.T., Wang Q.D., Perlman E.S., Donahue M.E., Schachter J., 1995, AJ 109, 1199
Taylor, et al., 1995, ApJS 88, 529
Treves A., Colpi M., 1991, A&A 241, 107
Trümper J., 1983, Adv. Space Res. 2, 241
Walter F.M., Wolk S.J., Neuhäuser R., 1996, Nat. 379, 233
Wang Q.D., 1994, in The Soft X-Ray Cosmos, E.M. Schlegel & R. Petre (eds.), AIP Conf. Proc. 313 (New York), 301
Wang Q.D., 1996, Nat. 379, 206

A Systematic Search for Supersoft X-Ray Sources in the ROSAT All-Sky Survey

J. Greiner

Max-Planck-Institut für Extraterrestrische Physik, 85740 Garching, Germany

Abstract. We have conducted a systematic search for supersoft X-ray sources using the ROSAT all-sky survey data. With the optical identification of the selected sources being almost complete, we discuss the statistics of the various source classes and their observability. Besides supersoft close binary sources this search also can be used to estimate the number of isolated neutron stars in the Galaxy, such as those described by Stocke et al. (1995) and Walter et al. (1996).

1 Introduction

Supersoft X-ray sources (SSS) are characterized by very soft X-ray radiation (most of the X-ray emission below 0.5 keV) of high luminosity. It is generally accepted that SSS involve steady nuclear burning on the surface of an accreting WD (van den Heuvel et al. 1992). ROSAT observations established SSS as a distinct class with somewhat more than 30 members known at present. While 16 sources belong to M 31, and a dozen to the Magellanic Clouds, only very few have been found in the Galaxy.

A simple scaling from the brightest supersoft X-ray sources in the Magellanic Clouds to galactic distances already shows that there are no such sources in our immediate neighbourhood. If CAL 83, the SSS prototype (Long et al. 1981), were at 1 kpc distance, we would expect a ROSAT PSPC countrate of 3000 cts/sec, much larger than the PSPC threshold intensity. Any such unabsorbed source in the solar neighbourhood would have forced the PSPC to switch off and it is safe to say that no such source has been detected in the ROSAT all-sky survey.

However, the intrinsic source luminosities are observed to vary from source to source by a factor of about 50, the mean temperatures of the WDs could be systematically lower in the Galaxy as compared to the Magellanic Clouds, and a moderate absorbing column could further dim the sources. We therefore have undertaken a systematic search of the ROSAT PSPC all-sky survey data for supersoft X-ray sources with emphasis on the galactic population.

2 Selection Criteria

We started with the list containing all detected X-ray sources of the all-sky survey in its version of March 1991 which contained about 100.000 sources (including detections of identical sources in overlapping strips). We applied a hardness ratio criterion $HR1 + \sigma_{HR1} \leq -0.80$ which is fulfilled by 304 sources. The hardness ratio HR1 is defined as the normalized count difference (N_{52-201} − $N_{11-41})/(N_{11-41} + N_{52-201}$), where N_{a-b} denotes the number of counts in the PSPC between channel a and channel b. After merging double and multiple detections from different strips at high ecliptic latitudes we are left with 165 sources. We have taken the results of each source from that strip in which the source has the largest distance to either edge of the strip. Finally, images in different energy bands were visually inspected to check for possible false detections, and to reduce spurious sources in bright supernova remnants (like Vela/Puppis). The final number of selected sources is 143.

This kind of hardness ratio selection has the implication of an implicit intensity selection. First, for decreasing brightness the error in the hardness ratio increases, and all sources with $\sigma_{HR1} \geq 0.2$ are exluded. Second, it also cuts away specific sources (like G or K stars) which are detected at high signal to noise ratio. This is due to the fact that these objects exhibit a faint hard component, and this is detected only for bright sources.

Finally we should note that the input list has been produced from source detections on the individual strips. At high ecliptic latitudes these strips overlap, and with the "survey II" processing which is done on adjacent sky fields instead of strips (and which will becaome available soon) a considerable improvement of the source detection and parameter estimation is possible.

3 Optical Identification and Source Statistics

A correlation with the Simbad database revealed probable source identifications for a total of 48 sources. Most of these identifications were single white dwarfs. Since we were not interested in discovering further WDs, and many of these WDs have been detected also in the extreme-ultraviolet sensitive Wide Field Camera (WFC), we have used the WFC source list to remove all sources from our master list which have been seen in the WFC (total number of 53). These sources were not considered for our own follow-up identification, but were left by purpose for the WFC consortium (and most of these were later identified as WDs).

The remaining sources have been observed optically by the author (mainly northern hemisphere) and the group of K. Beuermann (southern hemisphere). The majority of X-ray sources turned out to be magnetic cataclysmic variables, predominantly polars. The only galactic supersoft source (close binary) in this sample is RX J0019.8+2156 (Beuermann et al. 1995, Greiner & Wenzel 1995) while the other two are LMC/SMC sources (1E 0035.4–7230 and RX J0439.8–6809). Tab. 1 shows a breakdown of the sources on the individual source classes.

A few notes should be added to Tab. 1. The G and K star identifications (IDs) are probably systems with a WD in a close orbit, so after spectroscopic follow-up observations these two IDs will certainly be replaced. There is one additional object of presumably this type among the unidentified sources which also has a G star at the X-ray position. The only active galactic nuclei in our sample is the narrow-line Seyfert 1 (NLSy1) galaxy WPVS007, and many more are expected at somewhat harder hardness ratios (Greiner et al. 1996).

Table 1. The distribution among different classes of objects of supersoft X-ray sources which satisfy HR1 + $\sigma_{HR1} \leq -0.80$. PN means planetary nebula, the other abbreviations are explained in the text.

Class type	Simbad correlation	newly identified	total
SSS	0	3	3
PN	2	0	2
PG1159 stars	0	1	1
symbiotics	2	1	3
single WDs	32	66	98
magnetic CVs	5	12	17
NLSy1	1	0	1
B stars	4	0	4
G stars	1	0	1
K stars	1	0	1
unidentified			12

4 Discussion

In total, only seven out of the more than 30 known supersoft sources are members of our sample, namely the above listed RX J0019.8+2108, 1E 0035.4−7230 and RX J0439.8−6809 plus the planetary nebula RX J0058.6−7146 (N 67) plus the three symbiotics RX J0048.4−7332 (SMC3), RR Tel and AG Dra. This demonstrates the "conservative" approach of our selection criterium. In fact, it is more effective for low-temperature sources like single white dwarfs or PG1159 stars. The above listed sources are the softest among the known population of SSS. As soon as a source has a temperature higher than about 40 eV and the PSPC detects a significant number of photons above channel 41 (0.4 keV) than the HR1 increases and shifts the source out of our hardness ratio window. That is, our sampling of supersoft sources is complete only for temperatures below about 30 eV.

The brightest of the non-identified sources has a PSPC countrate of 0.37 cts/s. Above this intensity the identification is complete. This implies that any galactic source radiating at the Eddington rate with a temperature in the 20–40 eV range would have been detected above a galactic latitude bII>12°. Relaxing the luminosity constraint by a factor of 10 increases the latitude limit only slightly to bII>15°.

A different approach in searching for supersoft sources is to look specifically for strongly absorbed sources in the galactic plane. Their energy distribution then is expected to be sharply peaked around 0.5–0.7 keV with higher energy photons missing due to the spectrum of the source, and lower energy photons absorbed. Such a source would have a relatively hard HR1 but a very soft HR2. Indeed, these criteria have led to the discovery of RX J0925.7−4758 (Motch et al. 1994). From the results of the galactic plane survey identification these authors concluded that though sources radiating at the Eddington limit with a temperature below ~20 eV may be hidden in the galactic plane, sources at bII > 5–10° with temperatures above 40 eV and 4×10^{37} erg/s bolometric luminosity can be excluded. This supplements our finding for lower temperature sources.

Another possible class of soft X-ray sources are old, isolated neutron stars (IONS) accreting from the interstellar matter. Two sources have been proposed to be IONS, one of those (RX J1856.5–3754) is a bright source seen in the ROSAT all-sky survey (Walter et al. 1996). This source has been reported to have a best-fit blackbody temperature of below 60 eV. However, due to the distinct emission above 0.4 keV the hardness ratio is HR1=–0.29, far outside our selected range.

Such IONS are not expected to be easily visible in the optical range. All the unidentified sources of our sample have optical counterpart candidates brighter than 20th magnitude inside their corresponding error box. We therefore think that none of these sources might be another IONS candidate. This makes our search complete down to a limiting countrate of 0.1 cts/sec.

Since the temperature of the emitted radiation (assumed to be blackbody) is primarily determined by the ratio of accretion rate $\dot{M}$ to the fractional accreting area f ($kT = 20\,(\dot{M}/f)^{1/4}$ eV with $\dot{M}$ in units of 10^{10} g/s, Blaes & Madau 1993), we can determine the range of accretion rates for which our search for supersoft sources is sensitive. Assuming isotropic accretion for simplicity, sources with accretion rates above 50×10^{10} g/s are too hot for our search (have a hardness ratio greater than –0.8), while below 0.05×10^{10} g/s the temperature and luminosity get too low to be detectable beyond 20 pc.

Acknowledgement: JG is extremely grateful to T. Fleming for the frequent information on the status of single WD identifications. JG is supported by the Deutsche Agentur für Raumfahrtangelegenheiten (DARA) GmbH under contract FKZ 50 OR 9201. The *ROSAT* project is supported by the German Bundesministerium für Bildung, Forschung, Wissenschaft und Technologie (BMBW/DARA) and the Max-Planck-Society. This research has made use of the Simbad database, operated at CDS, Strasbourg, France.

References

Beuermann K., Reinsch K., Barwig H., Burwitz V., de Martino D., Mantel K.-H., Pakull M.W., Robinson E.L., Schwope A.D., Thomas H.-C., Trümper J., van Teeseling A., Zhang E., 1995, A&A 294, L1
Blaes O., Madau P., 1993, ApJ 403, 690
Greiner J., Wenzel W., 1995, A&A 294, L5
Greiner J., Danner R., Bade N., Richter G.A., Kroll P., Komossa S., 1996, A&A (in press)
Long K.S., Helfand D.J., Grabelsky D.A., 1981, ApJ 248, 925
Motch C., Hasinger G., Pietsch W., 1994, A&A 284, 827
Stocke J.T., Wang Q.D., Perlman E.S., Donahue M., Schachter J.F., 1995, ANJ 109, 1199
van den Heuvel E.P.J., Bhattacharya D., Nomoto K., Rappaport S.A., 1992, A&A 262, 97
Walter F.M., Wolk S.J., Neuhäuser R., 1996, Nat. 379, 233

A Search for Optical Counterparts to Supersoft X-Ray Sources in the ROSAT Pointed Database

C.M. Becker, R. Remillard, S.A. Rappaport

Department of Physics, Massachusetts Institute of Technology, Cambridge, MA 02139

Abstract. We present a progress report of our program to identify optical counterparts of supersoft X-ray sources detected during pointed ROSAT PSPC observations. We generate a parent sample of ∼4100 supersoft sources by starting with the WGA (White et al. 1994) and ROSAT SRC (Zimmermann 1994) catalogs, extracting X-ray detections on the basis of softness (ROSAT HR1 $\leq$ −0.5), condensing multiple detections, and combining sources seen in both catalogs. This unwieldy sample is divided into four (occasionally overlapping) subclasses: (1) a flux limited sample (WGA only) encompassing ∼120 square degrees of sky, (2) an all-sky "extremely soft" sample (HR1 $\leq$ −0.8), (3) an optically bright all-sky sample with candidates $V \leq 14.5$, and (4) an all-sky sample near the nuclear burning white dwarf (NBWD) track in the HR1-HR2 plane, as defined with Monte Carlo simulations.

This program has been underway for about one year and has included seven observing runs at the MDM observatory at Kitt Peak, Arizona. In addition, numerous sources have been identified via catalog and literature searches. More observing runs are currently scheduled. Although no new NBWDs have been discovered yet, we have found many new supersoft AGN and emission stars, and a few new white dwarfs and CVs. By the conclusion of our program we will be able to set constraints on models concerning the population of NBWD, to understand the proportions of objects that make up the class of "supersoft X-ray sources", to investigate models for the supersoft nature of AGN, and (hopefully) to discover some new NBWD systems.

1 Introduction

One of the interesting contributions of the ROSAT satellite is the discovery of celestial objects possessing such soft X-ray spectra that over 75% of the counts are detected between 0.1 and 0.4 keV. These "supersoft X-ray sources" (SXS) themselves do not form a uniform group, but rather a collection of astrophysical systems including hot white dwarfs, stars with active coronae, cataclysmic variables (CVs), white dwarfs undergoing surface nuclear burning of accreted matter (NBWD), active galactic nuclei (AGN), and galaxies with warm halos.

A large number of SXS were discovered in the ROSAT All-Sky Survey (RASS) conducted during the first 6 months of the mission. The exposure time for a typical portion of the sky is ∼500 seconds but increases to be comparable to the pointed studies within 5° of the ecliptic poles. Between that time and November 1994, three years of pointed guest observations were accumulated with ROSAT. Two groups, one in Germany and one at NASA — Goddard Space Flight Center, independently searched these data for serendipitous sources, with each group creating a catalog of their results. Each catalog contains about 50,000 detections, covers about ten percent of the sky, and yields a typical exposure time on any one source of about 11,000 seconds. Typical uncertainties in source location are 30″ (90% confidence error radius). This database provides a new SXS sample that is complementary to the RASS and more comprehensive, in the overlapping energy bands, than the EUVE or ROSAT WFC surveys (400-500 sources each).

We have undertaken a large program to find optical counterparts to SXS detected in these new databases. Upon completing the optical search, follow-up studies, and additional supporting projects, we will be able to: (1) determine in a statistically meaningful manner the composition of the general class "supersoft X-ray source" (2) test theoretical predictions regarding mass transfer in NBWD systems by modeling the orbital structure, (3) test theoretical predictions regarding populations of NBWD in the Galaxy, (4) discover new NBWD systems to a flux ten times fainter than that of the only such unobscured galactic system, (5) determine whether preliminary conclusions drawn during similar studies of the RASS sample still hold at deeper flux limits, and (6) provide a sound model (incorporating X-ray data) as to the nature of supersoft AGN.

This paper is a report on our ongoing program. In Sect. 2 we discuss the selection of targets for optical observation, in Sect. 3 we provide the current status of the program, and in Sect. 4 we present preliminary results of our work.

2 Program Definition

The Goddard (WGA) and German (SRC) ROSAT catalogs report not only the position of each source, but also spectral information. We use the following spectral bands:

$$A = 11 - 39 \text{ ADU} \approx 0.1 - 0.4 \text{ keV}$$
$$B = 40 - 85 \text{ ADU} \approx 0.4 - 0.9 \text{ keV}$$
$$C = 86 - 200 \text{ ADU} \approx 0.9 - 2.0 \text{ keV},$$

and form two hardness ratios

$$\text{HR1} = \frac{(C + B - A)}{(C + B + A)} \qquad \text{HR2} = \frac{(C - B)}{(C + B)}.$$

We select all detections with HR1 ≤ -0.5 from both catalogs. After this selection only ~5% of the original catalog remains of interest. This list is further condensed by looking for multiple detections of the same object. In the end, 4101 individual SXS remain to be identified; we define this as the "parent sample".

A literature search was conducted for each object in the parent sample, both through the SIMBAD database and catalogs compiled at MIT. In the end, about 500 sources could be identified through such literature searches[1], leaving over 3500 sources to be identified with optical counterparts. It is not feasible to carry

---
[1] The literature search is an on-going process. We are continually seeking and searching additional catalogs.

out so many observations in a reasonable amount of time. To effectively achieve our goals, we have defined four (occasionally overlapping) subsamples.

Our goal is to observe 90% of the sources in each subsample and determine whether or not they are NBWD. The exception is the flux limited sample where we demand an optical counterpart to the SXS independent of the system type.

2.1 Flux Limited Sample

In order to determine what proportion of the SXS are of various classes (CVs, AGN, etc.) in a statistically meaningful way, we have defined a flux limited sample of SXS. In this sample we restrict ourselves to the WGA catalog because of the careful consideration of point spread function effects for sources far off-axis (Haberl et al. 1994). Care has been taken to select objects with similar exposure times and interstellar absorption characteristics such that the ROSAT PSPC count rate can be used as a measure of flux. The resultant log(dN/dS)-log(S) plot is shown in Fig. 1. We have chosen a (conservative) lower flux limit of 0.014 cnts/s; Fig. 1 indicates that the X-ray data are complete to even lower flux values. This sample covers ≈120 square degrees of sky and contains 177 sources.

In addition to determining the constituency of the SXS, the flux limited sample will serve as a basis for studying SXS AGN. Table 2 (presented in Section 4) indicates that we will identify a large number SXS AGN from which we will be able to determine a luminosity function. We will look for signs of evolution in this distribution and compare it to other luminosity functions of general X-ray selected AGN.

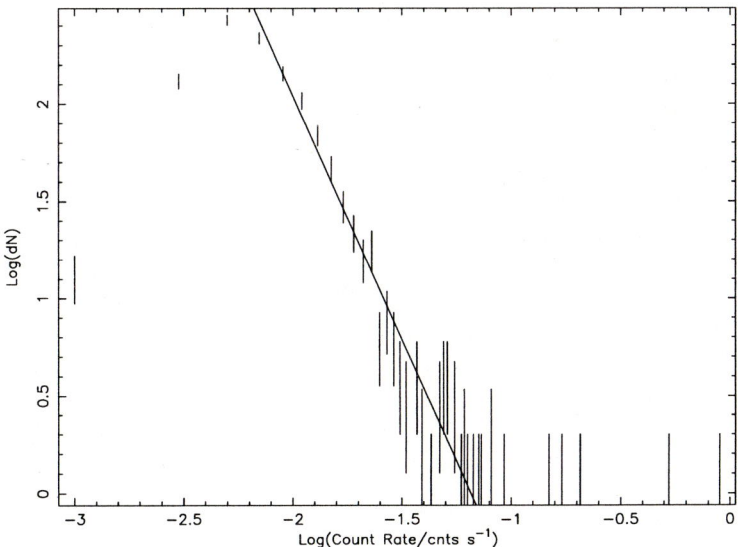

Fig. 1. Log(dN/dS)-Log(S) plot for the flux limited sample. A -5/2 power law is shown. We take a conservative completeness limit of 0.014 counts/s.

2.2 Extremely Soft Sample

To be classified as "supersoft" in this work, a source requires an HR1 value of ≤ -0.5 (75% of the counts in band A); however, there are a number of reasons to consider the very softest sources with HR1 ≤ -0.8 (90% of the counts in band A) distributed over the entire sky. To begin, NBWD systems are very soft, and it is a stated goal of this program to find more of these systems. In fact many of the known NBWD systems have been found in this range. Next the 0.1 – 0.4 keV region is being explored with great sensitivity for the first time with the ROSAT satellite. Therefore the identification of sources in this band is potentially rich in new phenomena. This sample includes 355 SXS.

2.3 Optically Bright Sample

Due to the very soft X-ray spectrum of a NBWD, any such object in the Milky Way detected by ROSAT is probably quite local because interstellar attenuation of soft X-rays would make more distant sources undetectable. At the same time, NBWD can be very luminous ($\sim 10^{38}$ erg/s). The combination of these properties might well result in an optically bright counterpart. For example RX J0019.8+2156, the only unobscured galactic NBWD known, has $m_V \sim 12.8$ (Beuermann et al. 1995). The opportunities that would arise from the discovery and study of a new NBWD system have prompted us to define a sample of all stars brighter than $V \sim 14.5$ within 1.3 error radii of the X-ray position (at 1.3 error radii the optical counterpart should be enclosed with a probability of 96%). A comparison of the parent sample to the Space Telescope Guide Star Catalog (ST-GSC) yields 694 stars meeting these criteria, distributed among 640 SXS. Although the ST-GSC is generally complete to 14.5 magnitude, this limit can can be significantly worse for dense fields. In these cases we attempt to include stars neglected in the ST-GSC via direct inspection of the finding charts produced with the Digital Sky Survey.

2.4 NBWD Track Sample

We also take another approach to finding galactic NBWDs among the 80,000 detections in the WGA and SRC catalogs, by generating a simulated catalog with only a population of NBWD distributed throughout the Milky Way. Using a simple "disk + bulge" model for interstellar gas, the column density to any point in the simulated galaxy can be analytically determined. The NBWD are placed in a disk population with a space density that decays exponentially with height above the midplane. The scale heights for the NBWDs and gas are taken to have the ratio $z_{\rm NBWD}/z_{\rm gas} = 0.5$ (following Di Stefano & Rappaport 1994), and their spectra (taken to be simple blackbodies) are selected from a luminosity-temperature distribution determined by (Rappaport & Di Stefano 1994). The probability of choosing any given (L,T) was weighted by the expected lifetime of that system.

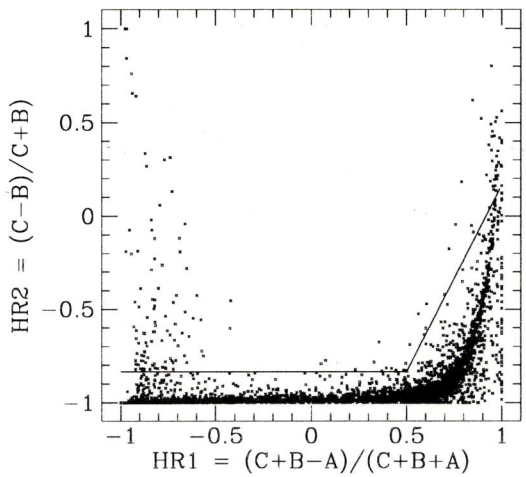

Fig. 2. Distribution of simulated galactic NBWD in the HR1-HR2 plane. 5000 "detections" from 46,464 seeded systems are plotted.

The intrinsic spectrum from each NBWD is modified by interstellar absorption, then "observed" by folding the incident spectrum through the ROSAT PSPC response matrix.

In order to be consistent with the actual catalogs, after including the effects of background and Poisson noise, we demand a signal-to-noise ratio ≥ 5.0 to be secure in a detection. Only the count rate and the number of counts in channels A, B and C are accumulated. The goal is to find relations among these spectral parameters which constrain the parameter space that is likely to contain Galactic NBWD.

Fig. 2 shows a very clear relation between HR1 and HR2 which agrees very well with known nuclear burning white dwarf systems. Based on this result, we are currently defining a fourth sample (the area below the line in Fig. 2) to optimize our search for NBWD[2].

3 Current Status

At the time this article was written we were about half way through the observing program. In about 50 nights of observing time we have taken 939 spectra, all but 5 on the Michigan - Dartmouth - MIT (MDM) 1.3 meter observatory atop Kitt Peak, Arizona. An additional 20 nights are scheduled in the first half of 1996.

Table 1 shows our progress on each subsample. The optically bright sample is virtually complete. The flux limited sample needs the most work; however, this sample was least conducive to study with the MDM 1.3m. Upcoming time on the MDM 2.4m should be highly productive in this regard. Note we are taking a very strict approach to "new identification" at this time, requiring compelling evidence that the counterpart has been found (e.g. emission lines, optical flux, and/or spatial coincidence with the X-ray position) Most notably we have not

---

[2] Other researchers have independently drawn similar conclusions utilizing other methods (e.g., Motch et al. 1994 and Kahabka 1996).

Table 1. Current status of the sub-samples

Sample	Total Sources	To Be Observed	Total IDs	New IDs	Literature IDs
Flux Limited	177	148	30	10	20
Extremely Soft	355	203	150	5	145
Optically Bright	694	134	147	13	134
Other	3035	2746	289	18	271

finished analyzing the ordinary stars, over 200 of which are probably optical identifications. This uncertainty is also reflected in Table 2 below.

4 Preliminary Results

A brief summary of our results to date is shown below in Tab. 2; general trends can be extracted from this information. The trends found in the RASS sample of SXS toward large numbers of SXS AGN (Greiner et al. 1996) and a predominance of SXS CVs being of the AM Her type (Beuermann & Burwitz 1995) continue to hold in the pointed survey.

Each sub-sample tends to have its own "signature class". The flux limited sample is dominated by SXS AGN (20 of 30 identified systems), The extremely soft sample is predominately white dwarfs and AM Her CVs, and the optically bright sample consists mainly of active coronae stars with notable numbers of white dwarfs. Note many of the sources in the optically bright sample have not tabulated at the moment and many of these are exepected to be active coronae stars.

New identifications are being made at a steady pace, mainly among AGN and emission stars. The fact that we have not made more firm identifications stems from a few things. Most of our observations have taken place on the MDM 1.3m during bright time. Next, the CCD used in most of the observations does not have sufficient blue response to examine the CaII H and K lines ($\lambda\lambda 3933, 3969$) which are excellent indicators of active coronae. As a result we have focused on the optically bright sample (see Table 1) which could be most easily carried out under these conditions. The Spring 1996 runs have been optimized to explore the other samples where actual identifications are the goal.

An example of the type of spectra we are accumulating during this project is shown for SXS AGN in Fig. 4. The spectra of four newly identified sources (at a variety of redshifts) are plotted in the figure. In addition to the obvious emission lines, iron emission complexes are noted on either side of the [OIII] doublet. These features are found in most of our SXS AGN and may signify the presence of dense, cool regions in the system, perhaps forming in parallel with the warm absorber. However, note there is nothing obviously different on the optical spectra of SXS AGN compared to the diverse varieties (e.g. emission

Table 2. Current SXS identifications by class and subsample

Class	Total (New)	Flux	Soft	Optical	Other
AGN (QSO)	78 (9)	12	5	1	60
AGN (Seyferts)	31 (14)	6	0	4	21
AGN (Other)	8 (0)	2	0	0	6
White Dwarfs	94 (5)	2	86	35	3
Emission Stars	44 (22)	3	12	20	15
CV (AM-Her)	19 (0)	3	16	0	3
CV (DQ-Her)	2 (0)	1	1	1	0
CV (Nova Like)	1 (0)	1	0	1	0
CV (Unclassified)	1 (1)	0	0	0	1
CSPN	5 (0)	0	3	2	2
Symbiotic Stars	3 (0)	0	3	2	0
Pulsars	3 (0)	0	1	0	2
Novae	2 (0)	0	1	1	1
Nuclear Burning WD	2 (0)	0	1	1	1
Early Stars (O1-F5)	>36 (?)	0	6	22	5
Late Stars (F6-M9)	>89 (?)	0	15	57	23

line ratios) in hard X-ray selected samples. The combination of optical and X-ray spectra will help us determine the physical mechanisms responsible for the 0.1–0.4 keV emission. A supporting SAX proposal has been submitted.

Finally, we note that we have yet to find a new NBWD; however, we will be looking in the southern hemisphere for the first time during spring, 1996. This, in combination with the new "NBWD Track selected sample" improves our chances of finding new systems of this type.

5 Summary

The WGA and SRC catalogs provide a complementary data set to the RASS, trading sky coverage for a factor of ~20 in average exposure time. We are halfway through a two-year program to identify SXS in the pointed catalogs. At this point, we can confirm that the RASS results regarding AM Her CVs (this subclass dominates all SXS CVs) and AGN (there are more SXS AGN than previously thought) still hold in our work. Although no new NBWDs have been found, we have made steady progress identifying new optical counterparts to SXS, most of which are AGN and emission stars.

Acknowledgements: This work is supported by funds of NASA ADP grant NAG5-3011 and NASA ROSAT guest observer grant NAG5-1784. In addition special thanks to the MDM observatories for time and technical support, the MPE ROSAT group for the SRC catalog, the White group at GSFC for the WGA catalog, and the CDS at Strausbourg, France for access to the SIMBAD database.

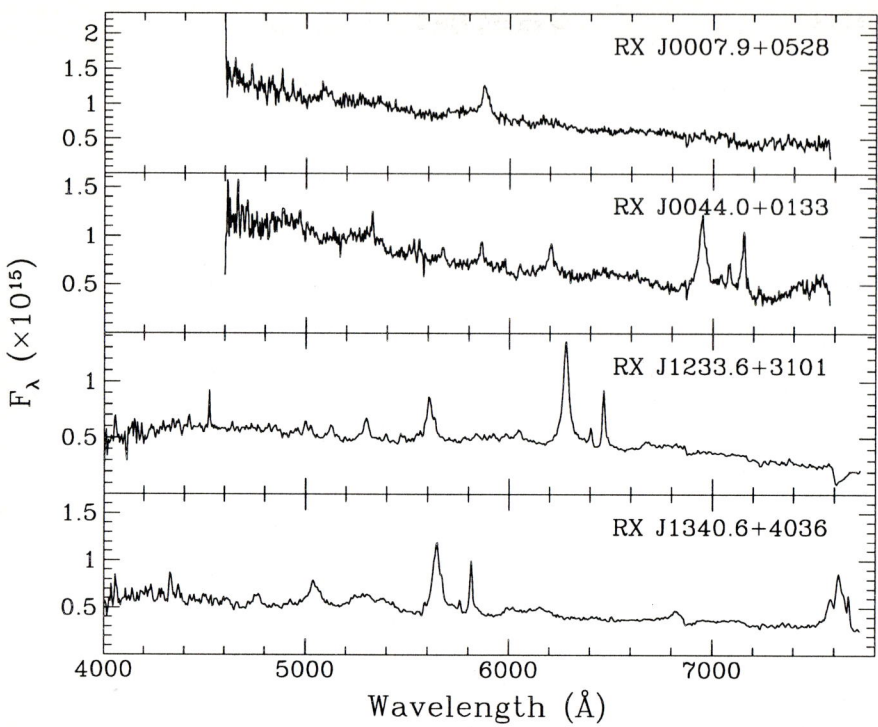

Fig. 3. Optical spectra of four newly identified SXS AGN. The redshifts (top to bottom) are 1.1, 0.44, 0.29, and 0.16.

References

Beuermann K., Reinsch K., Barwig H., et al. 1995, A&A 294, L1
Beuermann K., Burwitz V., 1995, in Cape Workshop on Magnetic Cataclysmic Variables, ASP Conf. Proceedings 85, eds. D.A.H Buckley & B. Warner
Di Stefano R., Rappaport S., 1994, ApJ 437, 733
Greiner J., Danner R., Bade N. et al. 1996, A&A, (in press)
Haberl F., Pietsch W., Voges W., 1994, "Differences in the Two ROSAT Catalogs of Pointed PSPC Observations" WWW Document, *http://rosat_svc.mpe-garching.mpg.de/rosat_svc/archive/sourcecat/wga_rosatsrc.html* (26 March 1996)
Kahabka P., 1996, this volume p. 215
Motch C., Hasinger G., Pietsch W., 1994, A&A 284, 827
Rappaport S., Di Stefano R., Smith J.D., 1994, ApJ 426, 692
White N., Giommi P., Angellini L., 1994, IAUC 6100
Zimmerman H.-U., 1994, IAUC 6102

Part VII

Catalog of Supersoft X-Ray Sources

Catalog of Luminous Supersoft X-Ray Sources

J. Greiner

Max-Planck-Institut für Extraterrestrische Physik, 85740 Garching, Germany

1 Introduction

This catalog comprises an up-to-date list of luminous ($>10^{36}$ erg/s) supersoft X-ray sources. Since most of the new sources are X-ray discoveries, the final inclusion in the group of luminous close binary supersoft sources has to await the optical identification. Only then a distinction is possible among the various and quite different types of objects which show a supersoft X-ray spectrum (i.e. emission only below 0.5 keV): single (i.e. non-interacting) white dwarfs, central stars of planetary nebulae, PG 1159 stars, symbiotic variables, magnetic cataclysmic variables, active galactic nuclei. An exception are F and G stars which also have supersoft X-ray spectra but which can readily recognised due to their optical brightness (low L_x/L_{opt} ratio). Due to this fact of necessary follow-up optical observations, it can well happen that a source is included in the catalog but later turns out to be of a different type. A rather recent example is RX J0122.9–7521 which has long been thought to be a SMC supersoft source (Kahabka et al. 1994), but has been identified as a galactic PG 1159 star (Cowley et al. 1995, Werner et al. 1996).

We include in this catalog accreting binary sources of high luminosity which are thought to be in a state of (steady or recurrent) hydrogen burning. Since CAL 83, the prototype, is known to have an ionisation nebula (Pakull and Motch 1989), and further supersoft binaries are expected to also have one, we include also sources associated with very luminous planetary nebulae. Not included are PG 1159 stars which reach similar magnitudes but form a rather distinct class of presently 27 members (Dreizler et al. 1995). Excluded are also supersoft active galactic nuclei which reach luminosities up to 10^{45} erg/s.

The first two supersoft X-ray sources, CAL 83 and CAL 87 (Long et al. 1981), have been discovered with Einstein satellite observations. However, ROSAT observations established these sources as a distinct class, and the majority of the X-ray measurements have been performed with the ROSAT position sensitive proportional counter (PSPC). The PSPC with its spectral resolution of about 50% below 1 keV has been used in nearly all cases to discover the supersoft X-ray spectrum. During the last years the high-resolution imager (HRI) has been used to improve the coordinates of the newly detected sources and to monitor their X-ray intensity. At these soft energies, the HRI countrates are typically a factor of 7.5–8 smaller than those of the PSPC (David et al. 1994, Greiner et al. 1996a).

2 Organisation

The catalog is organised as follows:

- The objects appear in order of increasing right-ascension (for equinox 2000.0) though this mixes SMC and M31 sources. Tab. 1 lists all sources according to their host galaxy with some important parameters for quick comparison.
- The original name of the source is written at the top left, and widely known secondary names on the right.
- The first two lines for each source contain the coordinates, the references for these coordinates and the discovery (in case these are different) and the type classification if the source is optically identified. If a source is not optically identified, we mark it as unidentified below the type identification, and the R.A./Dec. numbers are the X-ray positions with uncertainties as given in the above section of ROSAT related issues. Otherwise, the optical positions are given with typical uncertainties of $1''$. All coordinates are equinox 2000.0. The reference in the "Coordinates" field refers to the most accurate position, i.e. for identified objects the optical position overrides the X-ray position. The "Discovery" reference refers to the first paper which realises the luminous supersoft X-ray emission. Some of the sources have been known already for decades at this time, so the "Coordinates" reference can be much earlier than the "Discovery" reference.
- The next block contains general data which are not specific to any wavelength range. For sources supposed to belong to an external galaxy, the galaxy name is given instead of the presently known distance. This is motivated by recent changes in the distance determination of the LMC which reduced from the standard 55 kpc to 47.3±0.8 kpc (Gould 1995). This in turn also affects the distance to M31 because the distance ratio of LMC and M31 is more accurately known than the corresponding absolute distances. The most recent SMC distance is 57.5 kpc (van den Bergh 1989). The galactic absorbing column (N_H^{gal}, Dickey and Lockman 1990) is given for comparison with the values derived from the X-ray fits given in the next block.
- Two blocks follow for data derived from X-ray and optical/UV measurements. Except the brightness estimates, all numbers are referenced. The temperature derived from the X-ray spectra is given as a range including the error estimates and specifying the model used: bb for the blackbody model and wd for white dwarf atmosphere models. It should be noted that for some sources the values for the best-fit absorption are for the blackbody model while the temperature is from the white dwarf model, i.e. this is not consistent (unfortunately the fitted absorbing column in white dwarf models is only rarely given in the literature).
- Finally, the references are given in full with all co-authors, and stating the important pieces of new information (this reflects purely the subjective view of the author, and you should contact me if something is missing or wrong.) The references are sorted in time of appearance, so that the numbers in the data blocks will not change when the catalog is updated.

Table 1. Summary of all known supersoft X-ray sources with luminosities above 10^{36} erg/s excluding PG 1159-type stars and supersoft AGN. Given are for each source the name (column 1), the best fit X-ray temperature with bb indicating blackbody and wd white dwarf atmosphere models (2), the bolometric luminosity (3), the position (optical position if counterpart is identified, otherwise X-ray position which has a typical error of $\pm 20''$) (4), the type of system (SSS = Cal 83 like supersoft source; PN = planetary nebula; Sy = symbiotic system; N = Nova) (5), the binary period (6), the mass of WD (7), the mass of the companion (8) and References (9).

Name	Countrate[1] (cts/s)	$T^{[2]}$ (eV)	L_{bol} (erg/s)	Type	Period
Large Magellanic Cloud					
RX J0439.8−6809	1.35	20–25 (wd)	$10–14 \times 10^{37}$	SSS	3.37 h
RX J0513.9−6951	<0.06–2.0	30–40 (bb)	$0.1–2 \times 10^{38}$	SSS	18.24 h
RX J0527.8−6954	0.004–0.25	18–45 (bb)	$1–10 \times 10^{37}$	SSS?	
RX J0537.7−7034	0.02	18–30 (bb)	$0.6–2 \times 10^{37}$		
CAL 83	0.98	20–50 (bb)	$1–10 \times 10^{38}$	SSS	1.04 d
CAL 87	0.09	65–75 (wd)	$6–20 \times 10^{37}$	SSS	10.6 h
RX J0550.0−7151	<0.02–0.9	25–40 (bb)			
Small Magellanic Cloud					
1E 0035.4−7230	0.33	40–50 (wd)	$0.8–2 \times 10^{37}$	SSS	4.1 h
RX J0048.4−7332	0.19	25–45 (wd)	$1–8 \times 10^{38}$	Sy	
RX J0058.6−7146	<0.001–0.7	15–70 (bb)	2×10^{36}		
1E 0056.8−7154	0.29	30–40 (wd)	2×10^{37}	PN	
Andromeda Galaxy (M31)					
RX J0037.4+4015	0.3×10^{-3}	43			
RX J0038.5+4014	0.8×10^{-3}	45			
RX J0038.6+4020	1.7×10^{-3}	43			
RX J0039.6+4054	0.4×10^{-3}	45			
RX J0040.4+4009	0.8×10^{-3}	42			
RX J0040.7+4015	1.3×10^{-3}	42			
RX J0041.5+4040	0.3×10^{-3}	40			
RX J0041.8+4059	0.5×10^{-3}	43			
RX J0042.4+4044	1.7×10^{-3}	43			
RX J0043.5+4207	2.2×10^{-3}	45			
RX J0044.0+4118	2.5×10^{-3}	42			
RX J0045.4+4154	$<10^{-5}$–0.03	70–90 (wd)	$5–10 \times 10^{37}$		
RX J0045.5+4206	3.1×10^{-3}	20–48 (bb)	7×10^{37}		
RX J0046.2+4144	2.1×10^{-3}	38			
RX J0046.2+4138	1.1×10^{-3}	40			
RX J0047.6+4205	1.0×10^{-3}	39			

Name	Countrate[1] (cts/s)	T[2] (eV)	L_{bol} (erg/s)	Type	Period
Galactic Sources					
RX J0019.8+2156	2.0	25–37 (wd)	$3-9\times 10^{36}$	SSS	15.85 h
RX J0925.7−4758	1.0	70–75 (wd)	$3-7\times 10^{35(3)}$	SSS	3.5-4 d
GQ Mus	0.1	25–35 (bb)	$1-2\times 10^{38}$	N	1.41 h
1E 1339.8+2837	0.01–1.1	20–45 (bb)	$0.12-10\times 10^{35}$		
AG Dra	1.0	10–15 (bb)	$1.4\times 10^{36(3)}$	Sy	554 d
RR Tel	0.18	12 (wd)	1.3×10^{37}	Sy	387 d
Nova Cyg 1992	0.03–76		2×10^{38}	N	1.95 h

[1] Countrates in the ROSAT PSPC corrected for vignetting, i.e. absorbed on-axis count rates. Count rates in the HRI have been converted to PSPC rates using a conversion factor of PSPC/HRI = 7.8 (Greiner et al. 1996a).

[2] Temperatures for the M31 sources are the maximum blackbody temperatures derived from the hardness ratios at the appropriate absorbing column (Greiner et al. 1996b).

[3] Luminosity for assumed distance of 1 kpc.

I would like to emphasise that every user of this catalog should spare no pains to consult the original papers in order to avoid propagation of my errors in the literature. I will keep this catalog updated, and would appreciate (1) being informed on any errors users might discover and (2) getting preprints on supersoft sources to be included in the next version. An electronic version of this catalog will be available on the Web shortly after this volume has appeared (http://www.rosat.mpe-garching.mpg.de/~jcg).

Acknowledgements: I appreciate the help of many collegues who sent preprints and reprints of their work. Special thanks to R. Di Stefano for her steady encouragement to produce this catalog and for extensive discussion on its content. I apologise to anyone whose paper slipped through the literature search. JG is supported by the Deutsche Agentur für Raumfahrtangelegenheiten (DARA) GmbH under contract FKZ 50 OR 9201. The *ROSAT* project is supported by the German Bundesministerium für Bildung, Forschung, Wissenschaft und Technik (BMBW/DARA) and the Max-Planck-Society. This research has made use of the Simbad database, operated at CDS, Strasbourg, France.

References

Cowley A.P., Schmidtke P.C., Hutchings J.B., Crampton D., 1995, PASP 107, 927
David L.P., Harnden F.R., Kearns K.E., Zombeck M.V., 1994, The ROSAT HRI manual, GSFC
Dickey J.M., Lockman F.J., 1990, Ann. Rev. Astron. Astrophys., 28, 215
Dreizler S., Werner K., Heber U., 1995, in *White Dwarfs*, eds. D. Koester and K. Werner, Lecture Notes in Physics 443, Springer, Berlin, p. 160

Gold A., 1995, ApJ 452, 189
Greiner J., Schwarz R., Hasinger G., Orio M., 1996a, A&A (in press)
Greiner J., Supper R., Magnier E.A., 1996b, this volume p. 75
Kahabka P., Pietsch W., Hasinger G., 1994, A&A 288, 538
Pakull M.W., Motch C., 1989, in Extranuclear Activity in Galaxies, ed. E.J.A. Meurs, R.A.E. Fosbury, (Garching, ESO), p. 285
van den Bergh S., 1989, A&A Rev. 1, 111
Werner K., Wolff B., Cowley A.P., Schmidtke P.C., Hutchings J.B., Crampton D., 1996, this volume p. 131

RX J0019.8+2156

R.A.: $00^h 19^m 50\overset{s}{.}0$ LII: $113\overset{\circ}{.}30$ Discovery: [1] SSS
Dec.: $+21°56'54''$ BII: $-40\overset{\circ}{.}33$ Coordinates: [3]

General Data
D [kpc]: 2 ± 1 [2] N_H^{gal} [10^{21} cm^{-2}]: 0.429
L_{bol} [erg/s]: $3-9\times10^{36}$ [2] Orbital Period: 15.85 h [2]
Mass of central object [$M_\odot$]: 1.0–1.35 [4] Mass of companion [$M_\odot$]:
 Spectral type:

X-ray Data
T [eV]: 25–37 (wd) N_H^{fit} [10^{21} cm^{-2}]: 0.4–1.1 [2]
Orbital Modulation: quasi-sinusoidal, 10% amplitude [2]
Variability: constant between 1990–1995 [2]

Optical/UV Data
Finding Chart: [3] m_B [mag]: 12.3 m_V [mag]: 12.2
Orbital Modulation: 0.5 mag quasi-sinusoidal [2],[7]
Opt. Spectrum: [2] UV Spectrum: [2],[8]
Opt. Variability: 3 different timescales: 2 hr, weeks-months, 40 yrs [2], [3]
UV Variability: 20% irregular [8]
Nebula [erg/s]: Wind mass loss: 1000 km/s [2]

References
[1] Reinsch K., Beuermann K., Thomas H.-C., 1993, Astron. Ges. Abstr. Ser. 9, 41: short discovery report including X-ray spectral characteristics, optical and IUE spectrum, orbital period
[2] Beuermann K., Reinsch K., Barwig H., Burwitz V., de Martino D., Mantel K.-H., Pakull M.W., Robinson E.L., Schwope A.D., Thomas H.-C., Trümper J., van Teeseling A., Zhang E., 1995, A&A 294, L1: discovery, X-ray and optical properties
[3] Greiner J., Wenzel W., 1995, A&A 294, L5: long-term optical variability, improved orbital period, finding chart
[4] Kahabka P., 1995, A&A 304, 227: WD mass from variability timescale
[5] Fender R.P., Bell Burnell S.J., 1996, in Röntgenstrahlung from the Universe, Eds. H.U. Zimmermann et al. , MPE Report 263, p. 135: infrared imaging (JHK)
[6] Meyer F., Meyer-Hofmeister E., 1996, this volume p. 153: modelling of optical long-term lightcurve
[7] Will T., Barwig H., 1996, this volume p. 99: 3 yrs optical photometry, improved period
[8] Gänsicke B.T., Beuermann K., de Martino D., 1996, this volume p. 107: phase-resolved UV spectroscopy

1E 0035.4−7230   RX J0037.3−7214

R.A.: $00^h37^m19\overset{s}{.}8$ LII: $304\overset{\circ}{.}45$ Discovery: [3]   SSS
Dec.: $-72°14'13''$ BII: $-44\overset{\circ}{.}85$ Coordinates: [5], [9]

General Data
D [kpc]: SMC N_H^{gal} [10^{21} cm^{-2}]: 0.649
L_{bol} [erg/s]: $0.8-2\times10^{37}$ [3], [4], [10]   Orbital Period: 4.1 h [6], [9]
Mass of central object [$M_\odot$]:   Mass of companion [$M_\odot$]:
                                         Spectral type:

X-ray Data
T [eV]: 40−50 (wd) N_H^{fit} [10^{21} cm^{-2}]: 0.35−0.8 [4]
Orbital Modulation: strong, nearly sinusoidal [11], [9]
Variability: amplitude of factor 3, orbital variation not excluded [4]

Optical/UV Data
Finding Chart: [9] m_B [mag]: 20.0 m_V [mag]: 20.2
Orbital Modulation: sinusoidal, 0.4 mag amplitude [9]
Opt. Spectrum: [9],[12]   UV Spectrum:
Opt. Variability:
UV Variability:
Nebula [erg/s]: $<10^{34.6}$ (OIII) [8]   Wind mass loss:

References
[1] Seward F.D., Mitchell M., 1981, ApJ 243, 736: X-ray discovery in *Einstein* data
[2] Jones L.R., Pye J.P., McHardy I.M., Fairall A.P., 1985, Space Sci. Rev. 40, 693: suggested optical counterpart (no position or finding chart)
[3] Wang Q., Wu X., 1992, ApJS 78, 391: *Einstein* SMC survey, discovery of soft and luminous X-ray spectrum
[4] Kahabka P., Pietsch W., Hasinger G., 1994, A&A 288, 538: ROSAT X-ray data on several SMC/LMC sources
[5] Orio M., Della Valle M., Massone G., Ögelman H., 1994, A&A 289, L11: X-ray position (*ROSAT* PSPC), optical ID, finding chart, optical spectrum
[6] Schmidtke P.C., Cowley A.P., McGrath T.K., Hutchings J.B., Crampton D., 1994, IAUC 6107: orbital period, colours
[7] Greiner J., 1995, (unpublished): X-ray position (*ROSAT* HRI)
[8] Remillard R.A., Rappaport S., Macri L.M., 1995, ApJ 439, 646: ionisation nebula limit
[9] Schmidtke P.C., Cowley A.P., McGrath T.K., Hutchings J.B., Crampton D., 1996, AJ 111, 788: optical data, finding chart
[10] Teeseling A. van, Heise J., Kahabka P., 1996, in Compact stars in binaries, IAU Symp. 165, Eds J. van Paradijs et al. , p. 445: WD atmosphere modelling
[11] Kahabka P., 1996, A&A (in press): orbital modulation of X-rays
[12] van Teeseling A., Reinsch K., Beuermann K., Thomas H.-C., Pakull M.W., 1996, this volume p. 115: phase-resolved spectroscopy, radial velocity

RX J0037.4+4015

R.A.: $00^h37^m25\overset{s}{.}3$ LII: $120\overset{\circ}{.}04$ Discovery: [1]
Dec.: $+40°15'16''$ BII: $-22\overset{\circ}{.}54$ Coordinates: [1]   unidentified

General Data
D [kpc]: M 31 N_H^{gal} [10^{21} cm^{-2}]: 0.62
L_{bol} [erg/s]:   Orbital Period:
Mass of central object [$M_\odot$]:   Mass of companion [$M_\odot$]:
    Spectral type:

X-ray Data
T [eV]: <43 (bb) [2] N_H^{fit} [10^{21} cm^{-2}]:
Orbital Modulation:
Variability: no [2]

Optical/UV Data
Finding Chart: m_B [mag]: m_V [mag]:
Orbital Modulation:
Opt. Spectrum: UV Spectrum:
Opt. Variability:
UV Variability:
Nebula [erg/s]:   Wind mass loss:

References
[1] Supper R., Hasinger G., Pietsch W., Trümper J., Jain A., Magnier E.A., Lewin W.H.G., van Paradijs J., 1996, A&A (in press): ROSAT data of M31, position
[2] Greiner J., Supper R., Magnier E.A., 1996, this volume p. 75: X-ray spectral data, position

RX J0038.5+4014

R.A.: $00^h 38^m 32^s.1$ **LII:** $120°.27$ Discovery: [1]
Dec.: $+40° 14' 39''$ **BII:** $-22°.56$ Coordinates: [1] |unidentified|

General Data
D [kpc]: M 31
L_{bol} [erg/s]:
Mass of central object [$M_\odot$]:

N_H^{gal} [10^{21} cm^{-2}]: 0.62
Orbital Period:
Mass of companion [$M_\odot$]:
Spectral type:

X-ray Data
T [eV]: <45 (bb) [2]
Orbital Modulation:
Variability: no [2]

N_H^{fit} [10^{21} cm^{-2}]:

Optical/UV Data
Finding Chart: m_B [mag]: m_V [mag]:
Orbital Modulation:
Opt. Spectrum: UV Spectrum:
Opt. Variability:
UV Variability:
Nebula [erg/s]:   Wind mass loss:

References
[1] Supper R., Hasinger G., Pietsch W., Trümper J., Jain A., Magnier E.A., Lewin W.H.G., van Paradijs J., 1996, A&A (in press): ROSAT data of M31, position
[2] Greiner J., Supper R., Magnier E.A., 1996, this volume p. 75: X-ray spectral data, position

RX J0038.6+4020

R.A.: $00^h 38^m 40\overset{s}{.}9$ **LII:** $120\overset{\circ}{.}30$ Discovery: [1]
Dec.: $+40°20'00''$ **BII:** $-22\overset{\circ}{.}47$ Coordinates: [1]   unidentified

General Data
D [kpc]: M 31 N_H^{gal} [10^{21} cm^{-2}]: 0.62
L_{bol} [erg/s]:   Orbital Period:
Mass of central object [$M_\odot$]:   Mass of companion [$M_\odot$]:
   Spectral type:

X-ray Data
T [eV]: <43 (bb) [2] N_H^{fit} [10^{21} cm^{-2}]:
Orbital Modulation:
Variability: no [2]

Optical/UV Data
Finding Chart: [2] m_B [mag]: m_V [mag]:
Orbital Modulation:
Opt. Spectrum: UV Spectrum:
Opt. Variability:
UV Variability:
Nebula [erg/s]:   Wind mass loss:

References
[1] Supper R., Hasinger G., Pietsch W., Trümper J., Jain A., Magnier E.A., Lewin W.H.G., van Paradijs J., 1996, A&A (in press): ROSAT data of M31, position
[2] Greiner J., Supper R., Magnier E.A., 1996, this volume p. 75: X-ray spectral data, position, finding chart

RX J0039.6+4054

R.A.: $00^h 39^m 38^s.5$ **LII:** $120°.53$ **Discovery:** [1]
Dec.: $+40°54'09''$ **BII:** $-21°.91$ **Coordinates:** [1] |unidentified|

General Data
D [kpc]: M 31
L_{bol} [erg/s]:
Mass of central object [$M_\odot$]:

N_H^{gal} [10^{21} cm^{-2}]: 0.62
Orbital Period:
Mass of companion [$M_\odot$]:
Spectral type:

X-ray Data
T [eV]: <45 (bb) [2]
Orbital Modulation:
Variability: no [2]

N_H^{fit} [10^{21} cm^{-2}]:

Optical/UV Data
Finding Chart: [2] m_B [mag]: m_V [mag]:
Orbital Modulation:
Opt. Spectrum: UV Spectrum:
Opt. Variability:
UV Variability:
Nebula [erg/s]: Wind mass loss:

References
[1] Supper R., Hasinger G., Pietsch W., Trümper J., Jain A., Magnier E.A., Lewin W.H.G., van Paradijs J., 1996, A&A (in press): ROSAT data of M31, position
[2] Greiner J., Supper R., Magnier E.A., 1996, this volume p. 75: X-ray spectral data, position, finding chart

RX J0040.4+4009

R.A.: $00^h 40^m 26^s\!.3$ **LII:** $120°\!.65$ Discovery: [1]
Dec.: $+40°09'01''$ **BII:** $-22°\!.67$ Coordinates: [1]   unidentified

General Data
D [kpc]: M 31
L_{bol} [erg/s]:
Mass of central object [$M_\odot$]:

N_H^{gal} [10^{21} cm^{-2}]: 0.62
Orbital Period:
Mass of companion [$M_\odot$]:
Spectral type:

X-ray Data
T [eV]: <42 (bb) [2]
Orbital Modulation:
Variability: no [2]

N_H^{fit} [10^{21} cm^{-2}]:

Optical/UV Data
Finding Chart: m_B [mag]: m_V [mag]:
Orbital Modulation:
Opt. Spectrum: UV Spectrum:
Opt. Variability:
UV Variability:
Nebula [erg/s]:   Wind mass loss:

References
[1] Supper R., Hasinger G., Pietsch W., Trümper J., Jain A., Magnier E.A., Lewin W.H.G., van Paradijs J., 1996, A&A (in press): ROSAT data of M31, position
[2] Greiner J., Supper R., Magnier E.A., 1996, this volume p. 75: X-ray spectral data, position

RX J0040.7+4015

R.A.: $00^h 40^m 43^s.2$ **LII:** $120°.72$ **Discovery:** [1]
Dec.: $+40° 15' 18''$ **BII:** $-22°.57$ **Coordinates:** [1] unidentified

General Data
D [kpc]: M 31
L_{bol} [erg/s]:
Mass of central object [$M_\odot$]:

N_H^{gal} [10^{21} cm^{-2}]: 0.62
Orbital Period:
Mass of companion [$M_\odot$]:
Spectral type:

X-ray Data
T [eV]: <42 (bb) [2]
Orbital Modulation:
Variability: no [2]

N_H^{fit} [10^{21} cm^{-2}]:

Optical/UV Data
Finding Chart: [2] m_B [mag]: m_V [mag]:
Orbital Modulation:
Opt. Spectrum: UV Spectrum:
Opt. Variability:
UV Variability:
Nebula [erg/s]:   Wind mass loss:

References
[1] Supper R., Hasinger G., Pietsch W., Trümper J., Jain A., Magnier E.A., Lewin W.H.G., van Paradijs J., 1996, A&A (in press): ROSAT data of M31, position
[2] Greiner J., Supper R., Magnier E.A., 1996, this volume p. 75: X-ray spectral data, position, finding chart

RX J0041.5+4040

R.A.: $00^h 41^m 30\overset{s}{.}2$ LII: $120\overset{\circ}{.}90$ Discovery: [1]
Dec.: $+40°40'04''$ BII: $-22\overset{\circ}{.}16$ Coordinates: [1]   unidentified

General Data
D [kpc]: M 31 N_H^{gal} [10^{21} cm^{-2}]: 0.62
L_{bol} [erg/s]:   Orbital Period:
Mass of central object [$M_\odot$]:   Mass of companion [$M_\odot$]:
    Spectral type:

X-ray Data
T [eV]: <40 (bb) [2] N_H^{fit} [10^{21} cm^{-2}]:
Orbital Modulation:
Variability: no [2]

Optical/UV Data
Finding Chart: [2] m_B [mag]: m_V [mag]:
Orbital Modulation:
Opt. Spectrum: UV Spectrum:
Opt. Variability:
UV Variability:
Nebula [erg/s]:   Wind mass loss:

References
[1] Supper R., Hasinger G., Pietsch W., Trümper J., Jain A., Magnier E.A., Lewin W.H.G., van Paradijs J., 1996, A&A (in press): ROSAT data of M31, position
[2] Greiner J., Supper R., Magnier E.A., 1996, this volume p. 75: X-ray spectral data, position, finding chart

RX J0041.8+4059

R.A.: $00^h41^m49\overset{s}{.}9$
Dec.: $+40°59'21''$

LII: $120°\!.98$
BII: $-21°\!.85$

Discovery: [1]
Coordinates: [1]   unidentified

General Data
D [kpc]: M 31
L_{bol} [erg/s]:
Mass of central object [$M_\odot$]:

N_H^{gal} [10^{21} cm^{-2}]: 0.62
Orbital Period:
Mass of companion [$M_\odot$]:
Spectral type:

X-ray Data
T [eV]: <43 (bb) [2]
Orbital Modulation:
Variability: no [2]

N_H^{fit} [10^{21} cm^{-2}]:

Optical/UV Data
Finding Chart: [2]
Orbital Modulation:
Opt. Spectrum:
Opt. Variability:
UV Variability:
Nebula [erg/s]:

m_B [mag]:

UV Spectrum:

Wind mass loss:

m_V [mag]:

References
[1] Supper R., Hasinger G., Pietsch W., Trümper J., Jain A., Magnier E.A., Lewin W.H.G., van Paradijs J., 1996, A&A (in press): ROSAT data of M31, position
[2] Greiner J., Supper R., Magnier E.A., 1996, this volume p. 75: X-ray spectral data, position, finding chart

RX J0042.4+4044

R.A.: $00^h 42^m 27^s\!.6$
Dec.: $+40°44'32''$
LII: $121°\!.10$
BII: $-22°\!.10$
Discovery: [1]
Coordinates: [1] unidentified

General Data
D [kpc]: M 31
L_{bol} [erg/s]:
Mass of central object [$M_\odot$]:

N_H^{gal} [10^{21} cm^{-2}]: 0.62
Orbital Period:
Mass of companion [$M_\odot$]:
Spectral type:

X-ray Data
T [eV]: <43 (bb) [2]
Orbital Modulation:
Variability: no [2]

N_H^{fit} [10^{21} cm^{-2}]:

Optical/UV Data
Finding Chart: [2]
Orbital Modulation:
Opt. Spectrum:
Opt. Variability:
UV Variability:
Nebula [erg/s]:

m_B [mag]:

UV Spectrum:

Wind mass loss:

m_V [mag]:

References
[1] Supper R., Hasinger G., Pietsch W., Trümper J., Jain A., Magnier E.A., Lewin W.H.G., van Paradijs J., 1996, A&A (in press): ROSAT data of M31, position
[2] Greiner J., Supper R., Magnier E.A., 1996, this volume p. 75: X-ray spectral data, position, finding chart

RX J0043.5+4207

R.A.: $00^h 43^m 35^s.9$
Dec.: $+42°07'30''$

LII: $121°.31$
BII: $-22°.72$

Discovery: [1]
Coordinates: [1]   unidentified

General Data
D [kpc]: M 31
L_{bol} [erg/s]:
Mass of central object [$M_\odot$]:

N_H^{gal} [10^{21} cm^{-2}]: 0.62
Orbital Period:
Mass of companion [$M_\odot$]:
Spectral type:

X-ray Data
T [eV]: <45 (bb) [2]
Orbital Modulation:
Variability: no [2]

N_H^{fit} [10^{21} cm^{-2}]:

Optical/UV Data
Finding Chart: m_B [mag]: m_V [mag]:
Orbital Modulation:
Opt. Spectrum: UV Spectrum:
Opt. Variability:
UV Variability:
Nebula [erg/s]:   Wind mass loss:

References
[1] Supper R., Hasinger G., Pietsch W., Trümper J., Jain A., Magnier E.A., Lewin W.H.G., van Paradijs J., 1996, A&A (in press): ROSAT data of M31, position
[2] Greiner J., Supper R., Magnier E.A., 1996, this volume p. 75: X-ray spectral data, position

RX J0044.0+4118

R.A.: $00^h 44^m 04\overset{s}{.}8$ **LII:** $121\overset{\circ}{.}45$ Discovery: [1]
Dec.: $+41°18'20''$ **BII:** $-21\overset{\circ}{.}54$ Coordinates: [1] unidentified

General Data
D [kpc]: M 31
L_{bol} [erg/s]:
Mass of central object [$M_\odot$]:

N_H^{gal} [10^{21} cm^{-2}]: 0.62
Orbital Period:
Mass of companion [$M_\odot$]:
Spectral type:

X-ray Data
T [eV]: <42 (bb) [2]
Orbital Modulation:
Variability: no [2]

N_H^{fit} [10^{21} cm^{-2}]:

Optical/UV Data
Finding Chart: [2] m_B [mag]: m_V [mag]:
Orbital Modulation:
Opt. Spectrum: UV Spectrum:
Opt. Variability:
UV Variability:
Nebula [erg/s]:   Wind mass loss:

References
[1] Supper R., Hasinger G., Pietsch W., Trümper J., Jain A., Magnier E.A., Lewin W.H.G., van Paradijs J., 1996, A&A (in press): ROSAT data of M31, position
[2] Greiner J., Supper R., Magnier E.A., 1996, this volume p. 75: X-ray spectral data, position, finding chart

RX J0045.4+4154

R.A.: $00^h 45^m 29{.}^s 0$ LII: $121{.}^\circ 75$ Discovery: [2]
Dec.: $+41^\circ 54' 08''$ BII: $-20{.}^\circ 96$ Coordinates: [2] |unidentified|

General Data
D [kpc]: M 31
L_{bol} [erg/s]: $5-10 \times 10^{37}$ [2]
Mass of central object [$M_\odot$]:

N_H^{gal} [10^{21} cm^{-2}]: 0.84
Orbital Period:
Mass of companion [$M_\odot$]:
Spectral type:

X-ray Data
T [eV]: <154 (wd) [2],[3]
Orbital Modulation:
Variability: transient [2]

N_H^{fit} [10^{21} cm^{-2}]:

Optical/UV Data
Finding Chart: [4] m_B [mag]: m_V [mag]:
Orbital Modulation:
Opt. Spectrum: UV Spectrum:
Opt. Variability:
UV Variability:
Nebula [erg/s]:   Wind mass loss:

References
[1] White N.E., Giommi P., Angelini L., Fantasia S., 1994, IAU Circ. 6064: X-ray discovery, position
[2] White N.E., Giommi P., Heise J., Angelini L., Fantasia S., 1995, ApJ 445, L125: X-ray discovery, position, atmosphere modelling of X-ray data
[3] Supper R., Hasinger G., Pietsch W., Trümper J., Jain A., Magnier E.A., Lewin W.H.G., van Paradijs J., 1996, A&A (in press): ROSAT data of M31, position
[4] Greiner J., Supper R., Magnier E.A., 1996, this volume p. 75: X-ray spectral data, position, finding chart

RX J0045.5+4206

R.A.: $00^h 45^m 32\overset{s}{.}3$ LII: $121\overset{\circ}{.}76$ Discovery: [1]
Dec.: $+42°06'59''$ BII: $-20\overset{\circ}{.}74$ Coordinates: [1]   unidentified

General Data
D [kpc]: M 31 N_H^{gal} [10^{21} cm^{-2}]: 0.84
L_{bol} [erg/s]: 7×10^{37} [2]   Orbital Period:
Mass of central object [$M_\odot$]:   Mass of companion [$M_\odot$]:
     Spectral type:

X-ray Data
T [eV]: 20–48 (bb) [2] N_H^{fit} [10^{21} cm^{-2}]:
Orbital Modulation:
Variability: no [2]

Optical/UV Data
Finding Chart: [2] m_B [mag]: m_V [mag]:
Orbital Modulation:
Opt. Spectrum: UV Spectrum:
Opt. Variability:
UV Variability:
Nebula [erg/s]:   Wind mass loss:

References
[1] Supper R., Hasinger G., Pietsch W., Trümper J., Jain A., Magnier E.A., Lewin W.H.G., van Paradijs J., 1996, A&A (in press): ROSAT data of M31, position
[2] Greiner J., Supper R., Magnier E.A., 1996, this volume p. 75: X-ray spectral data, position, finding chart

RX J0046.2+4144

R.A.: $00^h 46^m 15\overset{s}{.}6$ **LII:** $121\overset{\circ}{.}90$ **Discovery:** [1]
Dec.: $+41°44'36''$ **BII:** $-21\overset{\circ}{.}12$ **Coordinates:** [1] $\boxed{\text{unidentified}}$

General Data
D [kpc]: M 31
L_{bol} [erg/s]:
Mass of central object [$M_\odot$]:

N_H^{gal} [10^{21} cm^{-2}]: 0.84
Orbital Period:
Mass of companion [$M_\odot$]:
Spectral type:

X-ray Data
T [eV]: <38 (bb) [2]
Orbital Modulation:
Variability: no [2]

N_H^{fit} [10^{21} cm^{-2}]:

Optical/UV Data
Finding Chart: [2] m_B [mag]: m_V [mag]:
Orbital Modulation:
Opt. Spectrum: UV Spectrum:
Opt. Variability:
UV Variability:
Nebula [erg/s]:   Wind mass loss:

References
[1] Supper R., Hasinger G., Pietsch W., Trümper J., Jain A., Magnier E.A., Lewin W.H.G., van Paradijs J., 1996, A&A (in press): ROSAT data of M31, position
[2] Greiner J., Supper R., Magnier E.A., 1996, this volume p. 75: X-ray spectral data, position, finding chart

RX J0046.2+4138

R.A.: $00^h 46^m 17\overset{s}{.}8$ LII: $121\overset{\circ}{.}90$ Discovery: [1]
Dec.: $+41°38'48''$ BII: $-21\overset{\circ}{.}21$ Coordinates: [1]   unidentified

General Data
D [kpc]: M 31
L_{bol} [erg/s]:
Mass of central object [$M_\odot$]:
N_H^{gal} [10^{21} cm^{-2}]: 0.84
Orbital Period:
Mass of companion [$M_\odot$]:
Spectral type:

X-ray Data
T [eV]: <40 (bb) [2]
Orbital Modulation:
Variability: no [2]
N_H^{fit} [10^{21} cm^{-2}]:

Optical/UV Data
Finding Chart: m_B [mag]: m_V [mag]:
Orbital Modulation:
Opt. Spectrum: UV Spectrum:
Opt. Variability:
UV Variability:
Nebula [erg/s]:   Wind mass loss:

References
[1] Supper R., Hasinger G., Pietsch W., Trümper J., Jain A., Magnier E.A., Lewin W.H.G., van Paradijs J., 1996, A&A (in press): ROSAT data of M31, position
[2] Greiner J., Supper R., Magnier E.A., 1996, this volume p. 75: X-ray spectral data, position

RX J0047.6+4205

R.A.: $00^h 47^m 38^s.5$ LII: $122°.18$ Discovery: [1]
Dec.: $+42°05'07''$ BII: $-20°.78$ Coordinates: [1]   unidentified

General Data
D [kpc]: M 31
L_{bol} [erg/s]:
Mass of central object [$M_\odot$]:

N_H^{gal} [10^{21} cm^{-2}]: 0.84
Orbital Period:
Mass of companion [$M_\odot$]:
Spectral type:

X-ray Data
T [eV]: <39 (bb) [2]
Orbital Modulation:
Variability: no [2]

N_H^{fit} [10^{21} cm^{-2}]:

Optical/UV Data
Finding Chart: m_B [mag]: m_V [mag]:
Orbital Modulation:
Opt. Spectrum: UV Spectrum:
Opt. Variability:
UV Variability:
Nebula [erg/s]:   Wind mass loss:

References
[1] Supper R., Hasinger G., Pietsch W., Trümper J., Jain A., Magnier E.A., Lewin W.H.G., van Paradijs J., 1996, A&A (in press): ROSAT data of M31, position
[2] Greiner J., Supper R., Magnier E.A., 1996, this volume p. 75: X-ray spectral data, position

RX J0048.4−7332   SMC 3

R.A.: $00^h48^m20\overset{s}{.}8$ LII: $303\overset{\circ}{.}23$ Discovery: [2]   Sy
Dec.: $-73°31'53''$ BII: $-43\overset{\circ}{.}59$ Coordinates: [1]

General Data
D [kpc]: SMC N_H^{gal} $[10^{21}$ cm$^{-2}]$: 0.516
L_{bol} [erg/s]: $1-8\times10^{38}$ [5], [6]   Orbital Period:
Mass of central object [$M_\odot$]:   Mass of companion [$M_\odot$]:
                                          Spectral type:

X-ray Data
T [eV]: 25–45 (wd) [2], [6], [5] N_H^{fit} $[10^{21}$ cm$^{-2}]$: 1.2–7.0 [2]
Orbital Modulation:
Variability:

Optical/UV Data
Finding Chart: [1] m_B [mag]: m_V [mag]: 15.5
Orbital Modulation:
Opt. Spectrum: [1],[7]   UV Spectrum:
Opt. Variability: 1.5 mag outburst in 1981 [1]
UV Variability:
Nebula [erg/s]: $<10^{34.6}$ (OIII) [4]   Wind mass loss:

References
[1] Morgan D.H., 1992, MNRAS 258, 639: symbiotic nature and outburst, finding chart
[2] Kahabka P., Pietsch W., Hasinger G., 1994, A&A 288, 538: ROSAT X-ray data on several SMC/LMC sources
[3] Vogel M., Morgan D.H., 1994, A&A 288, 842: IUE spectrum
[4] Remillard R.A., Rappaport S., Macri L.M., 1995, ApJ 439, 646: ionisation nebula limit
[5] Teeseling A. van, Heise J., Kahabka P., 1996, in Compact stars in binaries, IAU Symp. 165, Eds J. van Paradijs et al., p. 445: WD atmosphere modelling
[6] Jordan S., Schmutz W., Wolff B., Werner K., Mürset U., 1996, A&A (in press): model atmosphere for optical, UV and X-ray emission including luminosity, mass loss and abundances
[7] Mürset U., Schild H., Vogel M., 1996, A&A (in press): optical spectrum

RX J0058.6−7146

R.A.: $00^h58^m35\overset{s}{.}8$　　**LII:** $302\overset{\circ}{.}14$　　Discovery: [1]
Dec.: $-71°46'02''$　　**BII:** $-45\overset{\circ}{.}35$　　Coordinates: [1]　　unidentified

General Data
D [kpc]: SMC
L_{bol} [erg/s]: 2×10^{36}
Mass of central object [$M_\odot$]:

N_H^{gal} [10^{21} cm^{-2}]: 0.748
Orbital Period:
Mass of companion [$M_\odot$]:
Spectral type:

X-ray Data
T [eV]: 15-70 (bb)[1]
Orbital Modulation:
Variability: turn on within 2 days [1]

N_H^{fit} [10^{21} cm^{-2}]: 0.3–1.5 [1]

Optical/UV Data
Finding Chart:　　　　　m_B [mag]:　　　　　m_V [mag]:
Orbital Modulation:
Opt. Spectrum:　　　　　　　　　　UV Spectrum:
Opt. Variability:
UV Variability:
Nebula [erg/s]: $<10^{34.6}$ (OIII) [2]　　Wind mass loss:

References
[1] Kahabka P., Pietsch W., Hasinger G., 1994, A&A 288, 538: ROSAT X-ray data on several SMC/LMC sources, X-ray turn on
[2] Remillard R.A., Rappaport S., Macri L.M., 1995, ApJ 439, 646: ionisation nebula limit

1E 0056.8−7154   SMC N67

R.A.: $00^h 58^m 37\rlap{.}^s 0$ LII: $302\rlap{.}°12$ Discovery: [7]   **PN**
Dec.: $-71°35'48''$ BII: $-45\rlap{.}°52$ Coordinates: [2]

General Data
D [kpc]: SMC N_H^{gal} [10^{21} cm^{-2}]: 0.507
L_{bol} [erg/s]: 2×10^{37} [11]   Orbital Period:
Mass of central object [$M_\odot$]: 0.9   Mass of companion [$M_\odot$]:
     Spectral type:

X-ray Data
T [eV]: 30-40 (wd) [11] N_H^{fit} [10^{21} cm^{-2}]: 0.5 [11]
Orbital Modulation:
Variability: <20–40% [10], [9]

Optical/UV Data
Finding Chart: m_B [mag]: m_V [mag]: 16.7
Orbital Modulation:
Opt. Spectrum: UV Spectrum:
Opt. Variability:
UV Variability:
Nebula [erg/s]: <$10^{34.6}$ (OIII) [3], [12]   Wind mass loss:

References
[1] Henize K.G., 1956, ApJS 2, 315: position
[2] Aller L.H., Keyes C.D., 1987, ApJ 320, 159: position, UV spectrum
[3] Wood P.R., Meatheringham S.J., Dopita M.A., Morgan D.H., 1987, ApJ 320, 178: optical diameter limit from Speckle imaging, OIII and Hβ flux
[4] Dopita M.A., Meatheringham S.J., 1991, ApJ 367, 115: photoionisation modelling of optical spectrum, nebular parameters (L, T_{eff}, R_{in}, R_{out}, M_{neb}, density)
[5] Meatheringham S.J., Dopita M.A., 1991, ApJ Suppl. 75, 407: optical line intensities, abundances, comparison with other PN
[6] Wang Q., 1991, MNRAS 252, 47p: Einstein X-ray data, spectrum, luminosity
[7] Wang Q., Wu X., 1992, ApJS 78, 391: *Einstein* SMC survey, discovery of soft and luminous X-ray spectrum
[8] Brown T., Cordova F., Ciardullo R., Thompson R., Bond H., 1994, ApJ 422, 118: reanalysis Einstein data
[9] Kahabka P., Pietsch W., Hasinger G., 1994, A&A 288, 538: ROSAT X-ray data on several SMC/LMC sources
[10] Hughes J.P., 1994, ApJ 427, L25: variability ROSAT/Einstein
[11] Heise J., van Teeseling A., Kahabka P., 1994, A&A 288, L45
[12] Remillard R.A., Rappaport S., Macri L.M., 1995, ApJ 439, 646: ionisation nebula limit
[13] Teeseling A. van, Heise J., Kahabka P., 1996, in Compact stars in binaries, IAU Symp. 165, Eds J. van Paradijs et al., p. 445: WD atmosphere modelling

RX J0439.8−6809

R.A.: $04^h39^m49{.}^s6$ LII: $279{.}^\circ87$ Discovery: [1]   **SSS**
Dec.: $-68^\circ09'02''$ BII: $-37{.}^\circ10$ Coordinates: [3],[5],[7]

General Data
D [kpc]: LMC
L_{bol} [erg/s]: $10-14 \times 10^{37}$
Mass of central object [$M_\odot$]:

N_H^{gal} [10^{21} cm^{-2}]: 0.447
Orbital Period: 3.37 h [6]
Mass of companion [$M_\odot$]:
Spectral type:

X-ray Data
T [eV]: 20-25 (wd) [7]
Orbital Modulation:
Variability: constant 1990–1994 [1]

N_H^{fit} [10^{21} cm^{-2}]: 0.25–0.4 [7]

Optical/UV Data
Finding Chart: [1],[5],[6],[7] m_B [mag]: 21.5 m_V [mag]: 21.7
Orbital Modulation: 0.15 mag [6]
Opt. Spectrum: [5],[7]   UV Spectrum:
Opt. Variability: marginal [5],[6],[7]
UV Variability:
Nebula [erg/s]: $<10^{34.6}$ (OIII) [2]   Wind mass loss:

References
[1] Greiner J., Hasinger G., Thomas H.-C., 1994, A&A 281, L61
[2] Remillard R.A., Rappaport S., Macri L.M., 1995, ApJ 439, 646: ionisation nebula limit
[3] Schmidtke P.C., Cowley A.P., 1995, IAU Circ. 6278: optical counterpart proposed (position and colour)
[4] Reinsch K., van Teeseling A., Beuermann K., Thomas H.-C., 1996, in Röntgenstrahlung from the Universe, Eds. H.U. Zimmermann et al. , p. 183: optical counterpart proposed (colour)
[5] van Teeseling A., Reinsch K., Beuermann K., Thomas H.-C., Pakull M.W., 1996, this volume p. 115: HRI X-ray position, optical identification, U + V finding chart, optical spectrum, short-term variability
[6] Schmidtke P.C., Cowley A.P., 1996, this volume p. 123: U + V finding chart, orbital period
[7] van Teeseling A., Reinsch K., Beuermann K., 1996, A&A (in press): HRI X-ray position, optical identification, U + V finding chart, optical spectrum, short-term variability

RX J0513.9−6951   HV 5682

R.A.: $05^h13^m50\overset{s}{.}8$ LII: $280\overset{\circ}{.}80$ Discovery: [1]   SSS
Dec.: $-69°51'47''$ BII: $-33\overset{\circ}{.}69$ Coordinates: [1], [3], [7]

General Data
D [kpc]: LMC N_H^{gal} $[10^{21}$ cm$^{-2}]$: 0.838
L_{bol} [erg/s]: $0.1-6 \times 10^{38}$ [1]   Orbital Period: 0.76 d [7],[10],[13]
Mass of central object [$M_\odot$]: 1.3–1.4 [6]   Mass of companion [$M_\odot$]: >0.8 [7]
                                                                Spectral type:

X-ray Data
T [eV]: 30-40 (bb) [1] N_H^{fit} $[10^{21}$ cm$^{-2}]$: 0.9 [1]
Orbital Modulation: no
Variability: transient with on/off time scale of 4 weeks [1],[9]

Optical/UV Data
Finding Chart: [2], [3] m_B [mag]: 16.6 m_V [mag]: 16.7
Orbital Modulation: 0.1 mag [7],[10],[13]
Opt. Spectrum: [2],[3],[7],[10],[11],[12]   UV Spectrum: [2]
Opt. Variability: semi-regular 1 mag drops of $\approx$30 d duration [2],[7],[10],[11],[12]
UV Variability:
Nebula [erg/s]: $<10^{34.6}$ (OIII) [4]   Wind mass loss: 3800 km/s [2], [7]

References
[1] Schaeidt S., Hasinger G., Trümper J., 1993, A&A 270, L9: *ROSAT* X-ray data
[2] Pakull M.W., Motch C., Bianchi L., Thomas H.-C., Guibert J., Beaulieu J.-P., Grison P., Schaeidt S., 1993, A&A 278, L39: opt. ID, finding chart, opt. + UV spectrum
[3] Cowley A.P., Schmidtke P.C., Hutchings J.B., Crampton D., McGrath T.K., 1993, ApJ 418, L63: finding chart, opt. spectrum
[4] Remillard R.A., Rappaport S., Macri L.M., 1995, ApJ 439, 646: ionisation nebula limit
[5] Greiner J., 1995, Abano-Padova Conf. on Cataclysmic variables, eds. A. Bianchini, M. Della Valle, M. Orio, ASSL 205, 443: X-ray and optical variability
[6] Kahabka P., 1995, A&A 304, 227: WD mass from variability timescale
[7] Crampton D., Hutchings J.B., Cowley A.P., Schmidtke P.C., McGrath T.K., O'-Donoghue D., Harrop-Allin M.K., 1996, ApJ 456, 320: opt. photometry and spectroscopy, orbit, period, bipolar outflows
[8] Cowley A.P., Schmidtke P.C., Crampton D., Hutchings J.B., 1996, in Compact stars in binaries, IAU Symp. 165, Eds J. van Paradijs et al. , p.439 : prelim. period
[9] Schaeidt S., 1996, this volume p. 159: ROSAT HRI monitoring, several on-states
[10] Southwell K.A., Livio M., Charles P.A., Sutherland W., Alcock C., Allsman R.A., Alves D., Axelrod T.S., Bennett D.P., Cook K.H., Freeman K.C., Griest K., Guern J., Lehner M.J., Marshall S.L., Peterson B.A., Pratt M.R., Quinn P.J., Rodgers A.W., Stubbs C.W., Welch D.L., 1996, this volume p. 165: 3 yrs optical lightcurve, semi-regular brightness drops, orbital period
[11] Reinsch K., van Teeseling A., Beuermann K., Thomas H.-C., 1996, this volume p. 173: photometric monitoring, optical drop
[12] Reinsch K., van Teeseling A., Beuermann K., Abbott T.M.C., 1996, A&A (in press): photometric monitoring, optical drop
[13] Motch C., Pakull M.W., 1996, this volume p. 127: orbital period

RX J0527.8−6954

R.A.: $05^h27^m49\overset{s}{.}9$ **LII:** $280\overset{\circ}{.}56$ Discovery: [1],[2]
Dec.: $-69°54'09''$ **BII:** $-32\overset{\circ}{.}50$ Coordinates: [9] unidentified

General Data
D [kpc]: N_H^{gal} [10^{21} cm^{-2}]: 0.622
L_{bol} [erg/s]: $1-10\times10^{37}$ Orbital Period:
Mass of central object [$M_\odot$]: 1.1–1.35 [8],[10] Mass of companion [$M_\odot$]:
Spectral type:

X-ray Data
T [eV]: 18-45 (bb) [2] N_H^{fit} [10^{21} cm^{-2}]: 0.7–1.0 [2]
Orbital Modulation:
Variability: steady decrease with 5 yr timescale, possibly periodic [9],[10]

Optical/UV Data
Finding Chart: [4],[10] m_B [mag]: >19 m_V [mag]:
Orbital Modulation:
Opt. Spectrum: UV Spectrum:
Opt. Variability:
UV Variability:
Nebula [erg/s]: $<10^{34.6}$ (OIII) [6] Wind mass loss:

References
[1] Trümper J., Hasinger G., Aschenbach B., Bräuninger H., Briel U.G., Burkert W., Fink H., Pfeffermann E., Pietsch W., Predehl P., Schmitt J.H.M.M., Voges W., Zimmermann U., Beuermann K., 1991, Nat 349, 579: first report on X-ray discovery
[2] Greiner J., Hasinger G., Kahabka P. 1991, A&A 246, L17: X-ray discovery, position, temperature
[3] Orio M., Ögelman H., 1993, A&A 273, L56: X-ray fading
[4] Cowley A.P., Schmidtke P.C., Hutchings J.B., Crampton D., McGrath T.K., 1993, ApJ 418, L63: finding chart, X-ray position
[5] Hasinger G., 1994, Reviews in Modern Astronomy 7, 129: prel. X-ray lightcurve
[6] Remillard R.A., Rappaport S., Macri L.M., 1995, ApJ 439, 646: ionisation nebula limit
[7] Greiner J., 1995, in Proc. of the Abano-Terme Conference on Cataclysmic Variables, eds. A. Bianchini, M. Della Valle, M. Orio, Kluwer, ASSL 205, p. 443: prel. X-ray lightcurve
[8] Kahabka P., 1995, A&A 304, 227: WD mass from variability timescale
[9] Greiner J., Schwarz R., Hasinger G., Orio M., 1996, A&A (in press): X-ray lightcurve, improved X-ray position (HRI)
[10] Greiner J., Schwarz R., Hasinger G., Orio M., 1996, this volume p. 145: X-ray lightcurve, improved X-ray position (HRI), finding chart

RX J0537.7−7034   RX J0537.6−7033

R.A.: $05^h 37^m 43^s.0$ LII: $281°.19$ Discovery: [1]
Dec.: $-70°34'15''$ BII: $-31°.57$ Coordinates: [1]   unidentified

General Data
D [kpc]:
L_{bol} [erg/s]: 0.6–2×10^{37} [1]
Mass of central object [$M_\odot$]:

N_H^{gal} [10^{21} cm^{-2}]: 0.637
Orbital Period:
Mass of companion [$M_\odot$]:
Spectral type:

X-ray Data
T [eV]: 18-30 (bb) [1]
Orbital Modulation:
Variability: amplitude of factor 10 [2]

N_H^{fit} [10^{21} cm^{-2}]:

Optical/UV Data
Finding Chart: m_B [mag]: m_V [mag]:
Orbital Modulation:
Opt. Spectrum: UV Spectrum:
Opt. Variability:
UV Variability:
Nebula [erg/s]:   Wind mass loss:

References
[1] Orio M., Ögelman H., 1993, A&A 273, L56: X-ray discovery
[2] Orio M., Della Valle M., Massone G., Ögelman H., 1996, in Proc. of Workshop on Cataclysmic Variables, Keele, June 1995 (in press): X-ray variability, optical counterpart suggested

CAL 83   LHG 83

R.A.: $05^h 43^m 33{.}^s 5$ LII: $278{.}^\circ 56$ Discovery: [1]   SSS
Dec.: $-68°22'23''$ BII: $-31{.}^\circ 31$ Coordinates: [2]

General Data
D [kpc]: LMC N_H^{gal} [10^{21} cm^{-2}]: 0.652
L_{bol} [erg/s]: $1\text{--}10 \times 10^{38}$ [7]   Orbital Period: 1.04 d [5]
Mass of central object [$M_\odot$]:   Mass of companion [$M_\odot$]:
    Spectral type:

X-ray Data
T [eV]: 20-50 (bb) N_H^{fit} [10^{21} cm^{-2}]: 0.7–0.85 [7]
Orbital Modulation: no
Variability:

Optical/UV Data
Finding Chart: [5] m_B [mag]: 16.2–17.3 m_V [mag]: 16.2–17.3
Orbital Modulation: 0.22 mag sinusoidal [5]
Opt. Spectrum: [3]   UV Spectrum: [3],[4]
Opt. Variability: erratic
UV Variability: 50% irregular [3],[4]
Nebula [erg/s]: OIII ($10^{35.6}$), Hα ($10^{35.4}$)[6][9]   Wind mass loss:

References
[1] Long K.S., Helfand D.J., Grabelsky D.A., 1981, ApJ 248, 925: X-ray discovery (Einstein)
[2] Pakull M.W., Ilovaisky S.A., Chevalier C., 1985, Space Sci Rev 40, 229: optical identification
[3] Crampton D., Cowley A.P., Hutchings J.B., Schmidtke P.C., Thompson I.B., Liebert J., 1987, ApJ 321, 745: optical and IUE spectra, orbital period
[4] Bianchi L., Pakull M.W., 1988, in A decade of UV Astronomy with IUE, ESA SP-281, vol. 1, p. 145: UV spectrum
[5] Smale A.P., Corbet R.H., Charles P.Q., Ilovaisky S.A., Mason K.O., Motch C., Mukai K., Naylor T., Parmer A.N., van der Klis M., van Paradijs J., 1988, MNRAS 233, 51: orbital period, finding chart, optical spectra
[6] Pakull M.W., Motch C., 1989, in Extranuclear Activity in Galaxies, ed. E.J.A. Meurs, R.A.E. Fosbury, (Garching, ESO), p. 285: nebula
[7] Greiner J., Hasinger G., Kahabka P., 1991, A&A 246, L17: ROSAT X-ray spectrum and luminosity, position, temperature
[8] Brown T., Cordova F., Ciardullo R., Thompson R., Bond H., 1994, ApJ 422, 118: reanalysis Einstein data
[9] Remillard R.A., Rappaport S., Macri L.M., 1995, ApJ 439, 646: ionisation nebula

CAL 87   LHG 87

R.A.: $05^h46^m52\overset{s}{.}3$ LII: $281\overset{\circ}{.}75$ Discovery: [1]   SSS
Dec.: $-71°08'38''$ BII: $-30\overset{\circ}{.}76$ Coordinates:

General Data
D [kpc]: LMC N_H^{gal} [10^{21} cm^{-2}]: 0.749
L_{bol} [erg/s]: $6-20\times10^{37}$ [12]   Orbital Period: 10.6 h [6]
Mass of central object [$M_\odot$]:   Mass of companion [$M_\odot$]:
                                                                         Spectral type:

X-ray Data
T [eV]: 65-75 (wd) [12] N_H^{fit} [10^{21} cm^{-2}]:
Orbital Modulation: eclipse with amplitude of factor 3 [8]
Variability:

Optical/UV Data
Finding Chart: [6] m_B [mag]: 19.0 m_V [mag]: 18.9
Orbital Modulation: 2 mag eclipse with broad wings [5], [6]
Opt. Spectrum: [5]   UV Spectrum: [11]
Opt. Variability:
UV Variability:
Nebula [erg/s]: $<10^{34.6}$ (OIII) [10]   Wind mass loss:

References
[1] Long K.S., Helfand D.J., Grabelsky D.A., 1981, ApJ 248, 925: X-ray discovery
[2] Pakull M.W., Beuermann K., Angebault L.P., Bianchi L., 1987, Ap&SS 131, 689: optical identification
[3] Pakull M.W., Beuermann K., van der Klis M., van Paradijs J., 1988, A&A 203, L27: eclipse lightcurve
[4] Callanan P.J., Machin G., Naylor T., Charles P.A., 1989, MNRAS 241, 37p: eclipse lightcurve, orbital period
[5] Cowley A.P., Schmidtke P.C., Crampton D., Hutchings J.B., 1990, ApJ 350, 288: orbital period, optical spectra
[6] Schmidtke P.C., McGrath T.K., Cowley A.P., Frattare L.M., 1993, PASP 105, 863: orbital eclipse lightcurve, finding chart, X-ray lightcurve
[7] Brown T., Cordova F., Ciardullo R., Thompson R., Bond H., 1994, ApJ 422, 118: reanalysis Einstein data
[8] Kahabka P., Pietsch W., Hasinger G., 1994, A&A 288, 538: ROSAT X-ray data on several SMC/LMC sources, X-ray eclipse in CAL 87
[9] Khruzina T.S., Cherepashchuk A.M., 1994, Astron. Zh. 71, 442: binary parameters
[10] Remillard R.A., Rappaport S., Macri L.M., 1995, ApJ 439, 646: ionisation nebula limit
[11] Hutchings J.B., Cowley A.P., Schmidtke P.C., Crampton D., 1995, AJ 110, 2394: UV spectra
[12] Teeseling A. van, Heise J., Kahabka P., 1996, in Compact stars in binaries, IAU Symp. 165, Eds J. van Paradijs et al. , p. 445: WD atmosphere modelling
[13] Schandl S., Meyer-Hofmeister E., Meyer F., 1996, this volume p. 53: modelling of orbital lightcurve, inclination
[14] Schandl S., Meyer-Hofmeister E., Meyer F., 1996, A&A (subm.): modelling of orbital lightcurve, inclination

RX J0550.0–7151

R.A.: $05^h50^m00^s\!.2$ LII: $282°\!.52$ Discovery: [1]
Dec.: $-71°52'09''$ BII: $-30°\!.47$ Coordinates: [6]   unidentified

General Data
D [kpc]:
L_{bol} [erg/s]:
Mass of central object [$M_\odot$]:

N_H^{gal} [10^{21} cm^{-2}]: 0.892
Orbital Period:
Mass of companion [$M_\odot$]:
Spectral type:

X-ray Data
T [eV]: 25–40 (bb) [3],[6]
Orbital Modulation:
Variability: turn-off [6]

N_H^{fit} [10^{21} cm^{-2}]: 2.0 [3]

Optical/UV Data
Finding Chart: [6] m_B [mag]: m_V [mag]: >19.5
Orbital Modulation:
Opt. Spectrum: UV Spectrum:
Opt. Variability:
UV Variability:
Nebula [erg/s]: $<10^{34.6}$ (OIII) [2]   Wind mass loss:

References
[1] Cowley A.P., Schmidtke P.C., Hutchings J.B., Crampton D., McGrath T.K., 1993, ApJ 418, L63: X-ray discovery, position, temperature
[2] Remillard R.A., Rappaport S., Macri L.M., 1995, ApJ 439, 646: ionisation nebula limit
[3] Schmidtke P.C., Cowley A.P., 1995, IAU Circ. 6278: ROSAT HRI position in off-state, optical position and colour
[4] Charles P.A., Southwell K.A., 1996, IAU Circ. 6305: possible symbiotic ID
[5] Schmidtke P.C., Cowley A.P., 1996, this volume p. 123: difference to $2.\!''9$ nearby source RX J0549.8–7150
[6] Reinsch K., van Teeseling A., Beuermann K., Thomas H.-C., 1996, this volume p. 173: X-ray turn off, different source in $2.\!''9$ distance during off-state pointing = dMe star which is coincident with optical star from [3],[4]

RX J0925.7−4758

R.A.: $09^h25^m46^s.2$ LII: $271°.36$ Discovery: [1]   SSS
Dec.: $-47°58'17''$ BII: $1°.88$ Coordinates: [1]

General Data
D [kpc]: 0.4–2 [1]
L_{bol} [erg/s]: $3-7\times10^{35}$ (D/1 kpc)^2 [4]
Mass of central object [$M_\odot$]:

N_H^{gal} [10^{21} cm^{-2}]: 13.0
Orbital Period: 3.55–4.03 d [1], [3]
Mass of companion [$M_\odot$]:
Spectral type:

X-ray Data
T [eV]: 70-75 (wd) [4]
Orbital Modulation: possible [3]
Variability: less than 50% [1]

N_H^{fit} [10^{21} cm^{-2}]: 20–24 [4]

Optical/UV Data
Finding Chart: [1] m_B [mag]: 19.2 m_V [mag]: 17.2
Orbital Modulation: 0.3 mag amplitude
Opt. Spectrum: [1]   UV Spectrum:
Opt. Variability:
UV Variability:
Nebula [erg/s]:   Wind mass loss:

References
[1] Motch C., Hasinger G., Pietsch W., 1994, A&A 284, 827: X-ray discovery, optical ID, optical spectrum, prel. period, distance, finding chart
[2] Ebisawa K., Asai K., Dotani T., Mukai K., Smale A., 1996, in Röntgenstrahlung from the Universe, Eds. H.U. Zimmermann et al. , MPE Report 263, p. 133: ASCA observation
[3] Motch C., 1996, this volume p. 83: optical photometry and spectroscopy, orbital period, ROSAT HRI monitoring
[4] Hartmann H.W., Heise J., 1996, this volume p. 25: high-gravity NLTE modelling

GQ Mus   Nova Muscae 1983

R.A.: $11^h 52^m 02\rlap{.}^s 5$ LII: $297\rlap{.}^\circ 21$ Discovery: [6],[8]   N
Dec.: $-67^\circ 12' 24''$ BII: $-5\rlap{.}^\circ 00$ Coordinates: [2]

General Data
D [kpc]: 4.7 ± 1.5 N_H^{gal} [10^{21} cm^{-2}]: 4.24
L_{bol} [erg/s]: $1-2 \times 10^{38}$ [8]   Orbital Period: 1.41 [7]
Mass of central object [$M_\odot$]: $<1.1-1.25$ [13]   Mass of companion [$M_\odot$]: <0.2 [8]
                                                            Spectral type:

X-ray Data
T [eV]: 25–35 (bb) [8] N_H^{fit} [10^{21} cm^{-2}]: 1.0–3.4 [8]
Orbital Modulation: quasi-sinusoidal, 50% amplitude [14]
Variability: decline by factor >30 [12]

Optical/UV Data
Finding Chart: [3], [4] m_B [mag]: 7–21 m_V [mag]: 7–21
Orbital Modulation: 0.2 mag [7]
Opt. Spectrum: [4],[7],[9],[11]   UV Spectrum: [4]
Opt. Variability: nova outburst and decline
UV Variability:
Nebula [erg/s]:   Wind mass loss:

References
[1] Liller W., 1983, IAUC 3764: optical discovery, coordinate
[2] Cragg T., Nikoloff I., Johnston J., 1983, IAUC 3766: improved coordinate
[3] Bateson, Morel, 1983, Charts Southern Var. 16: finding chart
[4] Krautter J., Beuermann K., Leitherer C., Oliva E., Moorwood A.F.M., Deul E., Wargau W., Klare G., Kohoutek L., Paradijs J. van, Wolf B., 1984, A&A 137, 307: finding chart, IR, optical and UV spectra, distance, luminosity
[5] Whitelock P.A., Carter B.S., Feast M.W., Glass I.S., Laney D., Menzies J.W., Walsh J., Williams P.M., 1984, MNRAS 211, 421: optical + IR photometry/spectroscopy
[6] Ögelman H., Beuermann K., Krautter J., 1984, ApJ 287, L31: EXOSAT X-ray detection 1 yr after maximum
[7] Diaz M.P., Steiner J.E., 1989, ApJ 339, L41: orbital period, optical spectrum
[8] Ögelman H., Orio M., Krautter J., Starrfield S., 1993, Nat. 361, 331: X-ray discovery (ROSAT)
[9] Pequignot D., Petitjean P., Boisson C., Krautter J., 1993, A&A 271, 219: optical spectra from 1984–1988
[10] Diaz M.P., Steiner J.E., 1994, ApJ 425, 252: magnetic nature, photometry, spectroscopy, Doppler tomography, accretion rate, distance
[11] Diaz M.P., Williams R.E., Phillips M.M., Hamuy M., 1995, MNRAS 277, 959: optical spectra, photometry, photoionisation modelling
[12] Shanley L., Ögelman H., Gallagher J.S., Orio M., Krautter J., 1995, ApJ 438, L95: X-ray decline
[13] Kahabka P., 1995, A&A 304, 227: WD mass from variability timescale
[14] Kahabka P., 1996, A&A (in press): orbital modulation of X-rays

1E 1339.8+2837   RX J1342.1+2822

R.A.: $13^h42^m09\overset{s}{.}8$ LII: $42\overset{\circ}{.}23$ Discovery: [3]
Dec.: $28°22'45''$ BII: $78\overset{\circ}{.}71$ Coordinates: [3]   unidentified

General Data
D [kpc]: 10.4 kpc (M3) [2]
L_{bol} [erg/s]: $1.2 \times 10^{34} - 1.2 \times 10^{36}$ [3]
Mass of central object [$M_\odot$]:

N_H^{gal} [10^{21} cm^{-2}]: 0.11
Orbital Period:
Mass of companion [$M_\odot$]:
Spectral type:

X-ray Data
T [eV]: 20-45 (bb)
Orbital Modulation:
Variability: transient [3]

N_H^{fit} [10^{21} cm^{-2}]:

Optical/UV Data
Finding Chart: m_B [mag]: m_V [mag]:
Orbital Modulation:
Opt. Spectrum: UV Spectrum:
Opt. Variability:
UV Variability:
Nebula [erg/s]:   Wind mass loss:

References
[1] Hertz P., Grindlay J.E., 1983, ApJ 275, 105: Einstein data, position, flux
[2] Webbink R.F., 1985, in IAU Symp. 123, Dynamics of Star Clusters, ed. J. Goodman & P. Hut (Dordrecht:Reidel), p. 541: distance to M3
[3] Hertz P., Grindlay J.E., Bailyn C.D., 1993, ApJ 410, L87: X-ray transient discovery in ROSAT HRI data, position, flux
[4] Fender R.P., Bell Burnell S.J., 1996, in Röntgenstrahlung from the Universe, Eds. H.U. Zimmermann et al., MPE Report 263, p. 135: infrared imaging (JHK)

AG Dra   BD+67 922

R.A.: $16^h01^m40\overset{s}{.}9$ LII: $100\overset{\circ}{.}29$ Discovery: [10],[11]   Sy
Dec.: $+66°48'10''$ BII: $40\overset{\circ}{.}97$ Coordinates: [2]

General Data
D [kpc]: 0.7–4 kpc
L_{bol} [erg/s]: 1.4×10^{36} $(D/1\,kpc)^2$ [11]
Mass of central object [$M_\odot$]:

N_H^{gal} $[10^{21}$ cm^{-2}]: 0.315
Orbital Period: 554 d [5]
Mass of companion [$M_\odot$]:
Spectral type: K3III–K0Ib [3],[8]

X-ray Data
T [eV]: 10–15 (bb) [10],[11] N_H^{fit} $[10^{21}$ cm^{-2}]: 0.4 [11]
Orbital Modulation: not clear [10],[11]
Variability: major drop during optical outbursts [10],[11]

Optical/UV Data
Finding Chart: [1] m_B [mag]: 8–11.2 m_V [mag]: 8–9.8
Orbital Modulation: 0.5 mag in U [5]
Opt. Spectrum: [3]   UV Spectrum: [6],[7]
Opt. Variability: series of outbursts, possibly 15 yr period [4]
UV Variability: factor 2–5 during outburst [7],[10]
Nebula [erg/s]:   Wind mass loss: 100 km/s [7]

References
[1] Sharov A.S., 1954, Peremenn. Zvesdy 10, 55: discovery of optical variability and finding chart
[2] Wenzel W., Mitt. Veränd. Sterne No. 203: optical lightcurve, finding chart
[3] Boyarchuk A.A., 1966, Astrofizika 2, 101: spectral classification of cool component
[4] Robinson L.J., 1969, Peremennye Zvezdy 16, 507: optical long-term lightcurve
[5] Meinunger L., 1979, IVBS No. 1611: orbital period
[6] Viotti R., Ricciardi O., Ponz D., Giangrande A., Friedjung M., Cassatella A., Baratta G.B., Altamore A., 1983, A&A 119, 285: UV spectra, UV continuum evolution
[7] Viotti R., Altamore A., Baratta G.B., Cassatella A., Friedjung M., 1984, ApJ 283, 226: UV spectra, line profiles, ionised wind
[8] Huang C.C., Friedjung M., Zhou Z.X., 1994, A&AS 106, 413: near IR spectrum, spectral classification of cool component
[9] Mikolajewska J., Kenyon S.J., Mikolajewski M., Garcia M.R., Polidan R.S., 1995, AJ 109, 1289: optical and UV photometry and spectroscopy, orbital parameters, distance
[10] Greiner J., Bickert K., Luthardt R., Viotti R., Altamore A., González-Riestra R., 1996, this volume p. 267: ROSAT X-ray spectrum, luminosity, X-ray and UV lightcurve of 1994/1995 optical outburst, UV spectra
[11] Greiner J., Bickert K., Luthardt R., Viotti R., Altamore A., González-Riestra R., Stencel R.E., 1996, A&A (subm.): ROSAT X-ray spectrum, luminosity, X-ray and UV lightcurve of 1994/1995 optical outburst, UV spectra

RR Tel   HV 3181

R.A.: $20^h04^m18^s5$ LII: $342°16$ Discovery: [9]   Sy
Dec.: $-55°43'34''$ BII: $-32°24$ Coordinates: [3]

General Data
D [kpc]: 2.6 kpc [8] N_H^{gal} [10^{21} cm^{-2}]: 0.439
L_{bol} [erg/s]: 1.3×10^{37} [9] Orbital Period: 385–388 d [2],[4]
Mass of central object [$M_\odot$]: >0.9 [9] Mass of companion [$M_\odot$]:
                                             Spectral type: M-giant [7]

X-ray Data
T [eV]: 12 (wd) [9] N_H^{fit} [10^{21} cm^{-2}]: 0.17
Orbital Modulation:
Variability:

Optical/UV Data
Finding Chart: [3] m_B [mag]: 7–14 m_V [mag]: 7–14
Orbital Modulation: 2.5 mag amplitude [2]
Opt. Spectrum: [5]   UV Spectrum: [6]
Opt. Variability: outburst 1944/45 and decline [2],[3]
UV Variability: marginal
Nebula [erg/s]:   Wind mass loss:

References
[1] Pickering E.C., 1908, Harvard Coll. Obs. Circ. No. 143: discovery of variability by Mrs. Fleming
[2] Payne C.H., 1928, Bull. Harv. Coll. Obs. No. 861: long-term optical variations
[3] Mayall M.W., 1949, Harv. Coll. Obs. Bull. No. 919, p. 15: finding chart
[4] Gaposchkin S., 1945, Ann. Harv. Coll. Obs. No. 2: orbital period, long-term lightcurve
[5] Thackeray A.D., 1977, Mem. Roy. Astr. Soc. 83, 1: optical spectra 1951–1973, nebular evolution
[6] Penston M.V., Benvenuti P., Cassatella A., Heck A., Selvelli P., Macchetto F., Ponz D., Jordan C., Cramer N., Rufener F., Manfroid J., MNRAS 202, 833: IUE spectra and line identification, temperature and density
[7] Feast M.W., Whitelock P.A., Catchpole R.M., Robertson B.S.C., Carter B.S., 1983, MNRAS 202, 951: JHKL photometry 1972–1981, Mira identification
[8] Whitelock P.A., 1988, in The Symbiotic Phenomenon, IAU Coll. 103, ed. J. Mikolajewska, M. Friedjung, S.J. Kenyon, R. Viotti, Kluwer, p. 47: distance
[9] Jordan S., Mürset U., Werner K., 1994, A&A 283, 475: X-ray discovery, atmosphere modelling of X-ray, UV and optical data

V1974 Cyg   Nova Cyg 1992

R.A.: $20^h 30^m 31^s.2$ LII: $345°.92$ Discovery: [11]   N
Dec.: $-52°37'53''$ BII: $-36°.06$ Coordinates: [1]

General Data
D [kpc]: 1.8–3.2 [12] N_H^{gal} [10^{21} cm^{-2}]: 0.325
L_{bol} [erg/s]: 2×10^{38} [6]   Orbital Period: 1.95 h [9]
Mass of central object [$M_\odot$]: 0.75–1.1 [12],[13]   Mass of companion [$M_\odot$]:
    Spectral type:

X-ray Data
T [eV]: N_H^{fit} [10^{21} cm^{-2}]:
Orbital Modulation:
Variability: full outburst lightcurve [11]

Optical/UV Data
Finding Chart: [2] m_B [mag]: m_V [mag]: 4.5–21
Orbital Modulation: 0.2 mag [9]
Opt. Spectrum: [4],[10]   UV Spectrum: [5]
Opt. Variability:
UV Variability: [5],[7]
Nebula [erg/s]:   Wind mass loss:

References
[1] Collins P., Skiff B.A., 1992, IAUC 5454: optical discovery, coordinate
[2] A.A.V.S.O. Circ. No. 256, 1992: finding chart
[3] Shore S.N., Sonneborn G., Starrfield S., Gonzalez-Riestra R., Ake T.B., 1993, AJ 106, 2408: nebular evolution from HST + IUE spectra, HI limit from Lyα P Cyg profile, ejected mass
[4] Barger A.J., Gallagher J.S., Bjorkman K.S., Johansen K.A., Nordsieck K.H., 1993, ApJ 419: optical spectra during decline
[5] Shore S.N., Sonneborn G., Starrfield S., Gonzalez-Riestra R., Ake T.B., 1993, AJ 106, 2408: UV spectral evolution
[6] Shore S.N., Sonneborn G., Starrfield S., Gonzalez-Riestra R., Polidan R.S., 1994, ApJ 421, 344: IUE observations, UV and V lightcurve during first year, bolometric evolution
[7] Taylor M., Bless R.C., Ögelman H., Elliot J.L., Gallagher J.S., Nelson M.J., Percival J.W., Robinson E.L., van Citters G.W., 1994, ApJ 424, L45: UV photometry
[8] Bjorkman K.S., Johansen K.A., Nordsieck K.H., Gallagher J.S., Barger A.J., 1994, ApJ 425, 247: spectropolarimetry
[9] DeYoung J.A., Schmidt R.E., 1994, ApJ 431, L47: orbital period
[10] Rafanelli P, Rosino L., Radovich M., 1995, A&A 294, 488: optical spectra during outburst
[11] Krautter J., Ögelman H., Starrfield S., Wichmann R., Pfeffermann E., 1996, ApJ 456, 788: X-ray rise and decline, X-ray spectrum
[12] Paresce F., Livio M., Hack W., Korista K., 1996, A&A 299, 823: nebula resolved with HST, UV spectra, abundances, distance WD mass
[13] Krautter J., 1996 (priv. comm.)

Part VIII

Appendix

Subject and Object Index

1E 0035.4−7230, 6, 53, 115, 119, 120, 122, 216, 286, 287, 301, 305
1E 0056.8−7154, 301, 324
1E 1339.8+2837, 33, 34, 301, 334
4 Dra, 234, 235
4U 1822−37, 55

A 0535−668, 45, 47
A 0620−00, 48
abundance, 25−32, 57, 140
accreting white dwarf, 184
accretion disk, 46, 53−58, 60, 61, 65−72, 112, 170, 171
accretion disk, massive, 45−49
accretion induced collapse, 187, 190, 215, 221
AE Ara, 238, 239
AG Dra, 225, 228, 236, 237, 252, 259−265, 267−277, 287, 301, 335
AG Peg, 248, 249, 252, 253, 258
AR Pav, 244, 245
AS 201, 232, 233
AS 210, 236, 237
AS 217, 236, 237
AS 221, 238, 239
AS 239, 240, 241
AS 241, 240, 241
AS 245, 240, 241
AS 255, 240, 241
AS 269, 242, 243
AS 270, 242, 243
AS 276, 242, 243
AS 281, 242, 243
AS 282, 242, 243
AS 289, 242, 243
AS 293, 242, 243

AS 295B, 244, 245
AS 296, 244, 245
AS 297, 242, 243
AS 299, 244, 245
AS 302, 244, 245
AS 304, 244, 245
AS 313, 244, 245
AS 316, 244, 245
AS 323, 244, 245
AS 327, 246, 247
AS 329, 246, 247
AS 337, 246, 247
AS 338, 246, 247
AS 360, 246, 247
AS 373, 246, 247
AS 453, 248, 249
ASCA satellite, 31, 41, 43, 84, 86, 91
AX Per, 230, 231

BD+37 2318, 234, 235
BD+67 922, 236, 237, 335
BD−19 1821, 232, 233
BD−21 3873, 234, 235
BD−57 2768, 232, 233
BF Cyg, 246, 247
BI Cru, 234, 235
binary evolution, 3−13, 153, 184, 186, 187, 189, 193, 194, 196−199, 203, 205, 207, 210−212
black-hole candidate, 45
black hole, 5, 38, 47−49, 66
BL Tel, 246, 247
burning white dwarf, 15, 16, 18, 20, 21, 23, 24, 53, 54, 67, 70, 163, 170, 171
BX Mon, 232, 233

342 Subject and Object Index

CAL 83, 3, 5, 6, 45–47, 53, 54, 65–69, 72, 79, 81, 107, 113, 127, 130, 159, 170, 172, 200, 217, 276, 285, 299, 301, 329
CAL 87, 3, 5, 6, 45, 46, 53–55, 57, 59, 61, 62, 65–67, 72, 88, 112, 217, 299, 301, 330
CD-28 14567, 244, 245
CD-32 10517, 234, 235
CD-43 14304, 248, 249, 252
CD-64 665, 234, 235
Chandrasekhar limit, 24, 89, 183, 195–197, 200, 220, 221
CH Cyg, 246, 247, 252
CI Cyg, 246, 247
Cir X-1, 47
close-binary supersoft source, 3, 4, 6, 7, 13, 193–200, 203
CL Sco, 236, 237
CM Aql, 246, 247
common envelope, 3, 5, 7, 8, 10, 12, 186, 195–197, 199, 203, 207, 212
CPD-53 8091, 236, 237
CQ Dra, 234, 235
Crab, 49
CV, 4, 5, 38, 170, 184, 188, 195, 286, 289, 294, 295
Cyg X-3, 45

Draco C-1, 238, 239, 252

Eddington limit, 26, 46, 78, 150, 151, 170, 258
EG And, 230, 231, 252
Einstein satellite, 139, 225, 251, 259, 299
ER Del, 248, 249
EUVE satellite, 139, 290
evolution, 3, 15–17, 19, 22–24, 119, 127, 136, 207
EX Hya, 225, 228, 234, 235
EXOSAT satellite, 139, 225, 251, 259

FG Sge, 246, 247
FN Sgr, 246, 247
FR Sct, 244, 245

galaxies, 37, 193
gamma-ray burst, 49
GH Gem, 232, 233
GQ Mus, 19, 127, 171, 301, 333
GW Vir, 138
GX 1+4, 228, 238, 239, 252

H-R diagram, 15, 16, 24, 131
HD 149407, 236, 237
HD 161213, 240, 241
HD 161224, 240, 241
HD 177300, 246, 247
HD 182917, 246, 247
HD 214369, 248, 249
HD 316496, 240, 241
HD 319167, 244, 245
HD 330036, 234, 235
HD 34842, 230, 231
HD 4174, 230, 231
HD 59643, 232, 233
He 2-104, 234, 235
He 2-127, 234, 235
He 2-139, 236, 237
He 2-147, 236, 237
He 2-171, 236, 237
He 2-173, 236, 237
He 2-176, 236, 237
He 2-251, 238, 239
He 2-294, 240, 241
He 2-325, 242, 243
He 2-374, 244, 245
He 2-38, 232, 233
He 2-390, 244, 245
He 2-467, 248, 249
He 2-468, 248, 249
He 2-87, 234, 235
He 3-1092, 234, 235
He 3-1103, 234, 235
He 3-1213, 236, 237
He 3-1242, 236, 237
He 3-1341, 236, 237
He 3-1342, 238, 239
He 3-1383, 238, 239
He 3-1410, 238, 239
He 3-1591, 242, 243

Subject and Object Index   343

He 3-160, 232, 233
He 3-1674, 244, 245
He 3-1761, 246, 247
He 3-404, 232, 233
He 3-461, 232, 233
He 3-653, 232, 233
He 3-828, 234, 235
He 3-863, 234, 235
He 3-886, 234, 235
He 3-905, 234, 235
He 3-916, 234, 235
Hen 1591, 252, 254
Hen 1924, 248, 249
He nova, 15, 17, 23, 24
Her X-1, 56, 78, 88, 89
high-mass X-ray binary, 37–40, 47
HK Sco, 236, 237
HM Sge, 246, 247, 252
hot spot, 53
HST satellite, 137, 139
HV 10446, 232, 233
HV 107, 232, 233
HV 12671, 232, 233
HV 169, 242, 243
HV 3181, 246, 247, 336
HV 3372, 232, 233
HV 3376, 232, 233
HV 3519, 244, 245
HV 3655, 232, 233
HV 4035, 236, 237
HV 4493, 236, 237
HV 5491, 238, 239
HV 5682, 176, 326
HV 7199, 242, 243
HV 8771, 240, 241
HV 8827, 236, 237
HZ 43, 25

ionisation nebula, 6, 79, 138, 200
irradiation, 53, 55, 57–61, 63, 171
isolated old neutron star, 279–283, 285, 288
ISO satellite, 139
IUE satellite, 108, 110, 136, 139, 259, 262

IX Vel, 113

K 3-12, 244, 245
Kepler's SNR, 45
KM Vel, 232, 233
KW Sgr, 240, 241
KX TrA, 236, 237

LMC, 29, 30, 33, 57, 81, 123, 160, 165, 199, 200, 285, 300, 301
LMC1, 230, 231
LMC S63, 226, 232, 233
LMC X-3, 45, 47
LMC X-4, 45, 47
Ln358, 252
low-mass X-ray binaries, 33, 36
low-mass X-ray binary, 37–40, 45, 47, 48, 55
LT Del, 248, 249
LuSt-1, 238, 239
LW Cas, 230, 231

M 2-9, 47
M3, 33, 34
M31, 3, 7, 33, 37, 75–81, 199, 200, 217, 221, 285, 300, 301
MaC 1-17, 246, 247
MaC 1-3, 236, 237
MaC 1-9, 240, 241
MaC 2-4, 232, 233
MACHO project, 46, 163, 165, 167, 218
Maclaurin spheroid, 47
mass-donor star, 83, 86, 88, 89
massive accretion disk, 47
mass loss, 251
metal opacities, 139
metal opacity, 142
model atmospheres, 251
model atmospheres, LTE, 25, 26, 28, 29, 31, 139, 142
model atmospheres, NLTE, 25, 26, 28, 29, 31, 32, 94, 131–135, 138–142, 251, 256
multiwavelength observation, 259

MWC 265, 238, 239
MWC 295, 244, 245
MWC 315, 246, 247
MWC 400, 248, 249
MWC 411, 230, 231
MWC 412, 234, 235
MWC 413, 236, 237
MWC 414, 240, 241
MWC 415, 246, 247
MWC 416, 248, 249
MWC 560, 232, 233, 252
MWC 600, 244, 245
MWC 603, 244, 245
MWC 960, 244, 245

neutron star, 5, 43, 46, 48, 279
NGC 5272, 33
NGC 6341, 34
NGC 6397, 34
NGC 6544, 34
NGC 6626, 34
NGC 6642, 34
NGC 6656, 34
NGC 6752, 34
NGC 6888, 45
NGC 7099, 34
Nova Cyg 1978, 22
Nova Cyg 1992, 301, 337
Nova Mus 1983, 333
Nova Mus 1991, 48
NQ Gem, 232, 233
NSV 10142, 242, 243
NSV 10219, 242, 243
NSV 10241, 242, 243
NSV 12264, 246, 247
NSV 3539, 232, 233
NSV 4775, 232, 233
NSV 6160, 234, 235
NSV 6587, 234, 235
NSV 8226, 236, 237
NSV 8805, 238, 239
NSV 9601, 240, 241

o Cet, 230, 231

opacity, 15–17, 19, 23, 27, 45, 141, 142, 206, 207
orbital period, 83, 127
OX Cyg, 246, 247

PG 1159 star, 131, 132, 136–138, 299
planetary nebula, 4, 15, 16, 18, 19, 22, 24, 47, 53, 119, 131, 138, 200, 201, 226, 227
PN Ap 1-8, 242, 243
PN Ap 1-9, 242, 243
PN Ap 3-1, 246, 247
PN Bl 3-11, 240, 241
PN Bl 3-14, 240, 241
PN Bl 3-2, 240, 241
PN Bl 3-6, 240, 241
PN Bl L, 240, 241
PN H 1-36, 240, 241
PN H 2-19, 240, 241
PN H 2-28, 240, 241
PN H 2-34, 242, 243
PN H 2-38, 242, 243
PN H 2-5, 238, 239
PN M 1-21, 238, 239
PN Pe 2-16, 246, 247
PN Sa 3-142, 244, 245
PN Sa 3-43, 238, 239
PN Th 3-17, 238, 239
PN Th 3-18, 238, 239
PN Th 3-29, 238, 239
PN Th 3-30, 238, 239
PN Th 3-31, 238, 239
PN Th 3-7, 238, 239
PN Th 3-9, 238, 239
PN Th 4-4, 240, 241
post-AGB star, 117, 131, 136, 138
Pt-1, 238, 239
pulsar, 295
PU Vul, 246, 247, 252

QS Nor, 236, 237
QW Sge, 246, 247

R Aqr, 248, 249, 252
reprocessing, 54, 55, 57, 63, 65, 68–72, 153, 155

Roche lobe, 3–5, 7–9, 53, 54, 57, 58, 86, 88, 89, 99, 188, 194, 195, 207, 275, 277
ROSAT all-sky survey, 99, 107, 125, 126, 131, 159, 163, 173, 175, 178, 217, 225–228, 251, 270, 271, 279, 281, 285, 288, 289
ROSAT satellite, 33, 37, 41, 53, 75, 83, 86, 127, 139, 159, 173, 215, 225, 259, 285, 289, 299
RR Pic, 22
RR Tel, 246, 247, 251, 252, 255, 276, 287, 301, 336
RS Oph, 240, 241
RT Car, 232, 233
RT Ser, 238, 239
RW Hya, 234, 235, 252
RX J0019.8+2156, 6, 45, 46, 53, 54, 88, 99, 102–113, 153, 157, 215, 217, 220, 286, 287, 292, 301, 304
RX J0037.3–7214, 305
RX J0037.4+4015, 76, 301, 306
RX J0038.5+4014, 76, 301, 307
RX J0038.6+4020, 76, 80, 301, 308
RX J0039.6+4054, 76, 80, 301, 309
RX J0040.4+4009, 76, 79, 301, 310
RX J0040.7+4015, 76, 79, 80, 301, 311
RX J0041.5+4040, 76, 79, 80, 301, 312
RX J0041.8+4059, 76, 79, 80, 301, 313
RX J0042.4+4044, 76, 79, 80, 301, 314
RX J0043.5+4207, 76, 301, 315
RX J0044.0+4118, 76, 79, 80, 301, 316
RX J0045.4+4154, 76, 77, 80, 215, 217, 220, 221, 301, 317
RX J0045.5+4206, 76, 78–80, 301, 318
RX J0046.2+4138, 76, 301, 320
RX J0046.2+4144, 76, 80, 301, 319
RX J0047.6+4205, 76, 301, 321

RX J0048.4–7332, 251, 256–258, 287, 301, 322
RX J0058.6–7146, 287, 301, 323
RX J0059.2–7138, 41, 43, 45, 47
RX J0122.9–7521, 131–133, 135–138, 299
RX J0439.8–6809, 115–119, 123–125, 216, 286, 287, 301, 325
RX J0513.9–6951, 5, 6, 45–47, 53, 65–69, 72, 107, 113, 127, 128, 130, 151, 159–163, 165–173, 175–178, 215, 217–220, 301, 326
RX J0527.8–6954, 3, 145–152, 215, 220, 301, 327
RX J0537.7–7034, 301, 328
RX J0549.8–7150, 126, 331
RX J0550.0–7151, 123, 125, 173–175, 301, 331
RX J0925.7–4758, 6, 83, 84, 86, 88, 89, 91, 92, 97, 107, 127, 217, 220, 288, 301, 332
RX J1342.1+2822, 334
RX J1856.5–3754, 279, 280, 282, 288
RX J2117.1+3412, 131, 137
RX Pup, 232, 233, 252
RY Sct, 244, 245

S 147, 230, 231
S 190, 248, 249
S 308, 45
s 32, 230, 231
Sanduleak's star, 230, 231, 252
Sco X-1, 88
Sirius B, 25
SMC, 29, 30, 33, 41, 43, 81, 131, 199, 200, 285, 300, 301
SMC-N 60, 230, 231
SMC-N 73, 230, 231
SMC-S 18, 230, 231
SMC1, 230, 231
SMC2, 230, 231
SMC3, 230, 231, 251, 252, 255–258, 276, 287, 322
SMC Ln 358, 230, 231
SMC N67, 230, 231, 252, 287, 324

SMC X-1, 45, 47
spray, 53, 54, 56–61
SS 433, 46, 47
SS73 1, 232, 233
SS73 117, 242, 243
SS73 122, 242, 243
SS73 129, 242, 243
SS73 141, 242, 243
SS73 29, 232, 233
SS73 38, 234, 235
SS73 96, 238, 239
SSM 1, 240, 241
SS Sge, 246, 247
St 3-22, 234, 235
subsonic flow, 42
super-Eddington source, 45
supernova, 183, 193
supernova type Ia, 24, 45, 183, 193, 195, 205, 206, 212, 215, 221
supersoft AGN, 79, 81, 286, 289–291, 294–296, 299
supersonic flow, 42, 43
symbiotic nova, 37, 221, 251, 252, 258
symbiotic star, 4, 5, 183, 187–189, 194, 225–228, 251, 259, 295, 299
SY Mus, 232, 233
SZ 134, 236, 237

T CorB, 236, 237
TT Ari, 230, 231
TV Col, 230, 231
TX CVn, 234, 235

UKS Ce-1, 236, 237
U Sco, 21, 22
UU Ser, 240, 241
UV Aur, 230, 231

V1016 Cyg, 246, 247, 252
V1017 Sgr, 244, 245
V1148 Sgr, 242, 243
V1290 Aql, 246, 247
V1329 Cyg, 248, 249
V1974 Cyg, 171, 337

V1988 Sgr, 244, 245
V2051 Oph, 236, 237
V2110 Oph, 240, 241
V2116 Oph, 238, 239
V2416 Sgr, 240, 241
V2506 Sgr, 242, 243
V2601 Sgr, 244, 245
V2756 Sgr, 242, 243
V2905 Sgr, 244, 245
V345 Nor, 236, 237
V347 Nor, 236, 237
V352 Aql, 246, 247
V366 Car, 232, 233
V3804 Sgr, 244, 245
V3811 Sgr, 244, 245
V404 Cyg, 48
V4074 Sgr, 244, 245
V407 Cyg, 248, 249
V410 Cas, 230, 231
V4141 Sgr, 240, 241
V443 Her, 244, 245
V455 Sco, 236, 237
V503 Her, 238, 239
V603 Aql, 244, 245
V615 Sgr, 242, 243
V641 Cas, 230, 231
V704 Cen, 234, 235
V741 Per, 230, 231
V748 Cen, 234, 235
V835 Cen, 234, 235
V840 Cen, 234, 235
V852 Cen, 234, 235
V916 Sco, 240, 241
V917 Sco, 240, 241
V919 Sgr, 246, 247
variability, optical, 108, 116, 165, 167, 173, 176, 178, 179
variability, UV, 112, 259, 260, 262, 263
variability, X-ray, 43, 78, 108, 145, 146, 148, 150–152, 160–163, 173, 175, 176, 178, 179
Ve 2-57, 242, 243
VY Scl star, 171

W Cep, 248, 249
white dwarf, 25
wide-binary supersoft source, 3, 4, 7, 193, 194, 196, 199, 203
wind, 3–5, 8–13, 46–48, 200, 205, 206
wind, mass loss, 159, 169, 189, 208, 210, 211
wind, models, 251
wind, optically thick, 15–24, 208, 210
wind, radiation-driven, 9–13, 199, 205, 208
WPVS007, 79, 286
Wray 15-1470, 236, 237
Wray 15-157, 232, 233
Wray 16-202, 236, 237
Wray 16-312, 240, 241
Wray 16-377, 242, 243
WY Gem, 232, 233
WY Vel, 232, 233

X-ray heating, 121
X-ray pulsar, 41
X-ray spectrum, composite, 37–40

Y CrA, 242, 243
YY Her, 242, 243

Z And, 248, 249, 252

Author Index

Altamore A., 259, 267
Asai K., 91

Barwig H., 99
Becker C.M., 37, **289**
Beuermann K., 107, 115, 173
Bickert K.F., **225**, 267

Charles P.A., 165
Cowley A.P., 123, 131
Crampton D., 131

Davies M.B., 33
de Martino D., 107
Di Stefano R., 3, **33**, **37**, 65, **193**
Dotani T., 91

Ebisawa K., **91**

Fabbiano G., 37
Friedjung M., 259

Gänsicke B.T., **107**
González-Riestra R., 259, 267
Greiner J., 75, 145, 225, 259, **267**, **285**, **299**

Hachisu I., **205**
Hartmann H.W., **25**
Hasinger G., 145
Heise J., 25, 91
Hutchings J.B., 131

Jordan S., 251

Kahabka P., **215**
Kato M., **15**, 205

Kundt W., **45**
Kylafis N.D., **41**

Livio M., 165, **183**
Luthardt R., 267

Mürset U., **251**
MACHO Collaboration, 165
Maesano M., 259
Magnier E.A., 75
Mattei M., 259
Meyer-Hofmeister E., 53, 153
Meyer F., 53, **153**
Montagni F., 259
Motch C., **83**, **127**
Mukai K., 91

Nelson L.A., 3
Neuhäuser R., **279**
Nomoto K., 205

Orio M., 145

Pakull M.W., 115, 127, 131
Popham R. , **65**

Rappaport S.A., 289
Rauch T., **139**
Reinsch K., 115, **173**
Remillard R., 289

Schaeidt S.G., **159**
Schandl S., **53**
Schmidtke P.C., **123**, 131
Schwarz R., 145
Smale A., 91
Southwell K.A., **165**

Stencel R.E., 225
Supper R., 75
Sutherland W., 165

Thomas H.-C., 115, 173

van Teeseling A., **115**, 173
Viotti R., **259**, 267

W. Hartmann H.W., 91
Walter F.M., 279
Werner K., **131**
Will T., **99**
Wolff B., 131, 251
Wolk S.J., 279

Springer-Verlag and the Environment

We at Springer-Verlag firmly believe that an international science publisher has a special obligation to the environment, and our corporate policies consistently reflect this conviction.

We also expect our business partners – paper mills, printers, packaging manufacturers, etc. – to commit themselves to using environmentally friendly materials and production processes.

The paper in this book is made from low- or no-chlorine pulp and is acid free, in conformance with international standards for paper permanency.

Lecture Notes in Physics

For information about Vols. 1–439
please contact your bookseller or Springer-Verlag

Vol. 440: H. Latal, W. Schweiger (Eds.), Matter Under Extreme Conditions. Proceedings, 1994. IX, 243 pages. 1994.

Vol. 441: J. M. Arias, M. I. Gallardo, M. Lozano (Eds.), Response of the Nuclear System to External Forces. Proceedings, 1994, VIII. 293 pages. 1995.

Vol. 442: P. A. Bois, E. Dériat, R. Gatignol, A. Rigolot (Eds.), Asymptotic Modelling in Fluid Mechanics. Proceedings, 1994. XII, 307 pages. 1995.

Vol. 443: D. Koester, K. Werner (Eds.), White Dwarfs. Proceedings, 1994. XII, 348 pages. 1995.

Vol. 444: A. O. Benz, A. Krüger (Eds.), Coronal Magnetic Energy Releases. Proceedings, 1994. X, 293 pages. 1995.

Vol. 445: J. Brey, J. Marro, J. M. Rubí, M. San Miguel (Eds.), 25 Years of Non-Equilibrium Statistical Mechanics. Proceedings, 1994. XVII, 387 pages. 1995.

Vol. 446: V. Rivasseau (Ed.), Constructive Physics. Results in Field Theory, Statistical Mechanics and Condensed Matter Physics. Proceedings, 1994. X, 337 pages. 1995.

Vol. 447: G. Aktaş, C. Saçlıoğlu, M. Serdaroğlu (Eds.), Strings and Symmetries. Proceedings, 1994. XIV, 389 pages. 1995.

Vol. 448: P. L. Garrido, J. Marro (Eds.), Third Granada Lectures in Computational Physics. Proceedings, 1994. XIV, 346 pages. 1995.

Vol. 449: J. Buckmaster, T. Takeno (Eds.), Modeling in Combustion Science. Proceedings, 1994. X, 369 pages. 1995.

Vol. 450: M. F. Shlesinger, G. M. Zaslavsky, U. Frisch (Eds.), Lévy Flights and Related Topics in Physics. Proceedigs, 1994. XIV, 347 pages. 1995.

Vol. 451: P. Krée, W. Wedig (Eds.), Probabilistic Methods in Applied Physics. IX, 393 pages. 1995.

Vol. 452: A. M. Bernstein, B. R. Holstein (Eds.), Chiral Dynamics: Theory and Experiment. Proceedings, 1994. VIII, 351 pages. 1995.

Vol. 453: S. M. Deshpande, S. S. Desai, R. Narasimha (Eds.), Fourteenth International Conference on Numerical Methods in Fluid Dynamics. Proceedings, 1994. XIII, 589 pages. 1995.

Vol. 454: J. Greiner, H. W. Duerbeck, R. E. Gershberg (Eds.), Flares and Flashes, Germany 1994. XXII, 477 pages. 1995.

Vol. 455: F. Occhionero (Ed.), Birth of the Universe and Fundamental Physics. Proceedings, 1994. XV, 387 pages. 1995.

Vol. 456: H. B. Geyer (Ed.), Field Theory, Topology and Condensed Matter Physics. Proceedings, 1994. XII, 206 pages. 1995.

Vol. 457: P. Garbaczewski, M. Wolf, A. Weron (Eds.), Chaos – The Interplay Between Stochastic and Deterministic Behaviour. Proceedings, 1995. XII, 573 pages. 1995.

Vol. 458: I. W. Roxburgh, J.-L. Masnou (Eds.), Physical Processes in Astrophysics. Proceedings, 1993. XII, 249 pages. 1995.

Vol. 459: G. Winnewisser, G. C. Pelz (Eds.), The Physics and Chemistry of Interstellar Molecular Clouds. Proceedings, 1993. XV, 393 pages. 1995.

Vol. 460: S. Cotsakis, G. W. Gibbons (Eds.), Global Structure and Evolution in General Relativity. Proceedings, 1994. IX, 173 pages. 1996.

Vol. 461: R. López-Peña, R. Capovilla, R. García-Pelayo, H. Waelbroeck, F. Zertuche (Eds.), Complex Systems and Binary Networks. Lectures, México 1995. X, 223 pages. 1995.

Vol. 462: M. Meneguzzi, A. Pouquet, P.-L. Sulem (Eds.), Small-Scale Structures in Three-Dimensional Hydrodynamic and Magnetohydrodynamic Turbulence. Proceedings, 1995. IX, 421 pages. 1995.

Vol. 463: H. Hippelein, K. Meisenheimer, H.-J. Röser (Eds.), Galaxies in the Young Universe. Proceedings, 1994. XV, 314 pages. 1995.

Vol. 464: L. Ratke, H. U. Walter, B. Feuerbach (Eds.), Materials and Fluids Under Low Gravity. Proceedings, 1994. XVIII, 424 pages, 1996.

Vol. 465: S. Beckwith, J. Staude, A. Quetz, A. Natta (Eds.), Disks and Outflows Around Young Stars. Proceedings, 1994. XII, 361 pages, 1996.

Vol. 466: H. Ebert, G. Schütz (Eds.), Spin – Orbit-Influenced Spectroscopies of Magnetic Solids. Proceedings, 1995. VII, 287 pages, 1996.

Vol. 467: A. Steinchen (Ed.), Dynamics of Multiphase Flows Across Interfaces. 1994/1995. XII, 267 pages. 1996.

Vol. 468: C. Chiuderi, G. Einaudi (Eds.), Plasma Astrophysics. 1994. VII, 326 pages. 1996.

Vol. 469: H. Grosse, L. Pittner (Eds.), Low-Dimensional Models in Statistical Physics and Quantum Field Theory. Proceedings, 1995. XVII, 339 pages. 1996.

Vol. 470: E. Martínez-González, J. L. Sanz (Eds.), The Universe at High-z, Large-Scale Structure and the Cosmic Microwave Background. Proceedings, 1995. VIII, 254 pages. 1996.

Vol. 471: W. Kundt (Ed.), Jets from Stars and Galactic Nuclei. Proceedings, 1995. X, 290 pages. 1996.

Vol. 472: J. Greiner (Ed.), Supersoft X-Ray Sources. Proceedings, 1996. XIII, 350 pages. 1996.

New Series m: Monographs

Vol. m 1: H. Hora, Plasmas at High Temperature and Density. VIII, 442 pages. 1991.

Vol. m 2: P. Busch, P. J. Lahti, P. Mittelstaedt, The Quantum Theory of Measurement. XIII, 165 pages. 1991. Second Revised Edition: XIII, 181 pages. 1996.

Vol. m 3: A. Heck, J. M. Perdang (Eds.), Applying Fractals in Astronomy. IX, 210 pages. 1991.

Vol. m 4: R. K. Zeytounian, Mécanique des fluides fondamentale. XV, 615 pages, 1991.

Vol. m 5: R. K. Zeytounian, Meteorological Fluid Dynamics. XI, 346 pages. 1991.

Vol. m 6: N. M. J. Woodhouse, Special Relativity. VIII, 86 pages. 1992.

Vol. m 7: G. Morandi, The Role of Topology in Classical and Quantum Physics. XIII, 239 pages. 1992.

Vol. m 8: D. Funaro, Polynomial Approximation of Differential Equations. X, 305 pages. 1992.

Vol. m 9: M. Namiki, Stochastic Quantization. X, 217 pages. 1992.

Vol. m 10: J. Hoppe, Lectures on Integrable Systems. VII, 111 pages. 1992.

Vol. m 11: A. D. Yaghjian, Relativistic Dynamics of a Charged Sphere. XII, 115 pages. 1992.

Vol. m 12: G. Esposito, Quantum Gravity, Quantum Cosmology and Lorentzian Geometries. Second Corrected and Enlarged Edition. XVIII, 349 pages. 1994.

Vol. m 13: M. Klein, A. Knauf, Classical Planar Scattering by Coulombic Potentials. V, 142 pages. 1992.

Vol. m 14: A. Lerda, Anyons. XI, 138 pages. 1992.

Vol. m 15: N. Peters, B. Rogg (Eds.), Reduced Kinetic Mechanisms for Applications in Combustion Systems. X, 360 pages. 1993.

Vol. m 16: P. Christe, M. Henkel, Introduction to Conformal Invariance and Its Applications to Critical Phenomena. XV, 260 pages. 1993.

Vol. m 17: M. Schoen, Computer Simulation of Condensed Phases in Complex Geometries. X, 136 pages. 1993.

Vol. m 18: H. Carmichael, An Open Systems Approach to Quantum Optics. X, 179 pages. 1993.

Vol. m 19: S. D. Bogan, M. K. Hinders, Interface Effects in Elastic Wave Scattering. XII, 182 pages. 1994.

Vol. m 20: E. Abdalla, M. C. B. Abdalla, D. Dalmazi, A. Zadra, 2D-Gravity in Non-Critical Strings. IX, 319 pages. 1994.

Vol. m 21: G. P. Berman, E. N. Bulgakov, D. D. Holm, Crossover-Time in Quantum Boson and Spin Systems. XI, 268 pages. 1994.

Vol. m 22: M.-O. Hongler, Chaotic and Stochastic Behaviour in Automatic Production Lines. V, 85 pages. 1994.

Vol. m 23: V. S. Viswanath, G. Müller, The Recursion Method. X, 259 pages. 1994.

Vol. m 24: A. Ern, V. Giovangigli, Multicomponent Transport Algorithms. XIV, 427 pages. 1994.

Vol. m 25: A. V. Bogdanov, G. V. Dubrovskiy, M. P. Krutikov, D. V. Kulginov, V. M. Strelchenya, Interaction of Gases with Surfaces. XIV, 132 pages. 1995.

Vol. m 26: M. Dineykhan, G. V. Efimov, G. Ganbold, S. N. Nedelko, Oscillator Representation in Quantum Physics. IX, 279 pages. 1995.

Vol. m 27: J. T. Ottesen, Infinite Dimensional Groups and Algebras in Quantum Physics. IX, 218 pages. 1995.

Vol. m 28: O. Piguet, S. P. Sorella, Algebraic Renormalization. IX, 134 pages. 1995.

Vol. m 29: C. Bendjaballah, Introduction to Photon Communication. VII, 193 pages. 1995.

Vol. m 30: A. J. Greer, W. J. Kossler, Low Magnetic Fields in Anisotropic Superconductors. VII, 161 pages. 1995.

Vol. m 31: P. Busch, M. Grabowski, P. J. Lahti, Operational Quantum Physics. XI, 230 pages. 1995.

Vol. m 32: L. de Broglie, Diverses questions de mécanique et de thermodynamique classiques et relativistes. XII, 198 pages. 1995.

Vol. m 33: R. Alkofer, H. Reinhardt, Chiral Quark Dynamics. VIII, 115 pages. 1995.

Vol. m 34: R. Jost, Das Märchen vom Elfenbeinernen Turm. VIII, 286 pages. 1995.

Vol. m 35: E. Elizalde, Ten Physical Applications of Spectral Zeta Functions. XIV, 228 pages. 1995.

Vol. m 36: G. Dunne, Self-Dual Chern-Simons Theories. X, 217 pages. 1995.

Vol. m 37: S. Childress, A.D. Gilbert, Stretch, Twist, Fold: The Fast Dynamo. XI, 410 pages. 1995.

Vol. m 38: J. González, M. A. Martín-Delgado, G. Sierra, A. H. Vozmediano, Quantum Electron Liquids and High-T_c Superconductivity. X, 299 pages. 1995.

Vol. m 39: L. Pittner, Algebraic Foundations of Non-Commutative Differential Geometry and Quantum Groups. XII, 469 pages. 1996.

Vol. m 40: H.-J. Borchers, Translation Group and Particle Representations in Quantum Field Theory. VII, 131 pages. 1996.

Vol. m 41: B. K. Chakrabarti, A. Dutta, P. Sen, Quantum Ising Phases and Transitions in Transverse Ising Models. X, 204 pages. 1996.

Vol. m 42: P. Bouwknegt, J, McCarthy, K. Pilch, The W3 Algebra. Modules, Semi-infinite Cohomology and BV Algebras. IX, 204 pages. 1996.

PHYSICS LIBRARY